Shape-Me[illegible]ers and Multifunctional Composites

Edited by

Jinsong Leng · Shanyi Du

CRC Press is an imprint of the
Taylor & Francis Group, an **informa** business

CRC Press
Taylor & Francis Group
6000 Broken Sound Parkway NW, Suite 300
Boca Raton, FL 33487-2742

First issued in paperback 2019

ISBN-13: 978-1-4200-9019-2 (hbk)
ISBN-13: 978-0-367-38399-2 (pbk)

Library of Congress Cataloging-in-Publication Data

Shape-memory polymers and multifunctional composites / editors, Jinsong Leng, Shanyi Du.
p. cm.
"A CRC title."
Includes bibliographical references and index.
ISBN 978-1-4200-9019-2 (hardcover : alk. paper)
1. Composite materials. 2. Polymers--Thermomechanical properties. 3. Shape memory effect. I. Leng, Jinsong. II. Du, Shanyi. III. Title.

TA418.9.C6S457 2010
620.1'920429--dc22 2009042040

Visit the Taylor & Francis Web site at
http://www.taylorandfrancis.com

and the CRC Press Web site at
http://www.crcpress.com

Contents

Preface

Shape-memory polymers (SMPs) undergo significant macroscopic deformation upon the application of external stimuli (e.g., heat, electricity, magnetism, light, and moisture). Since the discovery of SMPs in the 1980s, international research interest in the shape-memory effect of polymers has grown rapidly in different fields. There has been a rapid development of SMPs and a number of achievements have been reported. It will not be an exaggeration to say that the appearance of the shape-memory effect in polymers is one of the revolutionary steps in the field of active materials research. Recently, the underlying behavior in response to an external stimulus, structures and applications of SMPs has increasingly drawn the attention of researchers. Due to their novel properties, SMPs can be used in a broad range of applications from outer space to automobiles, and to address challenges in advanced aerospace, and mechanical, bionic, and medical technologies.

This book aims to provide a comprehensive discussion of SMPs, covering the foundations and applications of all aspects of SMPs and their composites. Chapter 1 provides an overview of SMPs, where the different types of materials showing the shape-memory effect are described and some basic concepts are discussed. The definitions, architectures, and fundamental principles of different functions of SMPs are explained. Some examples, including direct and indirect triggering of SMPs, the light-induced dual shape effect, and triple-shape polymers are also presented. Chapter 2 presents the structural requirements and provides an overview of the variety of chemical compositions of SMPs. The synthesis, processing, and particular features of covalently cross-linked SMPs and thermoplastic SMPs are discussed in detail. Many practical applications require constitutive models that describe the three-dimensional finite deformation. Chapter 3 therefore presents thermo-mechanical constitutive models of SMPs based on viscoelasticity and phase transition. SMPs have been extensively studied in the last decade, especially their thermo-mechanical and electrical characterizations. Chapters 4 and 5 elaborate on these essential properties. Chapter 6 is devoted to an investigation of some actuation methods of SMPs, such as heat, electricity, light, magnetism, and moisture. These novel actuation approaches play a critical role in the development of multifunctional materials that not only exhibit the shape-memory effect but also perform particular functions. Chapter 7 elaborates on how SMP composites filled with particles or fibers may overcome the poor mechanical properties of pure SMP materials. Then, Chapters 8 through 12 present some typical potential applications of SMPs. In recent years, SMPs and SMP composites have been developed and qualified especially for deployable components and structures in aerospace. Chapter 8 describes the applications, which include hinges, trusses, booms, antennas, optical reflectors, and morphing skins. Chapter 9 discusses SMP foam technologies and identifies their potential applications in space, as well as their commercial and biomedical applications. Chapter 10 presents shape-memory fibers based on SMPs that can also be implemented to develop smart textiles that respond to thermal stimulus and may be used in future smart clothing. In recent times, a number of medical applications have been considered and investigated. SMP materials have been found to be biocompatible, nontoxic, and nonmutagenic. Chapter 11 describes SMP materials and their potential and existing medical applications. Finally, Chapter 12 presents some novel applications of SMPs, and proposes potential directions and applications of SMPs for future research and development.

Last but not least, we would like to take this opportunity to express our sincere gratitude to all the contributors for their hard work in preparing and revising the chapters. We are indebted to all members of the team as well as to those who helped with the preparation of this book. Finally, we wish to thank our families and friends for all their patience and support.

In addition, we hope that this book will be a useful reference for engineering researchers, and for senior and graduate students in their relevant fields.

Editors

Jinsong Leng is a Cheung Kong Chair Professor at the Centre for Composite Materials and Structures of Harbin Institute of Technology, Harbin, China. His research interests include smart materials and structures, sensors and actuators, fiber-optic sensors, shape-memory polymers, electro-active polymers, structural health monitoring, morphing aircrafts, and multifunctional nanocomposites. He has authored or coauthored over 180 scientific papers, 2 books, 12 issued patents, and has delivered more than 18 invited talks around the world. He also serves as the chairman and as a member of the scientific committees of international conferences. He served as the editor in chief of the *International Journal of Smart and Nano Materials* (Taylor & Francis Group) and as the associate editor of *Smart Materials and Structures* (IOP Publishing Ltd). He is the chairman of the Asia-Pacific Committee on Smart and Nano Materials. Professor Jinsong Leng has been elected as an SPIE Fellow in 2010.

Shanyi Du is a professor at the Harbin Institute of Technology, Harbin, China. He is an academician at the Chinese Academy of Engineering, China, and is an outstanding expert and educator in the field of mechanics and composite materials. He serves as the president of the Chinese Society of Composite Materials. He is one of the founders of the Centre for Composite Materials and Structures of Harbin Institute of Technology, Harbin, China. Previously, he has been the vice president of The Chinese Society of Theoretical and Applied Mechanics as well as a member of the Executive Council of the International Committee on Composite Materials. He has authored or coauthored over 200 journals and national/international papers, and 10 monographs or textbooks. He is a member of the editorial boards of famous international journals, such as *Composites Science and Technology* and the *International Journal of Computational Methods.* He also serves as the honorary editor of the *International Journal of Smart and Nano Materials* (Taylor & Francis Group).

Contributors

Marc Behl
Center for Biomaterial Development
Institute for Polymer Research
GKSS Research Center
Teltow, Germany

and

Berlin-Brandenburg Center for Regenerative Therapies
Berlin, Germany

Shanyi Du
Centre for Composite Materials and Structures
Harbin Institute of Technology
Harbin, P.R. China

Martin L. Dunn
Department of Mechanical Engineering
University of Colorado
Boulder, Colorado

Matthias Heuchel
Center for Biomaterial Development
Institute for Polymer Research
GKSS Research Center
Teltow, Germany

and

Berlin-Brandenburg Center for Regenerative Therapies
Berlin, Germany

Jinlian Hu
Institute of Textiles and Clothing
Hong Kong Polytechnic University
Hong Kong, China

Wei Min Huang
School of Mechanical and Aerospace Engineering
Nanyang Technological University
Singapore, Singapore

Hong-Yan Jiang
Mnemoscience GmbH
Übach-Palenberg, Germany

Karl Kratz
Center for Biomaterial Development
Institute for Polymer Research
GKSS Research Center
Teltow, Germany

and

Berlin-Brandenburg Center for Regenerative Therapies
Berlin, Germany

Xin Lan
Centre for Composite Materials and Structures
Harbin Institute of Technology
Harbin, P.R. China

Andreas Lendlein
Center for Biomaterial Development
Institute for Polymer Research
GKSS Research Center
Teltow, Germany

and

Berlin-Brandenburg Center for Regenerative Therapies
Berlin, Germany

Jinsong Leng
Centre for Composite Materials and Structures
Harbin Institute of Technology
Harbin, P.R. China

Yanju Liu
Department of Aerospace Science and Mechanics
Harbin Institute of Technology
Harbin, P.R. China

Haibao Lu
Centre for Composite Materials and Structures
Harbin Institute of Technology
Harbin, P.R. China

Hang Jerry Qi
Department of Mechanical Engineering
University of Colorado
Boulder, Colorado

Annette M. Schmidt
Institut für Organische Chemie und Makromolekulare Chemie
Heinrich-Heine-Universität Düsseldorf
Düsseldorf, Germany

Witold M. Sokolowski
Jet Propulsion Laboratory
California Institute of Technology
Pasadena, California

Wolfgang Wagermaier
Center for Biomaterial Development
Institute for Polymer Research
GKSS Research Center
Teltow, Germany

and

Berlin-Brandenburg Center for Regenerative Therapies
Berlin, Germany

Bin Yang
School of Mechanical and Aerospace Engineering
Nanyang Technological University
Singapore, Singapore

1

Overview of Shape-Memory Polymers

Marc Behl and Andreas Lendlein*

Center for Biomaterial Development, Institute for Polymer Research, GKSS Research Center, Teltow, Germany

CONTENTS

1.1 Introduction

The ability of polymers to respond to external stimuli such as heat or light is of high scientific and technological significance. Their stimuli-sensitive behavior enables such materials to change certain of their macroscopic properties such as shape, color, or refractive index when controlled by an external signal. The implementation of the capability to actively move into polymers has attracted the interest of researchers, especially in the last few years, and has been achieved in polymers as well as in gels. Sensitivity to heat, light, magnetic fields, and ion strength or pH value was also realized in gels [1]. In nonswollen polymers, active movement is stimulated by exposure to heat or light and could also be designed as a complex movement with more than two shapes.

* To whom correspondence should be addressed. E-mail: andreas.lendlein@gkss.de

Besides their scientific significance, such materials have a high innovation potential and can be found, e.g., in smart fabrics [2–4], heat-shrinkable tubes for electronics or films for packaging [5], self-deployable sun sails in spacecrafts [6], self-disassembling mobile phones [7], intelligent medical devices [8], and implants for minimally invasive surgery [9–11]. These are only examples and cover only a small region of potential applications. Actively moving polymers may even reshape the design of products [12]. In this chapter, different classes of actively moving materials are introduced with an emphasis on shape-memory polymers. The fundamental principles of the different functions are explained and examples for specific materials are given.

1.2 Definition of Actively Moving Polymers

Actively moving polymers are able to respond to a specific stimulus by changing their shape. In general, two types of functions have to be distinguished: the shape-memory and the shape-changing capability. In both cases, the basic molecular architecture is a polymer network while the mechanisms underlying the active movement differ [13,14]. Both polymer concepts contain either molecular switches or stimuli-sensitive domains. Upon exposure to a suitable stimulus, the switches are triggered resulting in the movement of the shaped body.

Most shape-memory polymers are dual-shape materials exhibiting two distinct shapes. They can be deformed from their original shape and temporarily assume another shape. This temporary shape is maintained until the shaped body is exposed to an appropriate stimulus. Shape recovery is predefined by a mechanical deformation leading to the temporary shape. So far, shape-memory polymers induced by heat or light have been reported. Furthermore, the concept of the thermally induced shape-memory effect has been extended by indirect actuation, e.g., irradiation with IR-light, application of electrical current, exposure to alternating magnetic fields, and immersion in water.

Besides exhibiting two distinct shapes, an important characteristic of shape-memory polymers is the stability of the temporary shape until the point of time of exposure to the suitable stimulus and the long-term stability of the (recovered) permanent shape, which stays unchanged even when not exposed to the stimulus anymore. Finally, different temporary shapes, substantially differing in their three-dimensional shape, can be created for the same permanent shape in subsequent cycles.

In contrast to shape-memory polymers, shape-changing polymers change their shape gradually, i.e., shrink or bend, as long as they are exposed to a suitable stimulus. Once the stimulus is terminated, they recover their original shape. This process of stimulated deformation and recovery can be repeated several times, while the geometry, i.e., of how a workpiece is moving, is determined by its original three-dimensional shape as the effect is based on a phase transition in a liquid crystalline elastomer network. Heat, light, and electromagnetic fields have been reported as suitable stimuli for shape-changing polymers.

1.3 Shape-Memory Polymer Architectures

The shape-memory effect is not an intrinsic material property, but occurs due to the combination of the polymer's molecular architecture and the resulting polymer morphology in combination with a tailored processing and programming technology for the creation of the temporary shape. To enable the shape-memory effect, a polymer architecture, which consists of netpoints and molecular switches that are sensitive to an external stimulus, is required.

The permanent shape in such a polymer network is determined by the netpoints that are cross-linked by chain segments (Figure 1.1). Netpoints can be realized by covalent bonds or intermolecular interactions; hence, they are either of a chemical or a physical nature. While chemical cross-linking can be realized by suitable cross-linking chemistry, physical cross-linking requires a polymer morphology consisting of at least two segregated domains. In such a morphology, the domains providing the second-highest thermal transition, T_{trans}, act as switching domains, and the associated segments of the multiphase polymers are therefore called "switching segments," while the

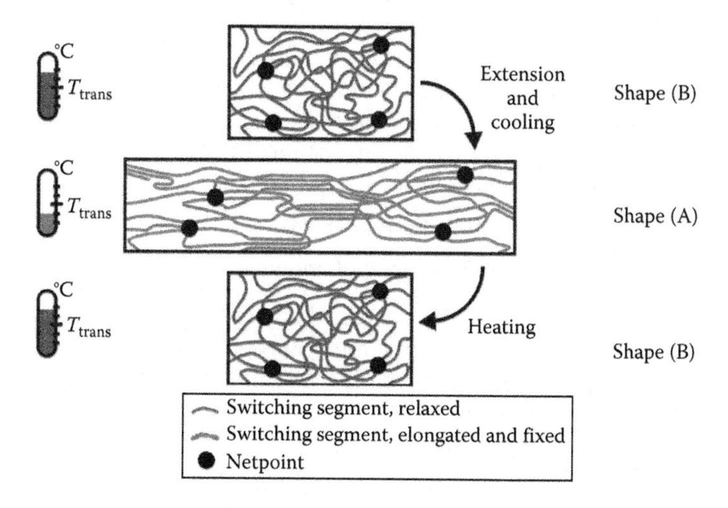

FIGURE 1.1
Molecular mechanism of the thermally induced shape-memory effect. T_{trans} is the thermal transition temperature of the switching phase. (Adapted from Lendlein, A. and Kelch, S., *Angew. Chem. Int. Ed.*, 41(12), 2034, 2002. With permission.)

domains associated-to-the highest thermal transition, T_{perm}, act as physical netpoints. The segments forming such hard domains are known as "hard segments." These switches must be able to fix the deformed shape temporarily under conditions relevant to the particular application. In addition to switching domains, they can be realized by functional groups that are able to reversibly form and cleave covalent cross-links. The thermal transition, T_{trans}, related to the switching domains can be a melting transition (T_m), or a glass transition (T_g). Accordingly, the temporary shape is fixed by a solidification of the switching domains by crystallization or vitrification. In suitable polymer architectures, these switching domains can be formed either by side chains that are only connected to one netpoint and do not contribute to the overall elasticity of the polymer, or by chain segments linking two netpoints and contributing to the overall elastic behavior. In both cases, the temporary stabilization is caused by the aggregation of the switching segments. Recently, blends of two thermoplastic polymers having shape-memory capability were presented, in which the segments forming the hard and the switching domains consisted of two different multiblock copolymers.

Functional groups that are able to reversibly form and cleave covalent bonds controlled by an external stimulus can be used as molecular switches providing chemical bonds. The introduction of functional groups that are able to undergo a photoreversible reaction, e.g., cinnamic acid (CA) groups, extends the shape-memory technology to light, which acts as a stimulus.

Shape-memory properties are quantified in cyclic stimuli-specific mechanical tests [15,16] in which each cycle consists of the programming of the test specimen and the recovery of its permanent shape. Different test protocols have been developed for the programming and recovery (see Chapter 3) from which the shape-memory properties are quantified by determining the shape-fixity ratio (R_f) for the programming and the shape-recovery ratio (R_r) for the recovery process. In thermally induced shape-memory polymers, the determination of the switching temperature, T_{sw}, characterizing the stress-free recovery process can be included in the test protocol.

1.3.1 Thermally Induced Dual-Shape Effect

1.3.1.1 Thermoplastic Shape-Memory Polymers

An important group of physically cross-linked shape-memory polymers are linear-block copolymers. Block copolymers with $T_{trans} = T_m$, and where polyurethanes and polyether-ester are prominent examples for such materials, are reviewed by Lendlein and Kelch [15]. In polyesterurethanes, oligourethane segments act as hard segments, while polyester, e.g., poly(ε-caprolactone) (T_m=44°C–55°C) forms the switching segment [17–19]. The phase separation and the domain orientation of poly(ε-caprolactone)-based polyesterurethanes could be determined by Raman spectroscopy using polarized light [20]. In polyesterurethanes where poly(hexylene adipate) provides the switching segment, and a hard segment is formed by the 4,4′-diphenyldiisocyanate and the 1,4-butanediol, the influence of the M_n of the switching segment as well as the hard segment content on the shape-memory properties were investigated [21]. The R_f increases with the increasing M_n of the switching segment but decreases with the increasing hard segment content. At the same time, the R_r decreases with the increasing M_n of the switching segment and the increasing hard segment content. Fibers from polyesterurethanes exert significantly higher recovery stress in the fiber axis when compared to the polymer films [22]. The exchange of the chain extender 1,4-butanediol with ethylenediamine can result in improved values of the R_f as urea-type bonding of the ethylene diamine can restrict the chain rotation and strengthen the physical interactions between the polyurethane segments [23]. Additionally, the shape-memory properties of polyurethanes can be enhanced by the addition of a second soft segment in small amounts so that segmented polyurethanes are obtained; e.g., 5 wt% of poly(ethylene glycol) can be added during synthesis to the poly(tetramethylene glycol) [24]. The addition of *N*-methyldiethanolamine as a cationomer in the hard segment of the segmented polyurethanes from poly(ε-caprolactone), 4,4′-diphenylmethane diisocyanate and 1,4-butanediol, simultaneously improved the R_f and the R_r. This effect is attributed to an improved switching segment crystallization [25]. A similar effect was found in copolyester-based ionomers obtained by the bulk polymerization of adipic acid and mixed monomers of bis(poly(oxyethylene)) sulfonated dimethyl fumarate and 1,4-butanediol [26]. The storage modulus of the rubbery plateau was significantly increased with increasing ionomer content and recovery rates of up to 95% were determined. Melt blending of an elastomeric ionomer based on the zinc salt of sulfonated poly{ethylene-r-propylene-r-(5-ethylidene-2-norbornene)} and low molecular mass fatty acids results in polymer networks in which the nanophase-separated ionomer provides the permanent network physically cross-linked by the zinc salt, and the fatty acids provide nanophases, whose melting triggers the shape recovery [27]. Polycarbonate segments containing polyurethanes were synthesized by the copolymerization of ethylene oxide in the presence of CO_2 catalyzed by a polymer-supported bimetallic catalyst, which yields an aliphatic polycarbonate diol. This macrodiol was further processed by the prepolymer method into a shape-memory polyurethane [28].

Thermoplastic multiblock copolymers with polydepsipeptide- and poly(ε-caprolactone)-segments providing shape-memory capability were synthesized via the coupling of oligodepsipeptide diol and oligo(ε-caprolactone) diol (PCL-diol) using a racemic mixture of 2,2,4- and 2,4,4-trimethylhexamethylene diisocyanate (TMDI). The multiblock copolymers were developed for biomedical applications and are supposed to degrade into less harmless degradation products than polyester-based materials. In the polymer molecules, the PCL block has the function of a switching segment forming the switching domains, that fix the temporary shape by crystallization [29].

Recently, binary polymer blends from two different multiblock copolymers with shape-memory capability were presented, whereby the first polymer component provided the segments forming the hard domains and the second, the segments forming the switching domains [30]. In both multiblock copolymers, a poly(alkylene adipate) mediator segment was incorporated to promote their miscibility as the hard segment poly(*p*-dioxanone) (PPDO) and the switching segment poly(ε-caprolactone) (PCL) are nonmiscible. All polymer blends investigated showed excellent shape-memory properties. The melting point associated with the PCL switching domains T_m, PCL, is almost independent of the weight ratio of the two blend components. At the same time, the mechanical properties can be varied systematically by the blend composition. In this way, a complex synthesis of new materials can be avoided. This binary blend system providing good biodegradability, a variability of mechanical properties, and a T_{sw} around the body temperature is thus an economically efficient, suitable candidate for diverse biomedical applications.

1.3.1.2 Covalently Cross-Linked Shape-Memory Polymers

Shape-memory polymer networks providing covalent netpoints can be obtained by the cross-linking of linear or branched polymers as well as by (co)polymerization/poly(co)condensation of one or several monomers, whereby one has to be at least trifunctional. Depending on the synthesis strategy, cross-links can be created during the synthesis or by postprocessing methods. Besides cross-linking by radiation (γ-radiation, neutrons, e-beam), the most common method for chemical cross-linking, after processing, is the addition of a radical initiator to polymers. An example of this is the addition of dicumyl peroxide to a semicrystalline polycyclooctene obtained by ring-opening methathesis polymerization containing unsaturated carbon bonds [31]. Here, the shape-memory effect is triggered by the melting of crystallites, which can be controlled by the *trans*-vinylene content. With increasing cross-linking density, the crystallinity of the material decreases. A melting temperature of 60°C was determined for pure polycyclooctene with a 81 wt% *trans*-vinylene content resulting in a shape recovery of these materials within 0.7 s at 70°C. The conversion to T_{trans}, when temperature is increased, can be monitored by the addition of a mechanochromic dye based on oligo(*p*-phenylene vinylene). Previously formed excimers of the dye are dissolved at this point and a pronounced change of their adsorption can be observed [32].

The other synthetic route to obtain polymer networks involves the copolymerization of monofunctional monomers with low molecular weight or oligomeric bifunctional cross-linkers. In a model system based on a styrene copolymer cross-linked with divinylbenzene, the influence of the degree of cross-linking on the thermomechanical properties has been investigated [33]. By increasing the amount of the cross-linker from 0 to 4 wt%, T_g increased from 55°C to 81°C accompanied by an increase of the gel content from 0% to 80%. The copolymerization of monofunctional monomers of low molecular weight with oligomeric bifunctional cross-linkers results in AB copolymer networks of increased toughness and elasticity at room temperature [34]. In nanoindentation studies of AB copolymer networks obtained from the copolymerization of diethyleneglycol with *t*-butyl acrylate, the increase of T_g and the rubbery modulus with increasing cross-linker addition could be confirmed [35]. From the copolymerization of various acrylates with amorphous poly[(L-lactide-ran-glycolide)]dimethacrylate, AB copolymer networks with T_{trans} based on T_g could be obtained, whose T_{sw} could be varied between 9°C and 45°C by the choice of the comonomer ratio. In copolymers with ethylacrylate as comonomer, values of R_f and R_r were higher than 97% and 98.5% while they were not influenced by the comonomer content. In similar AB copolymer networks prepared by copolymerization of n-butyl acrylate and semicrystalline oligo[(ε-hyrdroxycaproate)-co-glycolate]dimethacrylates, covalently cross-linked shape-memory polymer networks with $T_{trans} = T_m$ were obtained [36]. Here, the low T_g of butyl acrylate domains, as additional soft segment, contributes additional elasticity to the material at temperatures relevant for potential applications. The T_m of the oligo[(ε-hyrdroxycaproate)-co-glycolate] segments correlates with the T_{sw} and could be adjusted by the variation of the molecular weight and the glycolate content of the switching segment.

The photopolymerization of poly(ε-caprolactone) acrylate macromonomers with polyhedral oligosilsesquioxane (POSS) moieties located precisely in the middle of the network chains built polymer networks with $T_{trans} = T_m$ based on the T_m of the oligo(ε-caprolactone) segments [37]. In polymer networks with a POSS content of 47 wt%, a second rubbery plateau could be determined, which has been associated with the physical interactions of the POSS moieties.

Covalently cross-linked polymer can also be synthesized by polyaddition or polycondensation reactions. In polyurethane networks prepared by the prepolymer method using a diisocyanate and 1,1,1-trimethylol propane to provide covalent cross links and poly(tetrahydrofuran) to provide the switching segment, the T_{sw} could be controlled by the variation of the T_m of the precursor macrodiol, while the elastic properties were adjusted by the cross-link density [38]. High degrees of cross-linking were obtained by using hyperbranched polyesters providing the multiol component [39]. In the two-step process, the prepolymer was formed by the reaction of poly(butylene adipate) with diphenylmethane diisocyanate, which, afterward, reacted with the hyperbranched polyester Boltron H30 (hydroxyl number equal 470–500 mg KOH/g). In such polymer networks shape recovery rates of 96%–98% were determined.

Shape-memory polymer networks were also prepared using the addition reaction of oxiranes. Shape-memory properties were enabled by the cross-linking reaction of 3-amino-1,2,4-triazole with epoxidized natural rubber catalyzed by bisphenol-A [40]. In such a polymer network with $T_{trans} = T_g$, T_g could be controlled by the 3-amino-1,2,4-triazole content in the range between 29°C and 64°C.

A combination of polymerization and polyaddition enables covalently cross-linked interpenetrating polymer networks (IPN). Generally, such IPN are prepared first, through polyaddition and afterward by polymerization. The other sequence to enable IPN is from polyethyleneglycol dimethacrylate blended with star-shaped poly[(rac-lacide)-co-glycolide], which is first photopolymerized and later, the polyesterurethane network is formed using isophorone diisocyanate [41]. T_{trans} can be adjusted between −23°C and 63°C and the R_f and R_r are reported to be above 93%.

1.3.2 Indirect Triggering of Thermally Induced Dual-Shape Effect

The applicability of thermally induced shape-memory polymers has been broadened by indirect actuation. Two different strategies enable the indirect actuation of the shape-memory effect. The first strategy is by indirectly heating, e.g., by exposure to IR-radiation. In the second strategy, the T_{trans} is lowered so that the shape-memory effect is triggered but the sample temperature remains constant. The lowering of T_{trans} can be realized by the diffusion of a plasticizer, such as low molecular weight molecules, into the polymer.

Inducing the shape-memory effect by illumination with IR-light was demonstrated in a laser-activated polyurethane-based medical device [42,43]. Although the required energy is quite high, the principle could be extended to laser-activated shape-memory vascular stents and shape-memory polymer foams for aneurysm treatment [44,45] both of which require a light diffuser for the uniform application of light [46].

A better interconnection between the heat source and the shape-memory devices, resulting in a reduction of the required energy, can be reached by the incorporation of conductive fillers such as heat-conductive ceramics, carbon black, and carbon nanotubes (Figure 1.4) [47–49]. Besides a better heat transfer, the incorporation of particles also influences the mechanical properties: the incorporation of microscale particles results in increased stiffness and recoverable strain levels [50,51] and can be further enhanced by the incorporation of nanoscale particles [52,53]. But in the case of nanoscale particles, the molecular structure of the particles has to be considered to reach an enhanced photothermal effect and to improve the mechanical properties. Polyesterurethanes reinforced with carbon nanotubes, or carbon black, or silicon carbide of similar size display an inconsistent behavior in mechanical properties [54,55]. While carbon black–reinforced materials show limited shape recovery, i.e., around 25%–30%, in carbon nanotube–reinforced polymers, an R_{rs} of almost 100% can be observed [54]. This has been attributed to a synergism between the anisotropic carbon nanotubes and the crystallizing polyurethane-switching segments and can be confirmed by tensile tests.

When the filler is in the same size range of the soft segment lamellae, the soft segment crystallinity reduces drastically.

Energy conversion into heat can be reached by the incorporation of carbon nanotubes into shape-memory polyurethanes so that a certain level of conductivity can be reached [56–58]. Upon application of an electrical current, the sample temperature is increased, originating from the high ohmic resistance of the composite. The warming can be used to trigger the shape-memory effect. In case the conductivity of carbon nanotubes is not sufficiently high, conductivity can be increased by adding other conducting materials such as polypyrrole [57,58], or short carbon fibers [59], or nickel particle–forming chains [60,61]. While an addition of 2.5 wt% carbon nanotubes increases the modulus from 12 to 148 MPa, the addition of 5 wt% polypyrrole results in a modulus of 97 MPa. A composite with 2.5 wt% carbon nanotubes and 2.5 wt% polypyrrole displays a modulus of 112 MPa. The composites with nickel particles arranged in chains were prepared by curing the shape-memory resin containing small amounts of Ni powder in a weak magnetic field [60,61]. While the ohmic resistance is reduced so that a voltage of 20 V is sufficient for actuation, the storage modulus is increased and is therefore higher when compared to the pure shape-memory resin or the randomly distributed Ni.

Remote actuation of the thermally induced shape-memory effect in magnetic fields has been enabled by the incorporation of magnetic nanoparticles of iron(III)oxide cores in a silica matrix into shape-memory thermoplasts (Figure 1.2) [62]. When the nanocomposite is placed in an alternating magnetic field ($f = 258$ kHz, $H = 7–30$ kA m^{-1}), the sample temperature is increased by an inductive heating of the nanoparticles. Two different thermoplastic materials were selected as matrix. The first material was a biodegradable multiblock copolymer (PDC) with poly(*p*-dioxanone) as the hard segment and poly(ε-caprolactone) as the switching segment while the other was an aliphatic polyetherurethane (TFX) from methylene bis(*p*-cyclohexyl isocyanate), butanediol and polytetrahydrofuran. In contrast to the PDC, which has a crystalline-switching segment, the switching phase of TFX is amorphous. The usage of magnetite particles in the range of 9 μm allowed a reduction of the frequency and the magnetic field required for triggering

FIGURE 1.2
Magnetically induced shape-memory effect of a nanocomposite from magnetic nanoparticles and a polyetherurethane induced by an alternating magnetic field ($H = 30$ kA m^{-1}; $f = 258$ kHz) generated in an inductor. Upon stimulation, the nanocomposites transforms within 24 s from the rodlike temporary shape into the spiral-like permanent shape.

the shape-memory effect of composites with $T_{trans} = T_g$ from a thermoplastic polyurethane derived from diphenylmethane-4,4′-diisocyanate, adipic acid, ethylene oxide, propylene oxide, 1,4-butanediol and bisphenol A [64]. In these composites, an increase of the storage modulus with the increment of the magnetite volume fraction could be observed, which decreased more than 50 times at temperatures of $T_g + 20\,K$ [65]. Indirect magnetic actuation of thermosets has been reported by the incorporation of nickel zinc ferrite particles into a commercial ester-based thermoset polyurethane [66]. In contrast, composites from an epoxy thermoset, TEMBO-DP5.1 and Terefenol-D, a near single crystal metal alloy comprising of terbium, iron, and dysprosium-d of nominal composition $Tb_{0.3}Dy_{0.7}Fe_{1.92}$, enabled indirect heating through the use of the magnetoelectroelastic effect using radio frequency [67].

In commercially available polyurethanes and its composites with carbon nanotubes, the indirect actuation of the shape-memory effect by lowering the T_{trans} was enabled [68–71]. In all cases, the temporary shape was programmed by conventional methods for thermally induced shape-memory polymers. When immersed into water, moisture diffuses into the polymer sample, resulting in the shape recovery. Different processes have to be considered for this actuation. On the one hand, hydrogen bonds between the polyurea segments are broken [68–71]. On the other hand, water acts as a plasticizer [16]. Both effects result in the lowering of the T_g from 35°C to a temperature below ambient temperature, but the plasticizing effect is considered to be more important [72]. It could be shown, that the lowering of the glass transition temperature depends on the moisture uptake, which depends on the immersion time. As the maximum moisture uptake achieved after 240 h was around 4.5 wt%, this shape-memory polymer still needs to be considered as a polymer and not as a hydrogel. In a similar approach, the diffusion of DMF as small molecules in a cross-linked polymer network based on styrene based resin was reported. Here, the swelling was found to be significantly higher, i.e., 14.3 wt% after 120 min of immersion time. Therefore, a different mechanism, such as gel formation, most probably caused the shape-recovery.

Block copolymers derived from polyetherurethane polysilesquisiloxane possessing hydrophilic and hydrophobic domains display another possibility of realizing a water actuated movement [73]. Here, instead of lowering the T_{trans}, the domains providing the T_{trans} are dissolved. In the block copolymer, the domains formed by the hydrophobic polysilesquisiloxane provide the permanent shape, while the T_m of the domains of the hydrophilic polyether segment, in this case low molecular weight poly(ethylene glycol), is used to fix the temporary shape. When immersed in water, the poly(ethylene glycol) segment gets dissolved, resulting in the disappearance of the T_m and causing a movement of the material. A similar effect was enabled in a biopolymer-based system. Here chitosan was cross-linked with ethylene glycol diglycidyl ether, which was blended with polyethylene glycol. When immersed in water, the cross-linking polyethylene glycol segments get hydrated, resulting in a movement of the sample. When dried, the original shape is obtained [74]. Hence, this movement is mainly based on the swelling/de-swelling effect.

1.4 Light-Induced Dual-Shape Effect

Independency of the shape-memory effect from any temperature effect was realized by the development of light-induced shape-memory polymers [75]. Here, light of different wavelength ranges is used for the fixation of the temporary and the recovery of the permanent shape instead of increasing the sample's temperature. On the molecular level this has been realized by the incorporation of photosensitive molecular switches such as CA or cinnamyliden acetic acid (CAA). When irradiated with light of suitable wavelengths, the photosensitive functional groups form covalent cross-links with each other in a [2+2] cycloaddition reaction. Irradiation with light of different suitable wavelengths results in a cleavage of these bonds. In the course of programming, the polymer is deformed to ε_m and afterward irradiated with a UV-light of $\lambda > 260$ nm, so that new covalent bonds are created that fix the strained polymer chain segments in their uncoiled conformation. These newly formed covalent bonds can be cleaved and the permanent shape recovered, when the sample is irradiated with light of wavelengths $\lambda < 260$ nm.

The photosensitive groups have been incorporated in two alternative polymer network structures: a graft polymer network and an interpenetrating polymer network. In both cases, the permanent shape is determined by permanent netpoints, which cross-link the chain segments of the amorphous polymer networks.

The polymer networks with grafted CA molecules were synthesized by the copolymerization of n-butylacrylate, hydroxyethyl methacrylate, and ethylenegylcol-1-acrylate-2-CA with poly(propylene glycol)-dimethacrylate ($M_n = 560$ g mol^{-1}) as the cross-linker [75] while loading a permanent polymer network from butylacrylate and 3 wt% poly(propylene glycol)-dimethacrylate ($M_n = 1000$ g mol^{-1}) as the cross-linker with 20 wt% star-poly(ethylene glycol) end-capped with terminal CAA groups [75] results in a photosensitive interpenetrating network. The determination of shape-memory properties revealed an R_f of max. 52% and an R_r of max. 95% for the grafted polymer network in the fifth cycle, while an R_f of 33% and an R_r of 98% were measured in the third cycle for the interpenetrating network. From an application point of view, light-induced shape-memory polymers close the gap in the spectrum of available actively moving polymers where significant temperature changes are undesired as, e.g., in medical applications.

1.5 Triple-Shape Polymers

A recent development in actively moving materials is the development of materials having the capability to perform two subsequent shape changes [76,77]. Besides the triple-shape functionality, such materials allow by individual choice of the transition temperatures, two dual-shape effects, so that only one material needs to be designed, whose dual-shape capability can be easily adjusted to the application relevant requirements.

The triple-shape capability has been achieved on the molecular level by the integration of two switching segments into the polymer network that provide at least two segregated domains resulting in two transition temperatures: $T_{trans,A}$ and $T_{trans,B}$. While the original shape (C) is determined by the covalent cross-links during the polymer network formation, shapes (A) and (B) are fixed by additional physical cross-links created in a two-step thermomechanical programming process. The physical cross-links providing shape (B) are associated to the highest transition temperature $T_{trans,B}$, while the second-highest transition temperature, $T_{trans,A}$, determines shape (A).

In the course of programming, such a polymer network is heated to T_{high} where the material is in the elastic state and is deformed. When the material is cooled under external stress to T_{mid} ($T_{trans,A} < T_{mid} < T_{trans,B}$) physical cross-links related to $T_{trans,B}$ are formed. Shape B is obtained after the release of the external stress. Afterward, shape A is created by the subsequent deformation of shape (B) at T_{mid} and cooling to T_{low} under external stress and the subsequent release of the external stress. Reheating to T_{mid} recovers shape B and consequent heating to T_{high} results in shape C (Figure 1.3).

The generality of this approach has been shown by two independent polymer network architectures, which were prepared (Figure 1.4) by photo-induced copolymerization of a methacrylate-monomer and a poly(ε-caprolactone)dimethacrylate (PCLDMA) as the cross-linker. In a first system, called the MACL, cyclohexylmethacrylate (CHMA) and poly(ε-caprolactone) dimethacrylate (PCLDMA) were copolymerized so that the polymer network

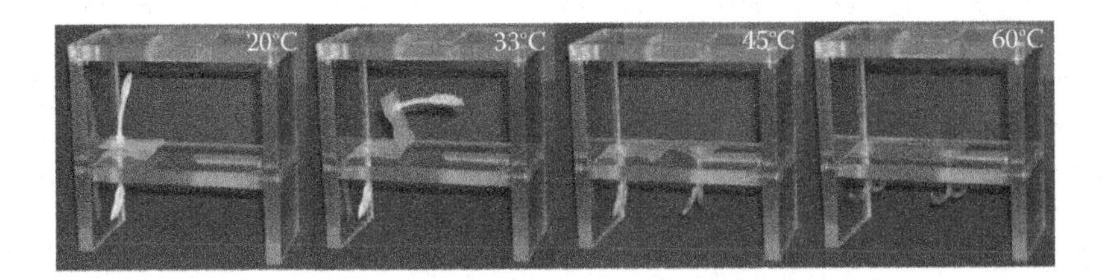

FIGURE 1.3
Series of photographs illustrating the triple-shape effect of a fastener device consisting of a plate with anchors as a demonstration object prepared from CL(50)EG. The picture series shows the recovery of shapes (B) and (C) by subsequent heating to 40°C and 60°C, beginning from shape (A), which was obtained as a result of the two-step programming process.

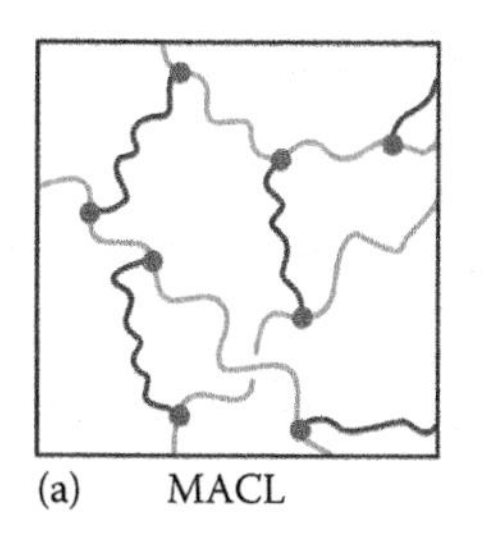

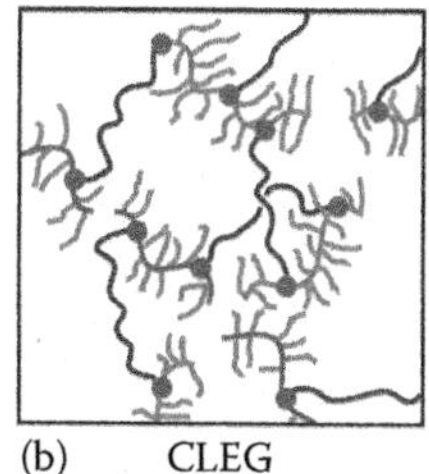

FIGURE 1.4
Polymer network architecture. (a) An MACL-network. (b) A CLEG-network. Light grey—PCHMA segments; black—PCL segments; grey—PEG side chains; dark grey circles—cross-links . (From Bellin, I. et al., *Proc. Natl. Acad. Sci. USA*, 103(48), 18043, 2006.)

structure is given by the PCLDMA and the polymerized CHMA segments. Here, both polymer chain segments contribute equally to the overall elasticity of the polymer network structure. In this polymer network, $T_{trans,A}$ is given by the T_g of the MACL and $T_{trans,B}$ by the T_m of the PCL segments. In the second polymer network architecture, provided by a system called the CLEG, poly(ε-caprolactone)dimethacrlyate (PCLDMA) was copolymerized with poly(ethyleneglycol)mono-methylethermethacrylate (PEGMA). In this polymer network architecture, the PCL-segments connecting two netpoints is what mainly determines the elasticity of the polymer network, while the poly(ethylene glycol)-segments (PEG-segments), which are introduced as side chains with one dangling end, do not contribute to the elasticity. In CLEG-networks, $T_{trans,B}$ and $T_{trans,A}$ are melting temperatures. A similar network architecture compared to the MACL network was achieved by the thermally induced post-polymerization of binary and ternary blends from ethylene-1-octene copolymers with varying degrees of branching [78]. Here, $T_{trans,A}$ and $T_{trans,B}$ are determined by the T_m of the different polyethylene crystal populations.

Besides the triple-shape capability, triple-shape materials can be used as dual-shape materials in which the T_{trans} can be varied according to the specific requirements. The dual shape experiments are performed either between T_{low} and T_{mid} using $T_{trans,A}$ or between T_{mid} and T_{high} using $T_{trans,B}$ [77]. Additionally, a combination of both switching phases is possible. Depending on the switching temperature in the CLEG networks, three mechanisms for dual-shape fixation can be differentiated (Figure 1.5). The capability of the PCL phase to act as the switching phase is determined by the temperature range between 40°C and 70°C (case I), while the temperature range between 0°C and 40°C (case II) allows the determination of the ability of the PEG phase to act as the switching phase. If the cyclic, thermomechanical experiments are conducted between 0°C and 70°C (case III) both crystallizable phases can support the fixation of a temporary shape. While in case I T_{sw} is in the range of $T_{m,PCL}$ values for T_{sw} are similar to the values determined for $T_{m,PEG}$ in case II. In case III, the crystallization of the PCL already fixes

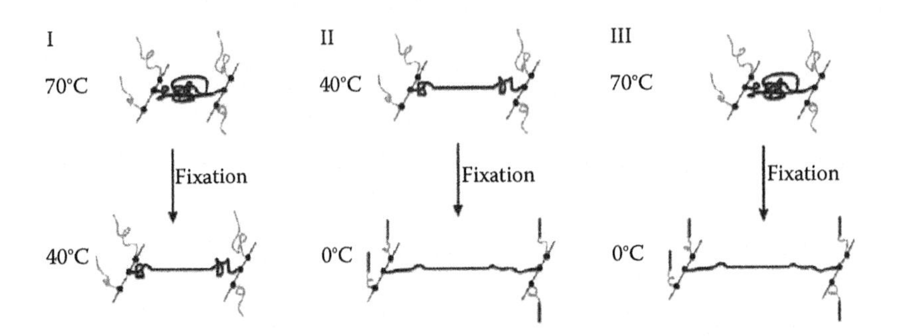

FIGURE 1.5
Molecular mechanism of three alternative ways of dual-shape programming for graft polymer networks. (a) Fixation of the temporary shape by the PCL phase (case I), (b) fixation of the temporary shape by the PEG phase (case II), (c) fixation of the temporary shape by the PCL and the PEG phases (case III). Straight black: crystalline PCL chain segments, coiled black: amorphous PCL chain segments, dark grey: crystalline PEG chain segments, grey: amorphous PEG chain segments, thin black lines: amorphous poly(methacrylate) chain segments. (From Bellin, I. et al., *J. Mater. Chem.*, 17, 2885, 2007. With permission.)

the deformation which leads to the orientation of the PCL chain segments. Therefore, T_{sw} is in the range of $T_{m,PCL}$ and no additional step in the recovery curve indicating a recovery triggered by melting of the PEG crystallites can be detected.

In a polymer network architecture in which both switching segments contribute to the overall elasticity such as in the MACL networks, the triple-shape capability is obtained by a one-step programming procedure [79]. Here, at $T_{high}=150°C$ both chain segments are flexible. At $T_{mid}=70°C$, the PCHMA chain segments are in the glassy state and upon further cooling to −10°C, the PCL chain segments form stiff flexible amorphous and rigid crystalline phases. A dual-shape programming with $T_{low}=-10°C$ and $T_{high}=150°C$ results in a triple-shape effect. The recovery curves after this one-step programming procedure are displayed in Figure 1.6 for $\varepsilon_m=30\%$, 50% and 100%. While for ε_m of up to 50%, around 54% of the temporary shape is fixed by the PCL segments. For larger deformations ($\varepsilon_m=100\%$) only 30% of the temporary shape is fixed by the PCL segments. This has been attributed to the polymer network architecture, where the PCL segments act as cross linkers and the PCHMA segments form the backbone and therefore can be uncoiled to a larger extent.

1.6 Outlook

There is rapid progress being made in the field of actively moving polymers [16,80]. While fundamental research focuses on new types or mechanisms

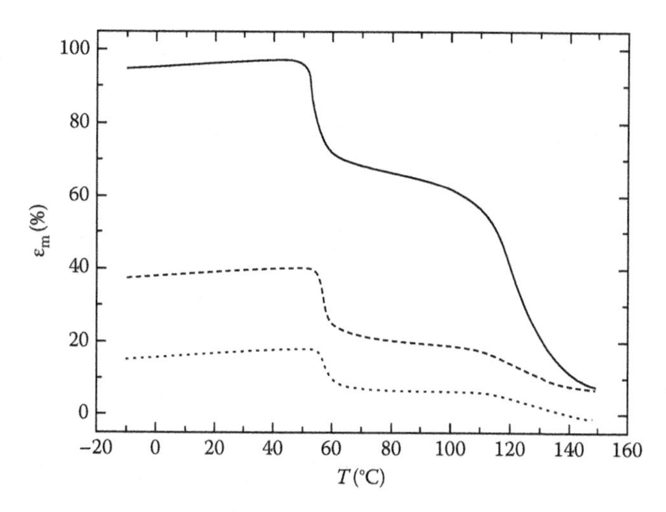

FIGURE 1.6
Recovery curves (second cycle) obtained from stress-controlled, thermomechanical experiments for PCL(50)CHM programmed in a one-step programming procedure with $T_{high} = 150°C$ and $T_{low} = -10°C$ and different mechanical deformations ε_m applied; solid line: $\varepsilon_m = 100\%$; dashed line: $\varepsilon_m = 50\%$; dotted line: $\varepsilon_m = 30\%$. (From Behl, M. et al., *Adv. Funct. Mater.*, 19, 102, 2009. With permission.)

of shape-memory effects, the technology platform of existing materials is moving toward some highly sophisticated applications. Nearly every area of everyday life bears application potential for such materials. Examples are switches, sensors, fabrics, intelligent packaging, or self-repairing autobodies. The implementation of stimuli such as alternating magnetic field or light will extend these application fields to noncontact operation. Here, the area of active medical devices, such as implants [81] or active prosthesis, is promising and potential applications have been demonstrated [42,43].

While thermally induced shape-memory polymers have meanwhile reached the mass market in the area of textiles [4], other promising applications for actively moving polymers are found mostly in niches, where their functionality is the key for enabling specific application. Such niches might be in aerospace or in the medical devices sector.

References

1. Gil, E. S. and Hudson, S. A. 2004. Stimuli-responsive polymers and their bioconjugates. *Progress in Polymer Science* 29:1173–1222.
2. Hu, J., Ding, X., Tao, X., and Yu, J. 2002. Influence of the surfactants on microstructure and the properties of wet coagulation polyurethane film. *Journal of Donghua University (English Edition)* 19:3–10.

3. Mondal, S. and Hu, J. L. 2006. Temperature stimulating shape memory polyurethane for smart clothing. *Indian Journal of Fibre & Textile Research* 31:66–71.
4. Hu, J. 2007. *Shape Memory Polymers and Textiles*, Woodhead Publishing Limited, Cambridge, U.K.
5. Charlesby, A.1960. *Polymer Physics*, Pergamon Press, Oxford, U.K.
6. Campbell, D., Lake, M. S., Scherbarth, M. R., Nelson, E., and Six, R. W. 2005. Elastic memory composite material: An enabling technology for future furable space structures. *Proceedings of the 46th AIAA/ASME/ASCE/AHS/ASC Structures, Structural Dynamics, and Materials Conference*, Austin, TX.
7. Hussein, H. and Harrison, D. 2004. Investigation into the use of engineering polymers as actuators to produce 'automatic disassembly' of electronic products. In *Design and Manufacture for Sustainable Development*, T. Bharma and B. Hon (Eds.), St Edmunds and London, U.K.: Professional Engineering Publishing Limited.
8. Wache, H. M., Tartakowska, D. J., Hentrich, A., and Wagner, M. H. 2003. Development of a polymer stent with shape memory effect as a drug delivery system. *Journal of Materials Science—Materials in Medicine* 14:109–112.
9. Lendlein, A. and Langer, R. 2002. Biodegradable, elastic shape-memory polymers for potential biomedical applications. *Science* 296:1673–1676.
10. Metcalfe, A., Desfaits, A. C., Salazkin, I., Yahia, L., Sokolowski, W. M., and Raymond, J. 2003. Cold hibernated elastic memory foams for endovascular interventions. *Biomaterials* 24:491–497.
11. Su, S. H. 2007. Mini review of the fully bioabsorbable polymeric stents. *Recent Patents on Engineering* 1:244–250.
12. Toensmeier, P. A. 2005. Compounders thwart counterfeiting with covert additive techniques. *Plastics Engineering* 61:10–15.
13. Behl, M. and Lendlein, A. 2007. Actively moving polymers. *Soft Matter* 3:58–67.
14. Gunes, I. S. and Jana, S. C. 2008. Shape memory polymers and their nanocomposites: A review of science and technology of new multifunctional materials. *Journal of Nanoscience and Nanotechnology* 8:1616–1637.
15. Lendlein, A. and Kelch, S. 2002. Shape-memory polymers. *Angewandte Chemie—International Edition* 41:2034–2057.
16. Behl, M. and Lendlein, A. 2007. Shape-memory polymers. *Materials Today* 10:20–28.
17. Kim, B. K., Lee, S. Y., and Xu, M. 1996. Polyurethanes having shape memory effects. *Polymer* 7:5781–5793.
18. Li, F. K., Hou, J. N., Zhu, W., Zhang, X., Xu, M., Luo, X. L., Ma, D. Z., and Kim, B. K. 1996. Crystallinity and morphology of segmented polyurethanes with different soft-segment length. *Journal of Applied Polymer Science* 62:631–638.
19. Ma, Z. L., Zhao, W. G., Liu, Y. F., and Shi, J. R. 1997. Intumescent polyurethane coatings with reduced flammability based on spirocyclic phosphate-containing polyols. *Journal of Applied Polymer Science* 63:1511–1514.
20. Liem, H. and Yeung, L. Y. 2007. Segment self-orientational behavior in shape memory polymer thin films probed by Raman spectroscopy. *Journal of Applied Polymer Science* 105:765–770.
21. Chen, S. J., Hu, J. L., Liu, Y. Q., Liem, H. M., Zhu, Y., and Liu, Y. J. 2007. Effect of SSL and HSC on morphology and properties of PHA based SMPU synthesized by bulk polymerization method. *Journal of Polymer Science Part B—Polymer Physics* 45:444–454.

22. Ji, F. L., Zhu, Y., Hu, J. L., Liu, Y., Yeung, L. Y., and Ye, G. D. 2006. Smart polymer fibers with shape memory effect. *Smart Materials and Structures* 15:1547–1554.
23. Chun, B. C., Cho, T. K., and Chung, Y. C. 2006. Enhanced mechanical and shape memory properties of polyurethane block copolymers chain-extended by ethylene diamine. *European Polymer Journal* 42:3367–3373.
24. Mondal, S. and Hu, J. L. 2007. Studies of shape memory property on thermoplastic segmented polyurethanes: Influence of PEG 3400. *Journal of Elastomers and Plastics* 39:81–91.
25. Zhu, Y., Hu, J., Yeung, K. W., Choi, K. F., Liu, Y. Q., and Liem, H. M. 2007. Effect of cationic group content on shape memory effect in segmented polyurethane cationomer. *Journal of Applied Polymer Science* 103:545–556.
26. Han, S. I., Gu, B. H., Nam, K. H., Im, S. J., Kim, S. C., and Im, S. S. 2007. Novel copolyester-based ionomer for a shape-memory biodegradable material. *Polymer* 48:1830–1834.
27. Weiss, R. A., Izzo, E., and Mandelbaum, S. 2008. New design of shape memory polymers: Mixtures of an elastomeric ionomer and low molar mass fatty acids and their salts. *Macromolecules* 41:2978–2980.
28. Xu, S. and Zhang, M. 2007. Synthesis and characterization of a novel polyurethane elastomer based on CO_2 copolymer. *Journal of Applied Polymer Science* 104:3818–3826.
29. Feng, Y. K., Behl, M., Kelch, S., and Lendlein, A. 2009. Biodegradable multiblock copolymers based on oligodepsipeptides with shape-memory properties. *Macromolecular Bioscience* 9:45–54.
30. Behl, M., Ridder, U., Feng, Y. K., Kelch, S., and Lendlein, A. 2009. Shape-memory capability of binary multiblock copolymer blends with hard and switching domains provided by different components. *Soft Matter* 5:676–684.
31. Liu, C. D., Chun, S. B., Mather, P. T., Zheng, L., Haley, E. H., and Coughlin, E. B. 2002. Chemically cross-linked polycyclooctene: Synthesis, characterization, and shape memory behavior. *Macromolecules* 35:9868–9874.
32. Kunzelman, J., Chung, T., Mather, P. T., and Weder, C. 2008. Shape memory polymers with built-in threshold temperature sensors. *Journal of Materials Chemistry* 18:1082–1086.
33. Zhang, D. W., Lan, X., Liu, Y., and Leng, J. 2007. Influence of cross-linking degree on shape memory effect of styrene copolymer. *Proceedings of SPIE—The International Society for Optical Engineering* 6526:65262W/1.
34. Kelch, S., Choi, N. Y., Wang, Z. G., and Lendlein, A. 2008. Amorphous, elastic AB copolymer networks from acrylates and poly[(L-lactide)-ran-glycolide] dimethacrylates. *Advanced Engineering Materials* 10:494–502.
35. Wornyo, E., Gall, K., Yang, F., and King, W. 2007. Nanoindentation of shape memory polymer networks. *Polymer* 48:3213–3225.
36. Kelch, S., Steuer, S., Schmidt, A. M., and Lendlein, A. 2007. Shape-memory polymer networks from oligo[(epsilon-hydroxycaproate)-co-glycolate]dimethacrylates and butyl acrylate with adjustable hydrolytic degradation rate. *Biomacromolecules* 8:1018–1027.
37. Lee, K. M., Knight, P. T., Chung, T., and Mather, P. T. 2008. Polycaprolactone-POSS chemical/physical double networks. *Macromolecules* 41:4730–4738.
38. Buckley, C. P., Prisacariu, C., and Caraculacu, A. 2007. Novel triol-crosslinked polyurethanes and their thermorheological characterization as shape-memory materials. *Polymer* 48:1388–1396.

39. Cao, Q. and Liu, P. 2006. Structure and mechanical properties of shape memory polyurethane based on hyperbranched polyesters. *Polymer Bulletin* 57:889–899.
40. Chang, Y. W., Mishra, J. K., Cheong, J. H., and Kim, D. K. 2007. Thermomechanical properties and shape memory effect of epoxidized natural rubber crosslinked by 3-amino-1,2,4-triazole. *Polymer International* 56:694–698.
41. Zhang, S. F., Feng, Y. K., Zhang, L., Sun, J. F., Xu, X. K., and Xu, Y. S. 2007. Novel interpenetrating networks with shape-memory properties. *Journal of Polymer Science Part A—Polymer Chemistry* 45:768–775.
42. Maitland, D. J., Metzger, M. F., Schumann, D., Lee, A., and Wilson, T. S. 2002. Photothermal properties of shape memory polymer micro-actuators for treating stroke. *Lasers in Surgery and Medicine* 30:1–11.
43. Small, W., Wilson, T. S., Benett, W. J., Loge, J. M., and Maitland, D. J. 2005. Laser-activated shape memory polymer intravascular thrombectomy device. *Optics Express* 13:8204–8213.
44. Baer, G. M., Small, W., Wilson, T. S., Benett, W. J., Matthews, D. L., Hartman, J., and Maitland, D. J. 2007. Fabrication and in vitro deployment of a laser-activated shape memory polymer vascular stent. *Biomedical Engineering Online* 6:43.
45. Maitland, D. J., Small, W., Ortega, J. M., Buckley, P. R., Rodriguez, J., Hartman, J., and Wilson, T. S. 2007. Prototype laser-activated shape memory polymer foam device for embolic treatment of aneurysms. *Journal of Biomedical Optics* 12:030504.
46. Small, W., Buckley, P. R., Wilson, T. S., Loge, J. M., Maitland, K. D., and Maitland, D. J. 2008. Fabrication and characterization of cylindrical light diffusers comprised of shape memory polymer. *Journal of Biomedical Optics* 13:024018.
47. Liu, C. D. and Mather, P. T. 2003. Thermomechanical characterization of blends of poly(vinyl acetate) with semicrystalline polymers for shape memory applications, *SPE ANTEC 2003 Proceedings*, Nashville, TN, 61(2):1962–1966.
48. Biercuk, M. J., Llaguno, M. C., Radosavljevic, M., Hyun, J. K., Johnson, A. T., and Fischer, J. E. 2002. Carbon nanotube composites for thermal management. *Applied Physics Letters* 80:2767–2769.
49. Li, F. K., Qi, L. Y., Yang, J. P., Xu, M., Luo, X. L., and Ma, D. Z. 2000. Polyurethane/conducting carbon black composites: Structure, electric conductivity, strain recovery behavior; and their relationships. *Journal of Applied Polymer Science* 75:68–77.
50. Liang, C., Rogers, C. A., and Malafeew, E. 1997. Investigation of shape memory polymers and their hybrid composites. *Journal of Intelligent Material Systems and Structures* 8:380–386.
51. Gall, K., Mikulas, M., Munshi, N. A., Beavers, F., and Tupper, M. 2000. Carbon fiber reinforced shape memory polymer composites. *Journal of Intelligent Material Systems and Structures* 11:877–886.
52. Ash, B. J., Stone, R., Rogers, D. F., Schadler, L. S., Siegel, R. W., Benicewicz, B. C., and Apple, T. 2001. Investigation into the thermal mechanical behavior of PMMA/alumina nanocomposites. In *Filled and Nanocomposite Polymer Materials*, Materials Research Society, Boston, KK2.10.1.
53. Bhattacharya, S. K. and Tummala, R. R. 2002. Epoxy nanocomposite capacitors for application as MCM-L compatible integral passives. *Journal of Electronic Packaging* 124:1–6.
54. Koerner, H., Price, G., Pearce, N. A., Alexander, M., and Vaia, R. A. 2004. Remotely actuated polymer nanocomposites—Stress-recovery of carbon-nanotube-filled thermoplastic elastomers. *Nature Materials* 3:115–120.

55. Gunes, I. S., Cao, F., and Jana, S. C. 2008. Evaluation of nanoparticulate fillers for development of shape memory polyurethane nanocomposites. *Polymer* 49:2223–2234.
56. Cho, J. W., Kim, J. W., Jung, Y. C., and Goo, N. S. 2005. Electroactive shape-memory polyurethane composites incorporating carbon nanotubes. *Macromolecular Rapid Communications* 26:412–416.
57. Sahoo, N. G., Jung, Y. C., and Cho, J. W. 2007. Electroactive shape memory effect of polyurethane composites filled with carbon nanotubes and conducting polymer. *Materials and Manufacturing Processes* 22:419–423.
58. Sahoo, N. G., Jung, Y. C., Yoo, H. J., and Cho, J. W. 2007. Influence of carbon nanotubes and polypyrrole on the thermal, mechanical and electroactive shape-memory properties of polyurethane nanocomposites. *Composites Science and Technology* 67:1920–1929.
59. Leng, J. S., Lv, H. B., Liu, Y. J., and Du, S. Y. 2007. Electroactivate shape-memory polymer filled with nanocarbon particles and short carbon fibers. *Applied Physics Letters* 91:144105.
60. Leng, J. S., Lan, X., Liu, Y. J., Du, S. Y., Huang, W. M., Liu, N., Phee, S. J., and Yuan, Q. 2008. Electrical conductivity of thermoresponsive shape-memory polymer with embedded micron sized Ni powder chains. *Applied Physics Letters* 92:014104.
61. Leng, J. S., Huang, W. M., Lan, X., Liu, Y. J., and Du, S. Y. 2008. Significantly reducing electrical resistivity by forming conductive Ni chains in a polyurethane shape-memory polymer/carbon-black composite. *Applied Physics Letters* 92:206101.
62. Mohr, R., Kratz, K., Weigel, T., Lucka-Gabor, M., Moneke, M., and Lendlein, A. 2006. Initiation of shape-memory effect by inductive heating of magnetic nanoparticles in thermoplastic polymers. *Proceedings of the National Academy of Sciences of the United States of America* 103:3540–3545.
63. Behl, M., Zotzmann, J., and Lendlein, A. Shape-memory polymers and shape-changing polymers, In *Advances in Polymer Science*, Eds. Lendlein, A., Springer-Verlag.
64. Razzaq, M. Y., Anhalt, M., Frormann, L., and Weidenfeller, B. 2007. Thermal, electrical and magnetic studies of magnetite filled polyurethane shape memory polymers. *Materials Science and Engineering A—Structural Materials Properties Microstructure and Processing* 444:227–235.
65. Razzaq, M. Y., Anhalt, M., Frormann, L., and Weidenfeller, B. 2007. Mechanical spectroscopy of magnetite filled polyurethane shape memory polymers. *Materials Science and Engineering A—Structural Materials Properties Microstructure and Processing* 471:57–62.
66. Buckley, P. R., McKinley, G. H., Wilson, T. S., Small, W., Benett, W. J., Bearinger, J. P., McElfresh, M. W., and Maitland, D. J. 2006. Inductively heated shape memory polymer for the magnetic actuation of medical devices. *IEEE Transactions on Biomedical Engineering* 53:2075–2083.
67. Hazelton, C. S., Arzberger, S. C., Lake, M. S., and Munshi, N. A. 2007. RF actuation of a thermoset shape memory polymer with embedded magnetoelectroelastic particles. *Journal of Advanced Materials* 39:35–39.
68. Yang, B., Huang, W. M., Li, C., Lee, C. M., and Li, L. 2004. On the effects of moisture in a polyurethane shape memory polymer. *Smart Materials and Structures* 13:191–195.

69. Huang, W. M., Yang, B., An, L., Li, C., and Chan, Y. S. 2005. Water-driven programmable polyurethane shape memory polymer: Demonstration and mechanism. *Applied Physics Letters* 86:114105.
70. Yang, B., Huang, W. M., Li, C., and Li, L. 2006. Effects of moisture on the thermomechanical properties of a polyurethane shape memory polymer. *Polymer* 47:1348–1356.
71. Yang, B., Huang, W. M., Li, C., Li, L., and Chor, J. H. 2005. Qualitative separation of the effects of carbon nano-powder and moisture on the glass transition temperature of polyurethane shape memory polymer. *Scripta Materialia* 53:105–107.
72. Leng, J. S., Lv, H. B., Liu, Y. J., and Du, S. Y. 2008. Comment on "water-driven programmable polyurethane shape memory polymer: Demonstration and mechanism" [*Applied Physics Letters* 86, 114105, (2005)]. *Applied Physics Letters* 92:206105.
73. Jung, Y. C., So, H. H., and Cho, J. W. 2006. Water-responsive shape memory polyurethane block copolymer modified with polyhedral oligomeric silsesquioxane. *Journal of Macromolecular Science Part B—Physics* 45:453–461.
74. Chen, M. C., Tsai, H. W., Chang, Y., Lai, W. Y., Mi, F. L., Liu, C. T., Wong, H. S., and Sung, H. W. 2007. Rapidly self-expandable polymeric stents with a shape-memory property. *Biomacromolecules* 8:2774–2780.
75. Lendlein, A., Jiang, H. Y., Jünger, O., and Langer, R. 2005. Light-induced shape-memory polymers. *Nature* 434:879–882.
76. Bellin, I., Kelch, S., Langer, R., and Lendlein, A. 2006. Polymeric triple-shape materials. *Proceedings of the National Academy of Sciences of the United States of America* 103:18043–18047.
77. Bellin, I., Kelch, S., and Lendlein, A. 2007. Dual-shape properties of triple-shape polymer networks with crystallizable network segments and grafted side chains. *Journal of Materials Chemistry* 17:2885–2891.
78. Kolesov, I. S. and Radusch, H. J. 2008. Multiple shape-memory behavior and thermal-mechanical properties of peroxide cross-linked blends of linear and short-chain branched polyethylenes. *Express Polymer Letters* 2:461–473.
79. Behl, M., Bellin, I., Kelch, S., and Lendlein, A. 2009. One-step process for creating triple-shape capability of AB polymer networks. *Advanced Functional Materials* 19:102–108.
80. Liu, C., Qin, H., and Mather, P. T. 2007. Review of progress in shape-memory polymers. *Journal of Materials Chemistry* 17:1543–1558.
81. El Feninat, F., Laroche, G., Fiset, M., and Mantovani, D. 2002. Shape memory materials for biomedical applications. *Advanced Engineering Materials* 4:91–104.

2

The Structural Variety of Shape-Memory Polymers

Hong-Yan Jiang
Mnemoscience GmbH, Übach-Palenberg, Germany

Annette M. Schmidt
Institut für Organische Chemie und Makromolekulare Chemie, Heinrich-Heine-Universität Düsseldorf, Düsseldorf, Germany

CONTENTS

The basic principle of the shape-memory effect in polymers is based on the physical concept of a frozen-in structural order rather than on a specific chemical interaction. Therefore, the ability to apparently "recall" a primary shape after a significant deformation that is entropically trapped to the material is reported to date for a number of different polymer systems and architectural structures. We present the structural requirements and give an overview of the variety of chemical compositions of shape-memory polymers (SMPs).

2.1 Structural Requirements for SMP

This chapter provides readers with a detailed description of the general structural features, the underlying principles of formation, and the specific chemical structures of shape-memory polymers (SMPs). Starting from the structural requirements in the molecular architecture and the material's microstructure, we give an overview of the various chemical structures that fulfill these requirements in the main part of this chapter.

SMP-based materials generally possess a three-dimensional structure that is determined by a network-like architecture. This network can be described by its cross-linking net points, and the polymer segments connecting them entropically to a given macroscopic shape.[1] The other basic requirement for an SMP is the ability to form strong reversible interactions (secondary cross-links) between the polymer segments, so that an external deformation of the macroscopic shape can be fixed ("frozen in"). The shape-memory effect is then achieved by applying a suitable trigger to detach the secondary cross-links and to release the inner stress from the material. The final shape is again determined by the network structure of the primary cross-linking points and the entropic state of the segments between them.

1. *Three-dimensional material architecture and primary cross-links*

   First of all, a shape-memory material obviously needs to show a distinct shape, meaning that the material has to be solid and show dimensional stability. In polymeric materials, such a structure is obtained by linking a set of polymeric chains strongly to each other, so that a three-dimensional network structure is obtained. These primary interchain cross-links may either be covalent bondage (chemical networks, thermosets), or physical and reversible in nature, like in thermoplastic polymeric materials that can be melt- and solution-processed (Figure 2.1).[2] In the latter materials, strong interchain interactions and chain cross-linking is achieved by a phase separation, followed by the formation of crystalline or glassy phases that

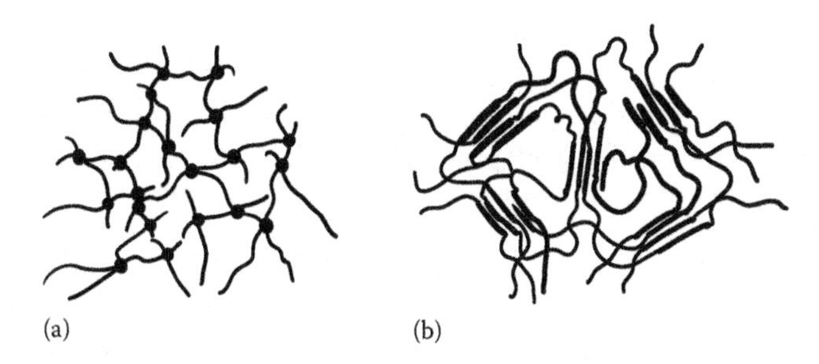

FIGURE 2.1
General structure of chemically and physically cross-linked polymeric elastomers: (a) covalent (chemical) network; (b) physical network from microphase-separation.

interconnect the individual polymer chains to form the network. More information on these types of SMPs can be found in Section 2.3. On the other hand, covalent networks characterized by chemical bonds between polymer chains are irreversible in nature, and can be obtained by chemical cross-linking during synthesis or processing. These materials are described in more detail in Section 2.2.

In either case, these cross-links determine the primary shape of the material, to which it will return after a deformation. The requirement of a cross-linked structure, together with easy elastic deformability, explains the fact that SMPs are in general more or less typical elastomers, with an additional option introduced via the polymeric segments between the primary cross-links.

2. *Polymer segments and secondary cross-links*

The second main parameter determining the actual shape of an SMP is the state of order possessed by the polymeric chains between the primary cross-links. At a fully amorphous state of the polymer chains, and above the glass-transition temperature, the entropically favored state of a polymer chain between two network points is a random coil. Thus, without the impact of an external force, the cross-linking points are pulled toward each other by the coiled chains, resulting in a rather compact state with a low degree of order. Upon deformation, the segments are stretched, resulting in an orientation of the polymer chains, and a dislocation of the network points. The fixation of this additional order in the network segments is the main principle underlying the shape-memory effect, and it can occur by effectively reducing the dynamics and flexibility of the segments while the network is being deformed. Usually, a thermal transition of the polymer segments (or part of them) is used for this "freezing in" of orientational order. This thermal transition is the entropic energy that can be released and used for shape recovery when the constraint of the secondary cross-links, like glassy or crystalline domains, is released.[3]

In this sense, the polymer segments in shape-memory networks can fulfill one or more of the different structural tasks:

The soft segments are flexible in nature, and highly contribute to the elastic mechanical properties of the SMPs. They are characterized by their amorphous structure, and low glass-transition temperatures.

In contrast, the *switching segments* possess a phase transition of a kind that is related to a significant mobility change within the polymer chains, e.g., a melting point or a glass transition, in the temperature range of interest. Both phase transitions involve the formation of rigid physical cross-links, either from glassy or crystalline domains that can be used to fix the secondary shape. Although the most-used transitions are thermal in nature, implying that the trigger for the shape-memory effect is basically temperature, the formation and release of the temporary cross-links can also occur by other means, e.g., by UV light[4] or electromagnetic irradiation.[5] More information on SMPs with stimuli other than temperature can be found in Chapters 5 and 6.

Hard segments, in contrast, are either glassy or crystalline over the whole temperature range of use. Thus, as described earlier, these physical cross-links with a high-transition temperature may act as the determiners of the primary shape.

The synthesis, processing, and the particular features of SMPs, starting from covalently cross-linked elastomers, and sorted by their parent material are introduced in the following sections.

2.2 SMPs with a Covalently Cross-Linked Primary Structure

Covalently cross-linked SMPs possess a three-dimensional, chemically interconnected structure that determines the material's primary macroscopic shape, while a thermal transition of the polymer segments is used as the shape-memory switch. Among the simplest types of SMPs is a cross-linked glassy polymer featuring a sharp T_g at the temperature of interest, and showing rubbery elasticity at temperatures above T_g derived from covalent cross-links. Alternatively, the segments may be semicrystalline in nature with a melting transition responsible for the shape-memory effect.

Polymer network-based SMPs can be synthesized by either adding a multifunctional cross-linker during the polymerization, or by a subsequent cross-linking of a linear or branched polymer. The networks can be based on many different polymer backbones, like polyolefines, polyurethanes (PUs), polyacrylates, and polystyrene.

In comparison with physically cross-linked SMPs, the polymer network-based systems show practically no creep, thus any irreversible deformation during the programming or the release progress is greatly diminished. As a consequence, polymer network-based SMPs show attractive characteristics, including excellent shape-recovery ratios, and a tunable work capacity during shape recovery, governed by a rubbery modulus that can be adjusted by the cross-link density.

2.2.1 SMP Networks Based on Polyolefines

An early example of a simple SMP[6–8] that is still in commercial use is heat-shrinkable foils and tubes for different industrial applications, which appeared on the market in the 1960s. They consist of covalently cross-linked and stretched low-density polyethylene (LDPE) and other semicrystalline polyolefines. The heat-shrink effect, which is employed for the tight enveloping of cables or packaging of goods, is based on the same principles as those of SMPs. Early works on the heat-shrinking properties of cross-linked polyolefines were mainly focused on the preparation of these new materials for various end-applications. Not much fundamental inquiry was involved, and little attempt was made to tune the thermomechanical properties. This changed not earlier than the 1980s, when a boost of scientific publications arose, resulting in the development of a great variety of chemical and architectural structures suitable for shape-memory actuation.

In general, heat-shrinking materials are based on a semicrystalline polyolefine with a melting transition between ambient and processing temperatures that can be exploited for the heat-shrinking effect. The basic principle for heat-shrinking materials is the same as discussed for shape-memory materials, with the difference being the deformation, which is mainly limited to expansion/shrinking processes in heat-shrinking applications.

The polyolefines are cross-linked to a three-dimensional structure after they are processed to the desired shape. Four different processes of cross-linking are known using either peroxides, silanes, azọcompounds, or γ-irradiation.[9] The cross-links determine the primary macroscopic shape of the material, and also improve the heat-stability and toughness. The fixing of the secondary stretched shape occurs by crystallization of the semicrystalline polyolefine.

The technological process for a heat-shrinkable specimen preparation after cross-linking the preform consists of exerting elongation stresses to the preform (through cross-sectional expansion), and then cooling it while still under stress. In this way, the stresses are "frozen" in the material upon cooling down. During product application, these stresses are released, thus causing a shrinking of the product.[7,10,11]

While the melting transition of the cross-linked PE segments is responsible for the shape transition, the crystallinity of the materials is found to be highly affected by cross-linking. In addition, the temperature of programming (stretching) is found to be in direct correlation to the actual transition temperature.[9]

Recently, a highly adaptable system was developed based on chemically cross-linked polycyclooctene.[12] The networks are obtained from a cyclooctene monomer with a ruthenium-based catalyst to allow control over the *cis/trans* double-bond ratio of the product. Curing of the obtained linear polymer is performed by the use of dicumyl peroxide. A fast and complete recovery process within 0.7 s was observed upon heating the sample in a water bath at 70°C.

2.2.2 SMP Networks Based on Polyurethanes

Because PUs are easy to design to fit a wide range of needs and properties in the broad parameter space of components and additives, the PU naturally becomes an important category of the SMP.[11,12] While the preparation and the performance of thermoplastic PU SMPs are described in detail in Section 2.3.1, we focus here mainly on additional features employed in chemically cross-linked SM-PUs, as the components involved in the formation are basically the same as for thermoplastic PUs, except for using a cross-linker like glycerine, trimethylol propane, or excess isocyanate during the PU synthesis.

The effects of cross-links in these materials are principally similar to the ones described for polyolefines: the thermal stability is improved, and the creep can considerably be reduced. The cross-links hereby not only act as an enhancer of the shape recovery ratio, but can also fully take over the role of the hard phase as the primary cross-links in the PU material. In case of PUs with crystallizable switching segments, the presence of cross-links reduces the crystallization of the switching segment. In addition, the materials are reported to show enhanced mechanical properties, and a sharper shape-transition temperature.[13]

These high definition properties make these materials especially attractive to the application in the medical area, where reliability and precise control is of special importance. This may be the reason for the observation that many of the reports in this section make use of biocompatible aliphatic polyether and polyester segments, and also of aliphatic isocyanates.

The wide range of segments available for the synthesis of the PU and the cross-linked PU enables the design of water-swellable PU networks by the introduction of poly(ethylene glycol) (PEG) segments.[14] The water uptake of these materials caused by the presence of hydrophilic PEG segments is assumed to be of positive impact on the biocompatibility of the materials. A microphase-separated hybrid network of this kind is realized by employing the strategy of interpenetrating networks (IPN).[15] Poly(tetrahydrofurane)-based PUs have been glycerol-cross-linked[16] and PEG-cross-linked,[17] and the results have been compared to the linear SM-PUs of a similar composition in order to gain insight into the impact of the polymer architecture on the materials properties (see Figure 2.2). Here, the chemical cross-links are shown to

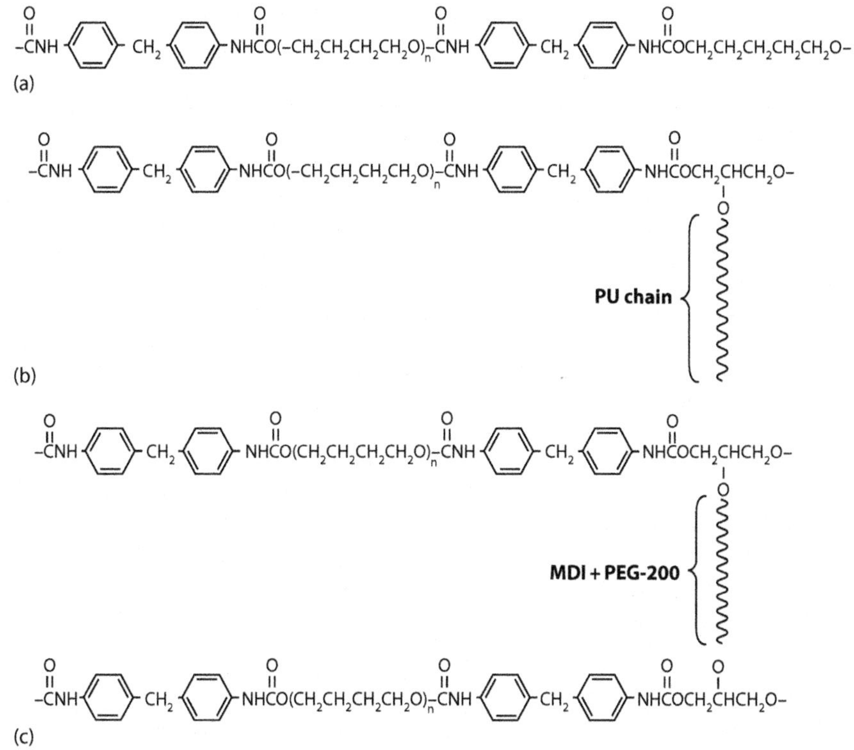

FIGURE 2.2
Structures of PU-SMPs based on PTHF, MDI, PEG, and propanetriol with (a) linear, (b) grafted, and (c) cross-linked structure. (Reproduced from Chun, B.C. et al., *J. Mater. Sci.*, 42, 9045, 2007. With permission.)

disrupt the ordered structure of the hard phase, as indicated by the x-ray diffraction (XRD). The flexible PEG segments in the latter case are reported to contribute to the elastic properties of the networks.

SMPs based on poly(carbonate urethanes) have been reported, and suggested for biostable implants.[18] An alternative based on biodegradable polyester segments is reported for poly(ester urethanes) from macrodiol- and -triols based on polylactide and poly(ε-caprolactone) (PCL), with mechanical properties tunable by the cross-linking density.[19–21] By using exclusively low molecular aliphatic amines, isocyanates, and polyols, optically clear polymer networks have been obtained with a tunable glass-transition temperature.[22]

2.2.3 SMP Networks Based on Poly((Meth)Acrylates)

The polymer that was first reported in patent literature to show a shape-memory effect was a polymethacrylate network, filed by the Vernon-Benshoff Company, Albany, New York in 1941[23] for its application as a denture material.

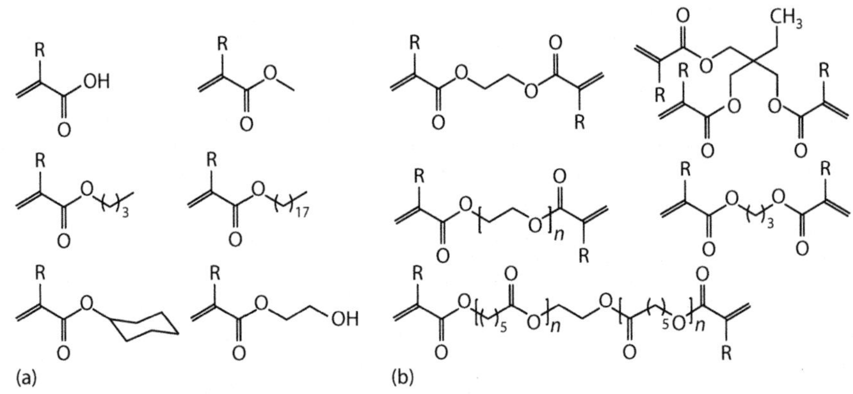

(a) (b)

FIGURE 2.3
(a) Some monomers and (b) cross-linkers employed in acrylic SMP networks.

Today, polyacrylates and polymethacrylates are of interest as shape-memory materials due to their ease of preparation, their simple phase behavior, and the possibility of easily tuning their key characteristics, like the thermal transition temperature and elastic modulus. In addition, some of these materials have excellent optical properties, like transparency and stainability.

(Meth)acrylates with a wide variety of ester functionalities are commercially available, and many of them can be copolymerized via radical polymerization. Cross-linking agents with a choice of functionality and architecture (see Figure 2.3) are employed. The synthesis is well-developed, and easy to perform, by free-radical polymerization in bulk or in solution.

In SMP networks, a handful of acrylic monomers are predominantly used, and these can be divided into three classes: (1) network segments, (2) a switching segment, and (3) a cross-linker.

As *network segments*, atactic polyacrylates and polymethacrylates are generally of a amorphous nature, and therefore serve as either glassy or rubbery segments. The glass-transition temperature of the respective homopolymers mainly depends on the nature of the side chain. For the aliphatic esters of (meth)acrylic acid the range is from 110°C for poly(methylmethacrylate) down to −55°C for poly(butylacrylate). For longer side chains, there is an increasing crystallization ability of the aliphatic side chains in the polymers, resulting in a melt transition that can also be used for the shape-memory effect.

As a *switching segment* in SMP networks, an add-on feature is the possibility of tuning the T_g or the T_m of the poly((meth)acrylates) by random copolymerization, to obtain the same single-phase materials with a single thermal transition. Such an example is a series of SMP networks based on poly(methylmethacrylate-*co*-butylmethacrylate) using tetra(ethylene glycol)dimethacrylate as the cross-linker.[24] Here, the comonomer composition (MMA/BMA) directly determines the glass-transition temperature, enabling the easy tuning of the shape-transition temperature (Figure 2.4).

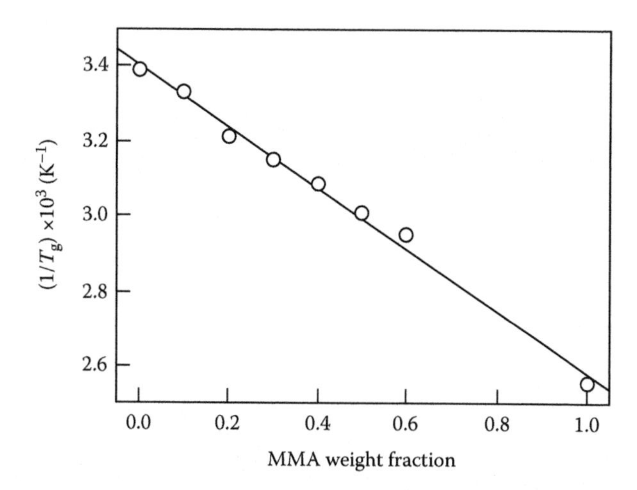

FIGURE 2.4
Dependence of T_g on copolymer composition expressed as T_g^{-1} vs. MMA weight fraction. Straight line is the Fox equation prediction. (Reproduced from Lui, C. and Mather, P.T., *J. Appl. Med. Polym.*, 6, 47, 2002. With permission.)

Apart from being built up by the main chain of poly((meth)acrylates), the switching segment can also be introduced as side chains or as long-chain cross-linkers by the use of (meth)acrylic ester macromonomers (see Figure 2.3). Such macromonomers based on oligo(ε-caprolactone),[25–27] PEG, and poly(lactide-*co*-glycolide)[28] were reported. In addition, tailored biodegradability can be introduced via these macromonomers.[26]

The low glass transition segments behave as rubbery, *softening domains* in the materials, preventing a brittle response of the materials at temperatures below the thermal transition of the switching segment.

As *cross-linkers*, multifunctional (meth)acrylates and -acrylamides are employed. Common small molecule cross-linkers include (meth)acrylates of polyols, like ethylene glycol, butylene glycol, glycerol, or trimethylolpropane (Figure 2.3). Cross-linkers can also be macromonomeric di(meth)acrylates, implying the possibility of introducing nonacrylic based polymer segments by using (oligoester)- or (oligoether)di(meth)acrylates.

As can be seen from the previous text, architectures in acrylic-based shape-memory networks range from simple, single phase materials cross-linked *in situ* by small molecules, to more sophisticated structures. For example, several acrylic networks with shape-memory properties based on side-chain physical cross-linking have been reported. By copolymerizing (meth)acrylates with aliphatic long side-chain comonomers, like stearyl acrylate[29] or octadecylvinylether,[30] polymer networks are obtained whose crystallizable side chains form the physical cross-link and fix the temporary shape (see Figure 2.5). In this example, the transition temperature can also be tuned by the comonomer ratio.

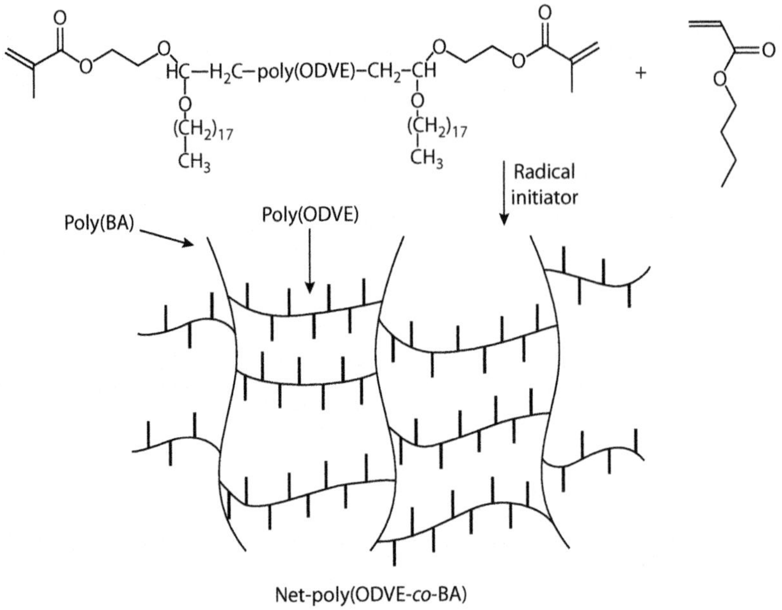

Radical initiator

Poly(BA)

Poly(ODVE)

Net-poly(ODVE-*co*-BA)

FIGURE 2.5
Polymer network structure obtained by free radical copolymerization of an octadecylvinylether-based bismacromonomer with BA. (Reproduced from Reintjens, W. et al., *Macromol. Rapid Commun.*, 20, 251, 1999. With permission.)

Next to side chain crystallization, the temporary cross-links can be formed by hydrogen bonding between accordingly designed side groups, as observed in SMP networks based on soft poly(butylacrylate), and carrying ureidopyrimidone side chains.[31] The cross-links can be reversed by heating the material up to a temperature above 60°C.

If a difunctional macromonomer is used in the copolymerization with (meth)acrylates, the macromonomer acts as the cross-linking segment, and thus an AB network structure is obtained.[27,28] In this way, PCL segments that are capable of forming a crystalline structure can be introduced to the network, to fix the secondary shape at a low temperature. Accordingly, the melting point of PCL segments controls the shape-recovery temperature, while the amorphous polybutylacrylate main chain with a low glass transition temperature ($T_g = -55°C$) results in the softening effect. The materials show excellent shape recovery and fixity ratios of more than 98%, while the mechanical properties can be widely adjusted from being soft to being flexible by varying the ratio of soft and switching segment (see Figure 2.6).[27]

By the combination of two immiscible segments with two distinct transition temperatures, SMPs are obtained that can hold two distinct shapes in memory.[32] A cross-linked copolymer formed from cyclohexyl methacrylate and PCL-dimethacrylate shows a glass transition at around 35°C originating

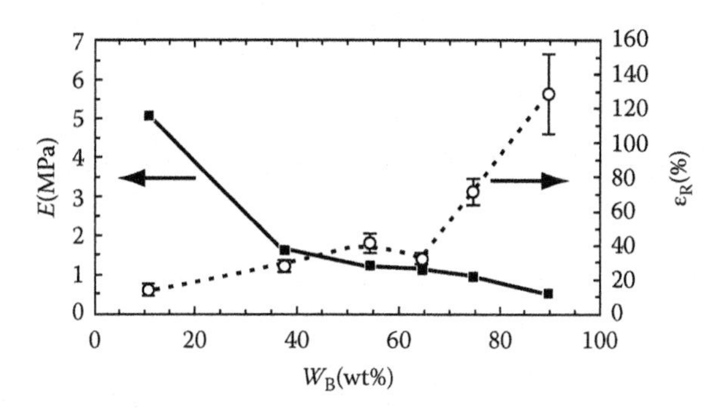

FIGURE 2.6
Tensile properties of copolymersates of PCLDMA2000 and butylacrylate as a function of comonomer content at room temperature. Filled squares: Young's modulus *E*, open circles: elongation at break ε_R. (Reproduced from Lendlein, A. et al., *PNAS*, 98, 842, 2001.)

from cyclohexyl methacrylate segments, and a melting temperature at around 50°C from PCL segments. In this case, the covalent polymer network is responsible for the memory of the permanent shape and the two distinct phases (amorphous and crystalline) determine the two different temporary shapes.

If such macrodi(meth)acrylates are used without the addition of a small comonomer, the resulting network structure will mainly contain the segments of the macromonomer main chain. SMP networks have been described that are based on the photoinduced homopolymerization of oligo(ε-caprolactone)dimethacrylates, resulting in rather low molecular oligoacrylates acting as polyfunctional, point-like junctions, and oligoester chains as the linking segments.[25] As a result, the melting temperature in these acrylic based networks is highly affected by the cross-linking density that is mainly determined by the length of the used macromonomer. An ambiguous shape-memory behavior is found in these networks, when macrodimethacrylates with a segment length above 3000 g mol^{-1} are employed.

For the formation of IPN and semi-IPN architectures, acrylic based polymers are mainly combined with PEG linear chains. By using poly(acrylic acid) main chains as the primary network, a complex formation is observed between the PEG and the PAA that can be thermally released to result in the recovery of the primary shape.[33] Alternatively, the melting transition of the incorporated PEG chains can be used for fixing an additional temporary shape. As a consequence, these materials are capable of holding two shapes in memory, similar to the AB-network structure from cyclohexyl methacrylate and PCL dimethacrylate described earlier.[34] Light microscopy confirms that a fibril-like structure is obtained in stretched samples of this material, while the original and recovered samples show no orientation (see Figure 2.7).

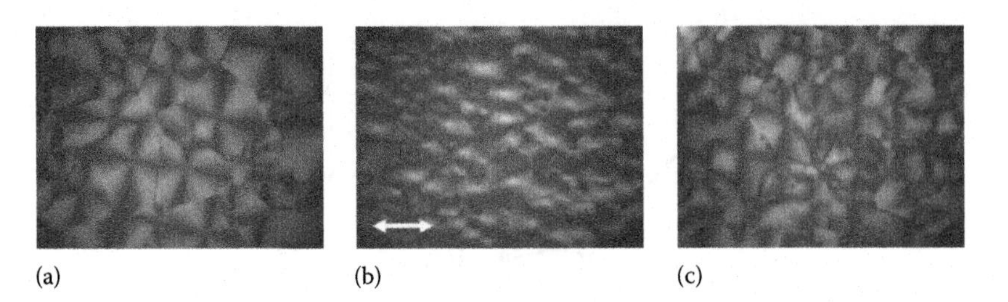

FIGURE 2.7
Structural changes in the surface texture of a PMMA–PEG semi-IPN at room temperature, using polarizing optical microscopy. The arrows indicate the stretch direction. (a) Original, (b) stretched with a draw ratio of 1.6; (c) recovered with a recovery ratio of 91%. (Reproduced from Liu, G. et al., *Macromol. Rapid Commun.*, 26, 649, 2005. With permission.)

Finally, acrylic based networks are reported to show light induced reversible cross-linking in polybutyl acrylate (PBA) or P(BA-*co*-HEA) networks containing cinnamoyl groups.[4] Here, a (2+2) cycloaddition between two pending cinnamoyl groups can be activated by UV irradiation at $\lambda > 260\,\text{nm}$. If the soft material is deformed at room temperature and irradiated, the freshly formed covalent cross-links between cinnamoyl moieties from two different polymer chains are able to fix up to 52% of the deformation. The primary shape in some materials can be almost completely (>98%) released upon the exposure of the material toward UV $\lambda < 260\,\text{nm}$ irradiation. The ability to show a shape fixation and recovery by the interaction with light of a distinct wave length is of great application potential in remote-controlled actuators that can be operated at ambient temperature. More shape-memory phenomena based on light irradiation and other stimuli are discussed in Chapters 5 and 6.

2.2.4 SMP Networks Based on Polystyrene

The variety of ways to polymerize styrene, and the wide availability of possible comonomers, make styrene-based materials highly interesting for shape-memory applications. Atactic polystyrene has a glass-transition temperature at around 100°C and is therefore rigid and brittle at ambient temperature. Copolymerization is the key to obtaining shape-memory materials with enhanced flexibility and a transition temperature in a useful range. Styrene can be polymerized by anionic, cationic, and free and controlled radical polymerization methods, so that by the proper choice of mechanism, initiator, and comonomers, a huge variety of network architectures can be designed.

By cationic polymerization, random copolymer networks, formed from renewable natural oils with a high degree of unsaturation, for example, soybean oils, copolymerized with styrene and divinylbenzene, are obtained.[35] On varying the comonomer ratio these networks show a tunable glass-transition temperature and mechanical properties. The shape-memory properties are shown to be closely related to the cross-linking density and the glass-transition temperature. In addition, the materials also show good reprogramming properties from the first cycle, and excellent shape fixity and recovery ratios. While being attractive in their unique composition, an unfavorably broad T_g might limit these materials to SMPs.

2.2.5 SMP Networks Based on Polysiloxanes

Polymethylsiloxanes are known for their high chain flexibility, resulting in high elongations and low glass-transition temperatures. In addition, polymethylsiloxanes are biocompatible and nontoxic. Therefore, they are promising materials as soft segments in SMP networks.

In search of a way to effectively cross-link PCL, which is widely used as a switching-segment in SMPs, a method was found to blend the same with polymethylvinylsiloxane and cross-link the resulting blend by γ-irradiation.[36] The appending vinyl double bond acts as a sensitizer for cross-linking. However, the gel content does not exceed 45%, and it remains unclear if a mixed network structure or rather an IPN morphology is achieved. The addition of the polysiloxane, followed by cross-linking, greatly improves the shape-memory properties. More information on blend-like materials can be found in Section 2.3.6.

SMPs based on liquid crystal elastomers (LCEs) have been reported, in which the order–disorder transition of the mesogens is used to fix the temporary shape.[37] The employed materials are main-chain smectic-C LCEs cross-linked in a polydomain state. It has been shown that these polymers can easily be deformed to about 300% tensile strains at a temperature above the smectic-C-to-isotropic phase transition, and that shape-fixation is realized by cooling to the smectic C phase. By adjusting the mesogen content in these polymethylsiloxane-based materials, the transition temperature can be tuned. These LCE-based SMPs are characterized by a pronounced soft actuation when compared to other SMPs, while still displaying excellent shape-fixity and shape-recovery ratios.

In general, chemically cross-linked SMPs can be realized on a number of chemical compositions, and can imply more sophisticated molecular architectures. Their shape-memory properties are often characterized by superior shape-recovery ratios due to their low tendency to creep. A drawback, however, is the need for a tailored processing, and the impossibility to reshape the primary form of a material after the cross-linking process.

2.3 Thermoplastic Shape-Memory Polymers

In contrast to the permanently cross-linked structure of the SMPs presented earlier, thermoplastic elastomers and shape-memory materials are constituted from linear polymer chains interconnected by physical cross-links that can be released either thermally or by the use of a solvent to allow for melt or solution processing. Thus, the interchain cross-links determining the three-dimensional structure of a given device need to be stable during the programming process, meaning that they need to possess a thermal-transition temperature that is well above the shape-transition temperature.

The principle of the thermally induced shape-memory effect of these materials is based on the formation of a phase-segregated morphology, with one phase acting as a molecular switch and another phase providing the physical cross-links. While the phase with the highest thermal transition on the one hand, provides the mechanical strength of the material by the formation of physical cross-links over the whole temperature range of the programming process, it is on the other hand, responsible for the fixation that determines the permanent primary shape. Actually, some single-phase thermoplastic polymers are reported to show shape-memory effect, in which molecular chain entanglements are suggested to determine the permanent shape (see Sections 2.3.3 and 2.3.5). In principal, such molecular entanglements can be seen as physical cross-links, however, the materials show strong creep effects (and thus low-shape recovery ratio) when the polymer is stretched to a large deformation or when keeping the deformation at a high temperature for a longer period time.

Among thermoplastic SMPs, PUs have received pronounced attention because of their versatile structure and mechanical properties. For this reason, shape-memory PUs are intensively considered. In addition, several other types of thermoplastic SMPs are introduced, including poly(ethylene oxide-ethylene terephthalate) copolymers, aliphatic polyester copolymers, polystyrene copolymers, and polynorbornene-type polymers. Finally, shape-memory materials made from polymer blends are described.

2.3.1 Shape-Memory Polyurethanes

Since the discovery of the shape-memory effect in PUs by Mitsubishi Heavy Industry (MHI) in the 1980s, extensive research on shape-memory PUs has been carried out, resulting in the most systematic knowledge on SMPs. The mechanism of forming shape-memory effect in PUs, the synthesis and properties, and the current development of these smart materials arepresented in the following text.

2.3.1.1 Principle of Formation

Although different building blocks were used, a two-phase heterogeneous structure consisting of a hard segment and soft segment is typical for PU shape-memory materials. The hard segments form the "netpoints" that link the soft segments, while the soft segments show a thermal transition at temperature T_{trans} and work as a reversible molecular switch.

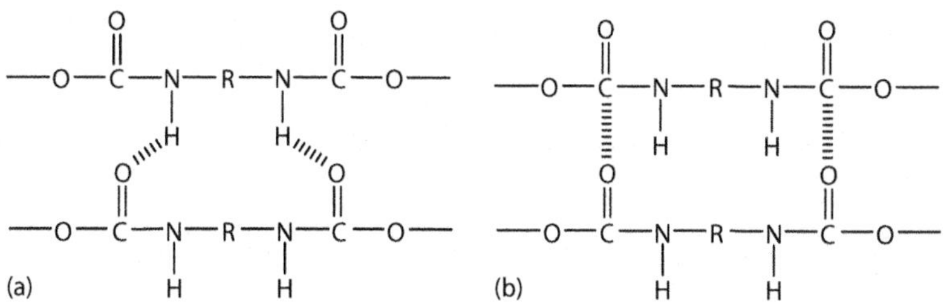

FIGURE 2.8
Possible interaction among hard segments.

The hard segments are formed by the reaction of diisocyanate with either diols or diamines, acting as low molecular weight chain-extenders to urethane-rich segments while, soft/switching segments are formed by long-chain polyether or polyester glycols. A key difference between the soft and hard segments is their polarity that may cause immiscibility of the segments. Together with the difference in molecule polarity, the strong hydrogen bonding function between carbonyl groups and amine groups in the urethane-containing hard segments (see Figure 2.8), dipole–dipole interactions, and/or segment crystallization can further enhance the incompatibility between the interconnected segments, which will cause the segregation of these two dissimilar blocks. The microphase separation is thus essential for the shape-memory properties as well as for the mechanical performance, as it results in the formation of hard segment-rich regions (domains) that act as cross-linking points for the soft/switching blocks.

2.3.1.2 Raw Materials

Shape-memory PUs differ from conventional PUs in that they exhibit a hard-segment phase and a soft-segment phase, forming a two-phase heterogeneous structure and morphology. In order to show good shape-memory properties, the hard segment content in the PUs must be high enough to inhibit the plastic flow of the chains by forming physical cross-links that are responsible for memorizing the primary shape. Hard segments can be formed either from diisocyanates and chain extenders or from a long chain macrodiol with a higher thermal transition temperature. Accordingly, shape-memory PUs can be classified into segmented PUs and multiblock PUs.

Segmented PUs are typically produced from three basic starting materials: (1) diisocyanate, (2) long-chain polyether or polyester macrodiols, and (3) short-chain chain extender, a diol, and/or a diamine. From the reaction between the diisocyanate and the chain extender, urethane-bond rich hard segments are obtained, while the long-chain polyether or polyester macrodiol forms the soft segments between them.

Scheme 2.1 shows the typical two-step prepolymer method that is usually employed for the production of thermoplastic PU elastomers. In this process, isocyanate-terminated prepolymers are obtained by the reaction of a long chain diol with an excess of a low molecular weight diisocyanate (Scheme 2.1, 1st reaction step). In the second step, the low molecular chain-extender is added to further couple these prepolymers. Linear, phase-segregated PU- or PU-urea block copolymers are obtained in this way, characterized by a microphase-segregated morphology that strongly depends on the chemical composition and the segment length.

Multiblock PUs are produced from three basic starting materials (1) a diisocyanate, (2) a long chain polyether, or polyester macrodiol with a high thermal transition temperature such as T_g or T_m as the hard segment, and (c) a long chain polyether, or polyester macrodiol with lower T_g or T_m as the switching segment (Scheme 2.2). In these materials, the hard segment content is given by the ratio of the two macrodiols rather than by the urethane bonds,

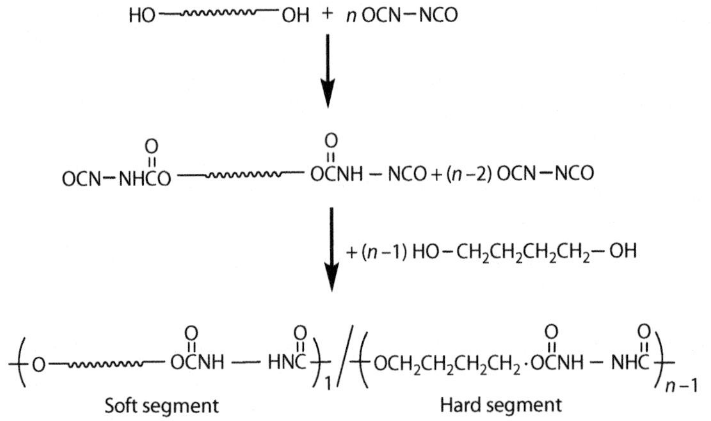

HO—∿∿∿—OH + n OCN—NCO

↓

OCN—NHC(=O)O—∿∿∿—OC(=O)NH — NCO + (n −2) OCN—NCO

↓ + (n −1) HO−$CH_2CH_2CH_2CH_2$−OH

$\left[O—∿∿∿—OC(=O)NH — HNC(=O) \right]_1 / \left[OCH_2CH_2CH_2CH_2 \cdot OC(=O)NH — NHC(=O) \right]_{n-1}$

Soft segment   Hard segment

SCHEME 2.1
Typical synthesis of segmented PUs.

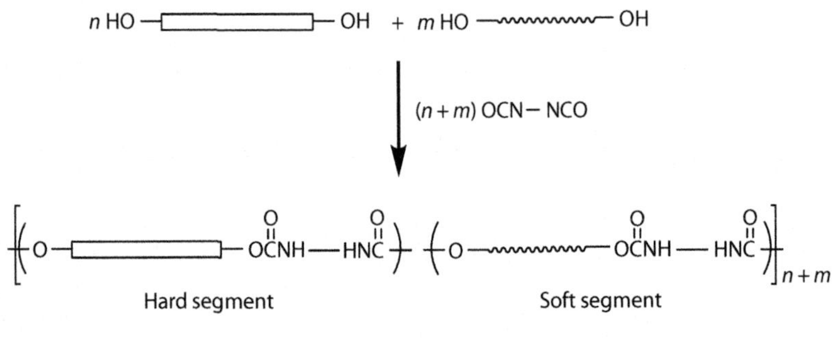

n HO—▭—OH + m HO—∿∿∿—OH

↓ (n + m) OCN— NCO

$\left[\left(O—▭—OC(=O)NH—HNC(=O) \right) \left(O—∿∿∿—OC(=O)NH — HNC(=O) \right) \right]_{n+m}$

Hard segment   Soft segment

SCHEME 2.2
Typical synthesis of multiblock PU copolymers.

and thus the fraction of urethane bonds along the chains can be diminished to 4%–10%. This may be an advantage for certain applications, like for the avoidance of harmful degradation products in biodegradable SMPs.

2.3.1.2.1 *Diisocyanates*

One of the key reactive materials required to produce PUs is a diisocyanate of choice. The basic structure of the diisocyanate can be either aromatic or aliphatic. Each type of diisocyanate has a different intrinsic ability to form a semi-crystalline hard segment. As the elasticity of the polymers depends on their degree of crystallinity and the degree of hard segment segregation, it is clear that the selection of the diisocyanate monomer will be one of the key parameters that influence PU mechanical characteristics. PUs synthesized under similar conditions, but with different types of diisocyanates, presented such different mechanical characteristics, with 1,4-phenylene diisocyanate (PPDI, see Table 2.1) showing the highest modulus (23.4 MPa), followed by MDI (13.1 MPa) and toluene diisocyanate (TDI, 2.1 MPa).[38] The differences can be attributed to the compact, rigid, and highly symmetric nature of PPDI, resulting in a high degree of crystallization.

The adequate selection of diisocyanate also involves the consideration of application. It should be considered that PUs made from aromatic diisocyanates tend to turn yellow upon exposure to light at ambient conditions as they form diquinones that act as chromophores. For the synthesis of medical-grade PUs, the MDI is often selected as the diisocyanate, because the resulting materials show strong mechanical properties.

Although diisocyanates in general are toxic compounds, once incorporated into the PUs, the isocyanate moieties are reacted to form urethane bonds, so that the resulting PU after hydrolytic degradation releases the corresponding diamine. Accordingly, the choice of the diisocyanates is largely governed by the toxicity of the corresponding diamine. PUs made from the MDI may release aromatic diamine as the biodegradation product, which is potentially carcinogenic. Alternatively, aliphatic diisocyanates are employed in multiblock copolymers. 1,6-hexanediisocyanate (HDI) and 1,4-butanediisocyanate (BDI) are the most investigated aliphatic diisocyanates in formulating biodegradable PUs, owing to the relatively nontoxic nature of the corresponding diamines, namely, 1,6-hexanediamine and 1,4-butanediamine. Table 2.1 lists the chemical structure and thermal properties of some typical diisocyanates for PU synthesis.

2.3.1.2.2 *Long Chain Macrodiols*

Long chain macrodiols for PU synthesis are usually a polyether or a polyester diol. Unlike diisocyanate compounds and chain extenders, which are usually of low molecular weight, a macrodiol is an oligomer with a molecular weight normally ranging from a few hundred to a few thousand. At room temperature, the macrodiols can be liquid or solid, depending on the molecular weight. Due to their aliphatic structure and intermolecular

TABLE 2.1

Some Diisocyanates Commonly Used in SM-PU Synthesis

Name	CAS Number	Structure	T_m (°C)	$T_{boiling}$ (°C mm Hg)	Producer
PPDI 1,4-phenylene diisocyanate	104-49-4	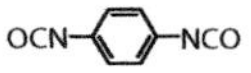 OCN– –NCO	94–96	—	Akzo
MDI 4,4′-diphenylmethane diisocyanate	101-68-8	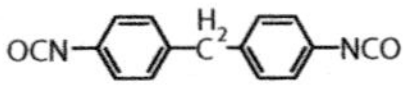 OCN– –H_2C– –NCO	39.5	—	BASF, Bayer, Dow, Enichem
TDI 2,4-toluene diisocyanate	584-84-9	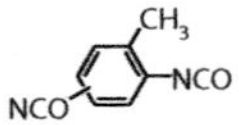 CH_3, NCO, NCO	20–21	—	BASF, Bayer, Dow, Enichem, Mitsui
HDI 1,6-hexane diisocyanate	822-06-0	$OCN(CH_2)6NCO$	—	127_{10} 255_{760}	Bayer Mitsui, Hüls, Lyondell
BDI 1,4-butane diisocyanate	4538-37-8	$OCN(CH_2)_4NCO$	—	$102–104_{14}$	DSM
TMDI 2,2(4),4-trimethylhexanediis ocyanate	34992-02-4	$OCNCH_2CH_2CH(CH_3)$ $C(CH_3)_2$ CH_2CH_2NCO	—	149_{760}	Hüls
CHDI *trans*-cyclohexane diisocyanate	2556-36-7	 OCN– –NCO	59–62	—	Akzo
IPDI isophorone diisocyanate	4098-71-9	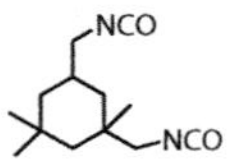 NCO, NCO	—	158_{10}	BASF, Bayer, Hüls, Lyondell
HMDI dicyclohexylmethane 4,4′-diisocyanate	5124-30-1	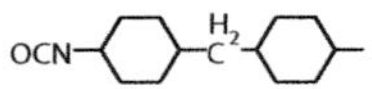 OCN– –H_2C– –	19–23	179_{10}	Bayer

interactions, particularly the abundant ether bonds, long chain macrodiol molecules show a high flexibility and a low glass or melting transition temperature, and are therefore soft materials. Consequently, the long chain macrodiol segments in the resulting SM-PUs are responsible for the formation of soft domains. Some typical long-chain macrodiols used for constructing shape-memory PUs are listed in Table 2.2. PCL diols and PEG diols are often used in the synthesis of SM-PUs with crystalline switching segments, while poly(tetramethylene) glycols (PTMG) and poly(ethylene adipate) diols (PEA) are used in the synthesis of PUs with amorphous switching segments (see Tables 2.3 and 2.4).

2.3.1.2.3 Chain Extenders

The direct reaction of a long chain diol with a diisocyanate produces a soft gum rubber with poor mechanical strength. The properties can be drastically improved by the addition of the chain extender. The role of the chain extender is to produce an "extended" sequence in the copolymer consisting of alternating chain-extenders and diisocyanates. These extended sequences, or hard segments, act both as filler particles and physical crosslink sites to increase mechanical strength. A PU-urea is obtained when a diamine is used and a PU results when a diol is employed. Commonly

TABLE 2.2

Some Typical Macrodiols Used in PU Synthesis

Macrodiols	Structure	T_m (°C)	T_g (°C)
PCL	$HO[(CH_2)_5CO]_nORO[CO(CH_2)_5]_nOH$	M_n=1,600: 46	−60
		M_n=2,000: 50	
		M_n=4,000: 54	
		M_n=7,000: 56	
		M_n=≥10,000: 59–64	
PEG	$HO[CH_2CH_2O]_nH$	M_n=600: 17–22	
		M_n=1,000: 39	
		M_n=1,500: 45	
		M_n=3,000: 54–58	
		M_n=4,000: 58–61	
		M_n=8,000: 62	
PTMG	$HO[CH_2CH_2CH_2CH_2O]_nH$	M_n=250: −8	
		M_n=650: 20	
		M_n=1,000: 24	
		M_n=2,000: 32	−100
		M_n=2,900: 46	
PEA diol	$HO[CH_2CH_2OOC(CH_2)_4COO]_nCH_2CH_2OH$	M_n=600: —	−46
Poly(buthylene adipate) diol	$HO[(CH_2)_4OOC(CH_2)_4COO]_n(CH_2)_4OH$	M_n=600: —	—
		M_n=1,000: 50–60	−70

TABLE 2.3

Overview of Shape-Memory PUs with T_g as Switching Temperature

Hard Segment	Soft Segment	T_{switch} (°C)	Synthesis/ Feature	Ref./Pub. Time
MDI/1,4-BD	PTMG M_n=250, 650	T_g=−13 to 54	First step bulk, second step in sol. polym.	[38]/1998
MDI/1,4-BD	PTMG M_n=250, 650, 1,000, 2,000, 2,900	T_g=−56 to 54	As above	[39]/1998
MDI/1,4-BD or MDI/EDA	PTMG M_n=2,000	T_g=−15 to −10 for 1,4-BD series T_g=15–22 for EDA series.	Bulk polym.	[47]/2006
MDI/1,4-BD	PEA M_n=300, 600, 1,000	T_g=−4.8 to −48.3	Bulk polym./ morphology	[48]/1996
MDI/1,4-BD	PEA M_n=600	T_g=50–60	Sol. polym./ fibers	[49]/2006
MDI/1,4-BD	PBA M_n=600	T_g=29–64	Sol. polym./ fibers	[50]/2006
MDI/1,6-HD	HDI/1,2-butane diol	T_g=41–53	Sol. polym.	[51]/2000
MDI, TDI or IPDI/1,4-BD	PLA M_n=3,200	T_g=48–63	Sol. polym.	[41]/2007
TDI/1,4-BD	Poly(L-lactide-*co*-ε-caprolactone) M_n=5,000	T_g=28–53	Sol. polym./ biodegradable	[52]/2007
PLLA M_n=2,000–10,000	Poly(glycolide-*co*-caprolactone) M_n=2,000–10,000	T_g=40–47	Sol. polym./ multi block, biodegradable	[60]/2005

used chain-extenders are 1,4-butanediol (1,4-BD), 1,6-hexanediol, ethylene diamine, and 4,4′-dihydroxy biphenyl (DHBP), which are shown in Figure 2.9.

Due to the aromatic and rigid nature of DHBP, some PUs using DHBP show greatly enhanced tensile modulus as compared to PUs using 1,4-BD as the chain extender.

2.3.1.2.4 Catalysts

Organometallic catalysts like stannous octoate (SnOct) and dibutyltin dilaurate (DBTL) are often used in PU synthesis to promote the reaction rate of isocyanates and polyols. While SnOct is employed mostly for flexible foam systems, DBTL is mainly used to produce PU elastomers.

2.3.1.3 Synthesis Procedures

SM-PUs are commonly produced by the same methods as conventional PUs. The polymerization involves a two-step prepolymer method either in

TABLE 2.4
Overview of Shape-Memory PUs with T_m as Switching Temperature

Hard Segment	Soft Segment	T_{switch} (°C)	Synthesis/ Feature	Ref./Pub. Time
MDI/1,4-BD	PCL M_n = 1,600, 4,000, 5,000, 7,000	T_m = 43–49	Sol. polym.	[42]/1997
MDI/1,4-BD	PCL M_n = 2,000, 4,000, 6,000	T_m = 50	Bulk polym.	[44]/1996
MDI/1,4-BD	PCL M_n = 4,000	T_m = 37–50	Sol. polym.	[43]/2007
HDI/DHBP	PCL M_n = 4,000	T_m = 38–59	Sol. polym./ liquid crystalline	[53]/2000
MDI/1,4-BD with DMPA	PCL M_n = 2,000, 4,000, 8,000	T_m = 50	Bulk polym./ ionomers	[54]/1998
HDI/MDI/ HMDA/PEG200 or DMPA	PCL M_n = 4,000	T_m = 49–51	Sol. polym./ WVP	[46]/2000
MDI/1,4-BD/ NMDA or BIN	PCL M_n = 10,000	T_m = 43–49	Sol. polym./ cationic ionomers	[55]/2007
Poly(*p*-dioxanone)	PCL	T_m = 40	Sol. polym./ multiblock, biodegradable	[58]/2002

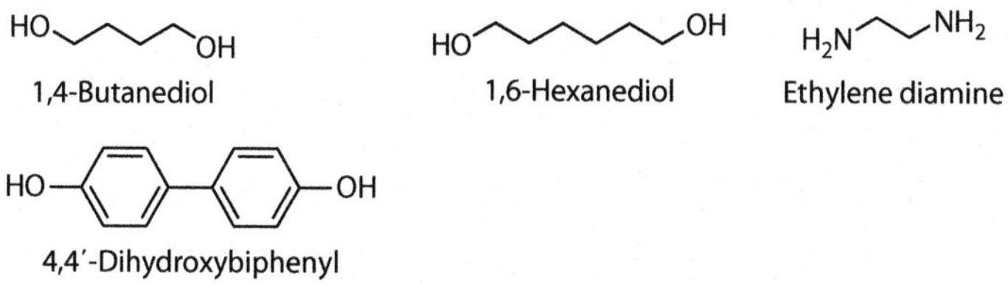

1,4-Butanediol

1,6-Hexanediol

Ethylene diamine

4,4′-Dihydroxybiphenyl

FIGURE 2.9
Chain extenders commonly used for SM-PU synthesis.

solution or in bulk as reported from the literature (see Tables 2.3 and 2.4). The general synthesis scheme is illustrated in Scheme 2.1. In the first step, isocyanate-terminated prepolymers are obtained by the reaction of a macrodiol with an excess of a low-molecular weight diisocyanate. Afterward, a chain extender is added to further couple these prepolymers. Linear, phase-segregated PU- or PU-urea block copolymers are obtained.

While the industrial processes for the production of PUs mostly rely on solvent free pathways, a solvent is often used in laboratory synthesis in

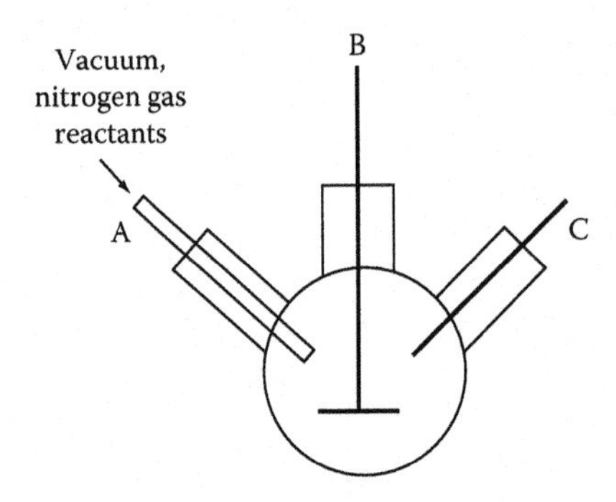

FIGURE 2.10
Set-up for the synthesis of PUs. A, Inlet; B, stirrer; C, thermometer.

order to reduce the viscosity and to promote the formation of high molecular weight polymers. For processes involving fiber spinning, a solvent polymerization method is often preferred, so that the resulting polymer solution can be directly used for fiber production.

2.3.1.3.1 Solution Synthesis

The laboratory synthesis of SM-PUs is usually carried out in a three-neck glass flask and a common setup is illustrated in Figure 2.10. The inlet has three functions: the connection to a vacuum line, the introduction of a nitrogen gas, and the addition of reactants. The reaction should be performed under a nitrogen atmosphere in order to protect the reaction mixture from moisture and oxygen. As a typical example, the procedure for the classic two-step solution synthesis of a PU composed of MDI, 1,4-BD, and PTMO or PCL is described in the following text.

1. The macrodiol is added to the reactor and dissolved in the solvent, e.g., DMF. The solution is heated to a predetermined temperature; here to 70°C.
2. The diisocyanate is added slowly to the reactor. Once the addition is completed, the reaction is maintained at 70°C–90°C with agitation for 2–3 h. A catalytic amount of dibutyltin dilaurate can be used to speed up the reaction. A predetermined amount of solvent is added to the reactor to reduce the viscosity of the PU, and to maintain effective agitation in the next chain-extending step.
3. The chain-extender is added slowly under vigorous agitation, and the reaction is kept at an elevated temperature until completion. At this stage, significant increase of viscosity and temperature will be noticed and efficient agitation is extremely important. The completion of the reaction is indicated by reaching constant viscosity or by following the residual isocyanate index. The reaction can be terminated by adding methanol.

2.3.1.3.2 *Bulk Polymerization*

The major advantage of solution synthesis is the relative ease in controlling the reaction. However, it is less frequently used in industrial production because of the high cost and inconvenience involved with the use of a solvent. For synthesis of shape-memory PUs, a two-step prepolymer bulk polymerization procedure is frequently employed (see Tables 2.3 and 2.4). The prepolymer (the product of isocyanate with long-chain polyether or polyester diol) is first prepared; then the chain extender is directly added to the reactor with vigorous agitation. When the viscosity of the product has reached a certain degree, it is poured out of the reactor and cured at an elevated temperature such as 120°C for 24h to complete the reaction.

2.3.1.3.3 *Modified Two-Step Prepolymer Method*

Another pathway employed to prepare shape-memory PUs[39,40] is a combined bulk and solution polymerization method. Here, excess isocyanate and macrodiol are mixed and heated in the bulk. Afterward, the NCO-terminated prepolymer is diluted with a solvent, and a dilute solution of the chain extender is added slowly under cooling. The reaction is carried out at an elevated temperature afterward. When reaching constant viscosity, the PU solution is cast onto a glass plate, and is dried under vacuum.

2.3.1.4 Structure–Property Relationships in PU-SMPs

As mentioned earlier, the properties of PUs can be significantly varied by the adjustment of the relative composition and the chemistry of a hard and soft/switching segment. These properties, and especially the shape-memory properties, can be greatly influenced by the processing and/or the programming procedure, which provides us a tool to control these properties according to specific applications. In addition, SM-PUs show some unique properties in the thermal transition temperature range, such as damping or permeability characteristics.

2.3.1.4.1 *Shape-Memory Properties*

SM-PUs typically present a microphase separated structure due to the thermodynamic incompatibility between the hard segment and the soft segment. The morphology of the phase separation, the phase composition, the microdomain size, and the phase distribution, have a significant impact on the mechanical performance and the shape-memory properties of PUs. Through regulating the relative content of the hard segment and the soft segment, as well as the type and molecular weight of the soft segment-forming long chain diols, and a suitable programming procedure (draw ratio, isotherm-crystallization etc.), the shape-memory properties can be greatly influenced and optimized.

Thermoplastic SM-PUs are often found to present a lower shape-recovery ratio than the thermoset counterpart (chemically cross-linked, Section 2.2).

This can be explained by the lower resistance toward mechanical stress in the physically cross-linked hard segment, resulting in a creep phenomenon during programming.

The switching temperature of PUs commonly originates either from a glass transition or a melting point of one of the polymer phases, which is determined, to a high extent, by the transition temperature of the long-chain diols. Table 2.2 lists the thermal properties of commonly used long-chain diols for PU synthesis. Generally speaking, in the case of a melting temperature, one observes a relatively sharp transition, while glass transitions always extend over a broad temperature range. Mixed glass-transition temperatures, $T_{g,mix}$, between the glass transition of the hard segment and the switching-determined soft blocks may occur in the cases where there is no sufficient phase-separation between the hard segment and the soft segment. Mixed glass-transition temperatures can also act as switching transitions for the thermally induced shape-memory effect.[3]

The actual switching-temperature of a specific SMP can be influenced not only by the composition of the material, but also by different aspects of the processing and programming procedures. The drawing temperature, the duration, and the draw-ratio on the transition temperature, as well as on the observed shape-recovery ratio also have an impact on the switching-temperature.

For example, it is reported[41] that a poly(L-lactide) (PLLA)-based PU rod was pressed into a disk at a temperature of about 5°C below its T_g and then cooled at a temperature of 25°C below its T_g within 5 s. Interestingly, by this processing procedure, all samples completed the shape-recovery at a temperature lower than their T_gs. This phenomenon could be explained by stating that the samples were deformed at a temperature below the T_g and stress was stored in the deformed sample during the programming process.

In a similar system,[42] the effect of the deformation temperature (T_D), the draw ratio, and the time period (t_D) at T_D after drawing was studied. The shape-memory behavior such as the shape recovery ratio R_r, the recovery temperature (T_r), and the recovery speed (V_r) remained almost unchanged when they were drawn at temperatures far above the T_m of the soft-segment crystals. However, specimens programmed (deformed) in the range of the T_m, exhibited a lower transition temperature, a lower recovery speed and slightly less final shape-recovery ratio. The shape-recovery ratio R_r decreased with an increasing draw-ratio and with an increasing holding time period t_D at draw-temperature indicating that the hard segment domains, acting as physical cross-links in segmented PUs, are not very stable and are very sensitive to the conditions of the specimen preparation. In contrast, the shape-fixity ratio R_f increased on raising the deformation amplitude,[43] suggesting that the increased draw-ratio enhanced the crystallization of the soft segment, thus improving the shape-fixity ratio. Consequently, good shape-memory properties can be observed only for draw-ratios within an optimized range.

2.3.1.4.2 Physical and Mechanical Property Changes at the Transition Temperature

Several physical properties of SMPs, other than the ability to memorize shape, are significantly altered in response to external changes in temperature and stress, particularly at the melting point or the glass-transition temperature of the soft/switching segment. These properties include the mechanical properties, like elastic modulus, hardness, and damping, and also diffusion properties, like water vapor permeability (WVP).

Figure 2.11 shows the typical changes of the storage modulus of SM-PUs and nonshape-memory materials when they were heated across the melting point.[44] PCL4000- and PCL8000-based PU polymers clearly show a sharp glass-rubber transition at around 50°C, corresponding to the T_m of the soft segment. On the other hand, the PCL2000-based PU (nonshape-memory material) does not show a sharp transition—instead a broad transition appears. This is due to the amorphous nature of the soft segment. Only PUs with a sufficient soft-segment length are able to crystallize, and thus to show a shape-memory behavior, in coincidence with a sharp storage modulus drop at the melting temperature in the DMTA experiments. It has been reported by several authors that, especially for PCL-based PUs, the crystallinity of the PCL is significantly depressed in segmented PUs, and a lower limit of PCL molecular weight is found to be around 2000–3000, below which the PCL segments are not able to crystallize at the usual processing conditions.[42,43]

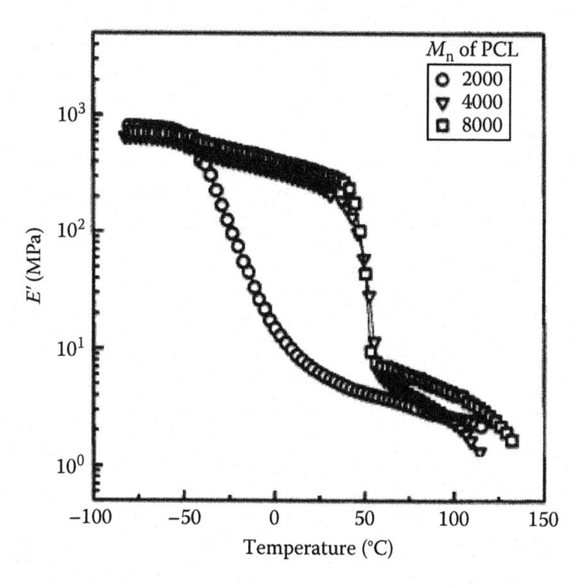

FIGURE 2.11
Effect of PCL molecular weight on the storage modulus of PU ionomers with 70% soft segment content. (Reproduced from Kim, B.K. et al., *Polymer*, 37(26), 5781, 1996. With permission.)

SM-PUs are also reported to have a discontinuous change in moisture permeability when crossing the T_g or T_m of the soft segments. Typically, the materials have low moisture permeability below T_g or T_m, showing a strong increase at the thermal transition temperature. The WVP of the SM-PUs based on an amorphous soft segment from the reaction of the HDI and 1,2-butanediol, and the crystalline hard segment from the MDI and 1,6-hexanediol, was studied.[45] In addition, the PUs were modified by hydrophilic segments, i.e., PEG or dimethylol propionic acid (DMPA, see Figure 2.14). Considerable thermo-sensitive WVP is reported for these and similar materials that can be of use in stimuli-sensitive membranes for sophisticated applications.[46]

2.3.1.5 Current State of Development

Since the discovery of the shape-memory effect in PUs in the 1980s, extensive research on segmented PUs has been carried out. According to the nature of the thermal transition of the switching segments, thermoplastic SM-PUs can be classified into two main categories—one with PUs with a crystallizable switching-segment showing a melting temperature T_m, and the other with PUs containing an amorphous segment and a glass transition temperature T_g.

SM-PUs with a crystallizable switching-segment often show a relatively sharp transition, while glass transitions always extend over a relatively broad temperature range. A complex phase behavior gives rise to the occurrence of mixed glass-transition temperatures between the glass transition of the hard segment and the switching segment, in the cases where there is no sufficient phase separation. On the other hand, mixed glass-transition temperatures can also act as switching transitions for the thermally induced shape-memory effect. In the following paragraph, we will demonstrate the current state of development in the area of SMPUs with amorphous and crystalline switching segments by selected examples. Due to their difference from conventional segmented PUs, multiblock PUs are introduced in a separate paragraph.

2.3.1.5.1 *Polyurethanes with Amorphous Switching Segment*

SM-PUs with a soft/switching-segment phase showing a glass transition in a useful temperature range, i.e., 20°C–60°C, are mostly obtained by using one of the suitable macrodiols from PTMG, PEA glycol, or copolyesters from glycolide, ε-caprolactone, and/or lactides. Table 2.3 gives an overview of the chemical composition, switching temperatures, and the synthesis method for the PUs with an amorphous switching segment.

The influence of the hard segment and the soft segment on the morphology and the shape-memory behavior using PTMG-based PUs was systematically studied.[39,40] These PUs consist of a soft/switching segment formed from PTMG, and a hard-segment phase formed from 4,4′-diphenylmethane diisocyanate (MDI) and 1,4-butanediol (1,4-BD) by using a modified prepolymer method. These polymers are typical materials with mixed T_gs, enabling the easy tuning of the switching temperatures by the hard-segment content and the molecular weight of the PTMG diols. The switching temperature value increased with increasing hard-segment content and decreased with an increasing molecular weight of PTMG diols. The shape-recovery ratio R_r increased when the amounts of the hard segment were increased, and reached values of up to 99%. By using ethylene diamine (EDA) as a chain extender, enhanced mechanical and shape-memory properties were obtained.[47] The urea-type bonding formed from EDA restricts the chain rotation in the hard segments, thus leading to strengthened PUs.

PUs containing a hard-segment-forming phase formed of MDI and 1,4-BD and a switching segment based on PEA glycols are an interesting class of SMPs with amorphous switching segment.[48] The glass-transition temperature of these PUs decreases from 48.3°C to −4.8°C with increasing PEA molecular weight, and in addition is strongly influenced by the hard-segment content.

Polarizing optical microscopy (POM) was utilized to investigate the morphological changes while increasing the hard segment (Figure 2.12). The size of spherulites decreases with the decrease of the hard-segment content (Figure 2.12b and c). When the hard-segment content decreases to 75 mol%, the spherulite structure is no longer observed in the POM micrograph.

Shape-memory fibers were spun from SM-PUs consisting of a hard segment made from MDI and 1,4-BD, and a soft/switching segment from PEA with M_n 600.[49] Compared with the PU polymers, the fibers showed a lower shape-fixity ratio (R_f=87%–92%) and a higher shape-recovery ratio (R_r=85%–90%), and the recovery force of the fibers was twice as much as those of the PU polymers. Similar shape-memory properties are observed in PUs containing poly(buthylene adipate) glycol with M_n 600 and fibers made of them.[50]

Segmented PUs with a crystalline hard segment formed from the reaction of MDI and 1,6-hexanediol (1,6-HD), and the amorphous soft segment from 1,6-hexane diisocyanate (HDI) and 1,2-butane diol were investigated.[45,51] All polymers show good shape-memory properties as characterized by thermo-mechanical cyclic tensile testing. Polymers with a lower molecular weight soft/switching segment have higher shape-recovery ratios, and the difference in the stress-strain curves between the first and second cycles decreases as the block lengths are decreased.

By systematically employing different diisocyanates in the synthesis of SMPUs, the influence of hard segments on the mechanical and shape-memory properties was examined.[52] With PLLA diol (M_n=3200 g mol^{-1}) as the soft/switching segment material, 1,4-BD and one of the three diisocyanates (DI) from MDI, 2,4-toluene diisocyanate (TDI) and isophorone diisocyanate (IPDI), the hard segment of the PUs was constructed. The resulting PUs were all of an amorphous nature, showing T_gs in the range of 48°C–63°C. With a similar hard-segment content, MDI-based PUs showed the highest T_gs, and relatively better mechanical properties than the other two series.

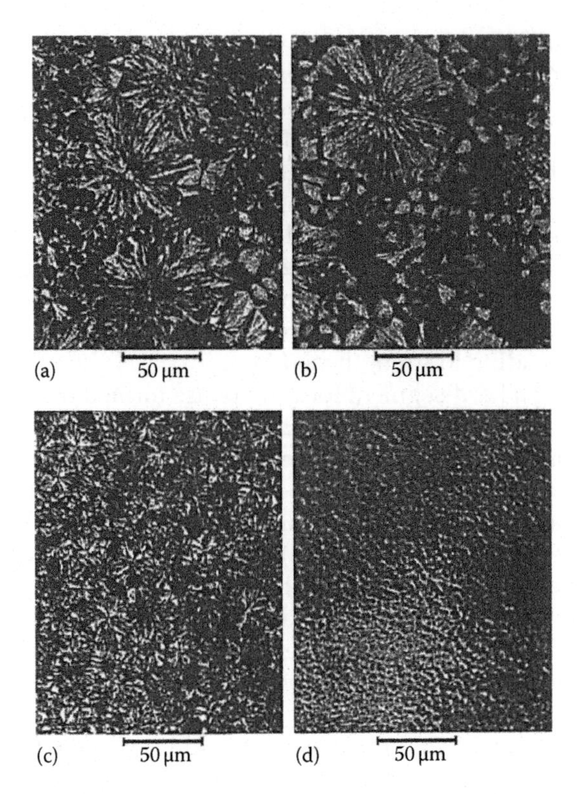

FIGURE 2.12
Polarizing optical micrographs of PUs utilizing PEA600 as soft segment with different hard-segment content: (a) 90 mol%, (b) 88 mol%, (c) 84 mol%, and (d) 75 mol%. (Reproduced from Takahashi, T. et al., *J. Appl. Polym. Sci.*, 60, 1061, 1996.)

By using copolyesters formed from L-lactide and ε-caprolactone as switching segments, biodegradable SMPUs can be obtained.[52] The T_gs and transition temperatures of these PUs are in the range of 28°C–53°C, depending on the comonomer ratio in the soft/switch segments, and the hard-segment content. All these PUs have a shape recovery ratio of more than 93% and a recovery force between 1.4 and 5.5 MPa depending on the composition.

From the previous text, we summarize that SM-PUs with amorphous switching segments are mostly synthesized from an MDI with 1,4-BD as the hard segment, a PTMG, and a PEA, and more recently, degradable polymers as the soft/switching segment. These polymers are basically synthesized by a two-step prepolymer method, either in solution or in bulk.

2.3.1.5.2 *Polyurethanes with Crystalline Switching Segment*

One of the advantages of PUs with T_m as the switching temperature is that the melting transition, in general, is much sharper than the glass transition, and therefore, SMPUs with a crystallizable switching segment possess a narrower shape-transition temperature range. In addition, the melting temperature is affected to a much lesser degree by the formation of mixed phases, and the main factor determining the shape-memory transition temperature is the chemical composition and the segment length of the switching segment. However, the crystallizability of the soft segment, important for a good shape-fixity ratio, is influenced by a number of factors like the hard-segment content and the processing conditions. An overview of such SM-PUs is shown in Table 2.4.

The shape-memory behavior of a series of PUs based on PCL as soft segment and an MDI and 1,4-BD as hard segments was investigated.[42,44] The influence of the molecular structure on the shape-memory effect of the segmented PUs was obvious. Only specimens of PUs with a soft-segment length of 2000 g mol^{-1} and a hard-segment content exceeding 10 wt% were excellent SMPs, with a high final shape recovery ratio of 93%–98%. Their response temperature, depending on the melting temperature of the soft segment crystals of the PUs, was in the range of 43°C–49°C, while the final shape-recovery ratio and the recovery speed were mainly related to the hard-segment content of the PUs.

The relationship between the morphology and the shape-memory behavior of segmented PUs from PCL, MDI and 1,4-BD was systematically studied by DSC and diffraction methods.[43] Depending on the hard-segment content, different phase morphologies were found, ranging from interconnected hard segments, isolated hard segments and no hard domains.

In the DSC test results, PUs with 20 wt% hard segments or less showed no endothermic peaks of the hard-segment phase. However, the TMA and SAXS tests showed that in PU-15 and PU-20, the crystallization of the hard segment was still detected owing to the higher sensitivity of the methods.[43]

The shape-memory behavior of these materials was quantified by thermomechanical cyclic tensile tests. In agreement with the findings on many other systems, the shape-fixity ratio decreased rapidly with the increase of the hard-segment content of more than 40 wt%. In order to understand the effect of crystallization on shape-fixity, three sets of WAXD tests were performed on the PU films before extension, after shape-fixing and after shape-recovery to trace the crystallinity changes that happen to these segmented PUs in the course of shape memorization. Before extension (Figure 2.13a), the intensity of the diffraction peaks that were attributed to the crystallization of the soft-segment phase decreased with an increase of the hard-segment content, until vanishing completely at a hard-segment content of 45 wt%–50 wt%.

As shown in Figure 2.13b, the diffraction traces indicate a substantial increase in the crystallinity after shape-fixing, and even PUs with a hard-segment content of 45% and 50% show weak diffraction peaks. The increase of crystallinity is attributed to the strain-induced crystallization that took place in the soft segments after extension and shape-fixing. After the shape-recovery process (Figure 2.13c), the PUs showed a little higher diffraction

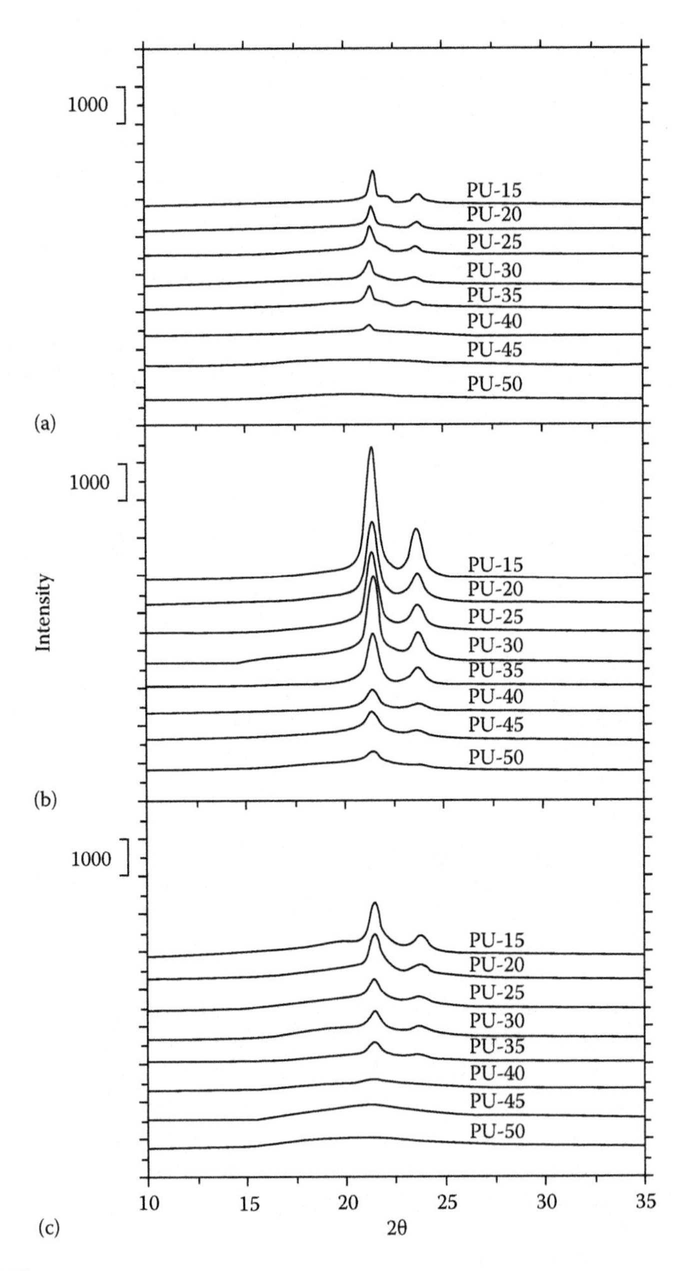

FIGURE 2.13
Tracing the structural changes taking place in the PUs in a cycle of shape memorization by the WAXD tests at room temperature: (a) before extension; (b) after shape fixing (being extended 100% and fixed in a temporary shape); (c) after shape recovery. PU-15 means PU with 15 wt% hard segment. (Reproduced from Li, F.L. et al., *Polymer*, 48, 5133, 2007. With permission.)

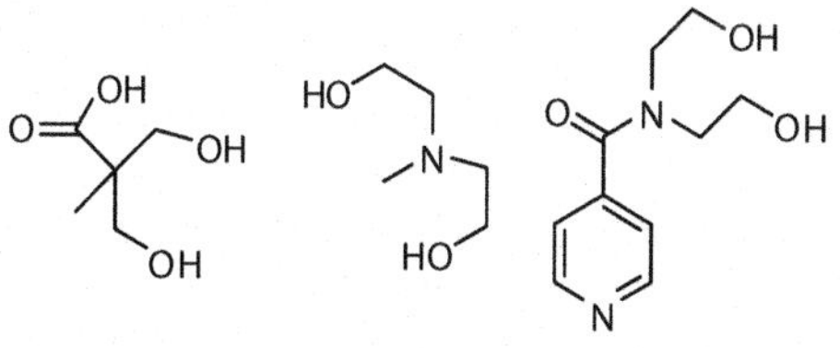

FIGURE 2.14
Dimethylolpionic acid (DMPA), *N*-methyldiethanolamine (NMDA), *N,N*-bis(2-hydroxyethyl) isonicotinamide (BIN).

intensity when compared to the PUs before deformation. This implies that after a cycle of shape-memorization the soft segments remain partially in a more oriented state, even though the PUs almost recover their original shape.

Further methods have been employed to influence the hard segments of PUs by incorporating an aromatic chain extender[53] or ionic groups (Figure 2.14).[46,54,55] By using DHBP as the chain extender, the tensile modulus of the resulting PUs is significantly enhanced when compared to those of corresponding PUs using 1,4-BD as the chain extender, due to the aromatic and rigid nature of the DHBP. The same group[54,55] studied the effect of the soft-segment content and lengths on the shape-memory effect of PUs based on PCL, MDI, 1,4-BD, and DMPA, and the results between PUs ionomers and the corresponding nonionomers were compared. It was found that ionomers gave higher hardness, modulus and strength than the nonionomers due to Coulombic forces resulting from an increased cohesion between the hard segments and the microphase separation, and the fatigue of the shape-memory effect was reduced by the incorporation of the ionic moieties into the hard segments.[55]

Using a chain-extender with a tertiary amine group, the effect of cationic groups on the shape-memory properties can be studied after protonating at a low pH. For this purpose, *N*-methyldiethanolamine (NMDA) and *N,N*-bis(2-hydroxyethyl)isonicotinamide (BIN, see Figure 2.14) were incorporated into PUs based on PCL, MDI, and 1,4-BD.[46] The shape-fixity ratio and the shape-recovery ratio of the NMDA series were improved simultaneously by the insertion of cationic groups within the hard segments. However, in the BIN series, the cationic pendant groups had less effect on the shape-memory effect.

These examples show that shape-memory PUs with crystalline switching segments are mostly synthesized from the MDI and 1,4-BD as the hard segment, and the PCL as the soft segment. The molecular weight of the PCL and the hard-segment content of the PUs have a particularly great influence on the mechanical performance and the shape-memory properties. Incorporation of an aromatic chain-extender or ionic groups into the hard segment will impart some additional effect on these properties.

2.3.1.5.3 *Multiblock Polyurethanes*

In conventional segmented PUs, the hard segment is formed from diisocyanates and chain extenders, and the urethane segment content is usually higher than 20 wt% in order to provide sufficient mechanical strength and/or to show the shape-memory effect. In contrast, multiblock PUs are typically produced from two long-chain macrodiols, one with a lower thermal transition as the switching segment, and the other with a higher transition temperature as the hard segment to provide the physical cross-links. These two macrodiols are coupled with a diisocyanate to form the multiblock copolymer. In multiblock PUs, the hard segment is formed from one of the two long chain macrodiols, and diisocyanate is used only as the coupling agent, thus significantly diminishing the content of urethane groups to 4 wt%–10 wt%.

The lower content of urethane groups in multiblock PUs is a clear advantage for medical applications, especially in absorbable implants. The nature and toxicity of the degradation products of PUs are of high concern for medical applications, because the hard segments are mostly produced with aromatic diisocyanates. The degradation of the hard segments produces aromatic diamines that are toxic and potentially carcinogenic to humans.[56] For durable PUs that show practically no or very slow degradation, the risk of toxic effects from the degradation products is correspondingly low. However, in case the PUs are designed to be biodegradable, the risk of relevant levels of toxic degradation products is much higher.

For this reason, diisocyanates such as MDI and TDI, that are commonly used in many industrial PU formulations, are not used in formulating biodegradable PUs. Biodegradable PUs have increasingly been synthesized with aliphatic diisocyanates, mainly 1,6-hexamethylene diisocyanate (HDI) and BDI, to overcome this potential safety problem (see Section 2.3.1.2).[56,57] Their choice is largely due to the relatively nontoxic nature of the corresponding degradation products—1,6-hexanediamine and 1,4-butanediamine. The symmetric molecular structure also leads to strong intermolecular attractions through hydrogen bonding resulting in elastomers with high strength.

An intensively studied system of such biodegradable multiblock SM-PUs is composed of oligo(ε-caprolactone)diol and oligo(*p*-dioxanone)diol coupled with 2,2(4),4-trimethylhexanediisocyanate (TMDI).[58,59] While segments from oligo(ε-caprolactone) form the switching phase, hard domains from crystalline oligo(*p*-dioxanone) with a melting temperature of 88°C–90°C form the physical cross-links for the primary shape. The materials show shape-transition temperatures at around 40°C, shape-fixity ratios of 98% and higher, and shape-recovery ratios of about 80% in the first cycle, and > 98%–99% in the third cycle. A potential application of these materials is suggested in surgical sutures.

Multiblock PUs using PLLA diols as hard segments have also been investigated, by systematically studying the impact of segment lengths and the hard-segment content on the mechanical and shape-memory properties of the multiblock copolymers.[60] Depending on the composition, these copolymers show a T_g between 41°C and 49°C and a T_m between 150°C and 160°C. By adjusting the PLLA content, the copolymers can be changed from rubbery materials when the PLLA content is less than 60% to as hard as the pure PLLA polymer when the PLLA content reaches 90%. In vitro degradation tests (phosphate buffer solution, pH 7.4 at 37°C) indicate a fast degradation of these multiblock copolymers, with a weight loss exceeding 40% at 30 weeks. The weight loss is accelerated by decreasing the molecular weight of both segments in the copolymers. The copolymers lose all mechanical strength after 5–7 weeks of in vitro degradation.

While PUs represent the majority of thermoplastic SMPs, other polymers and copolymers with a different chemistry were found to show analogous shape-memory behavior. The shape-memory mechanism of these polymers is based, either on the formation of a phase-segregated morphology, or on molecular chain entanglements. Table 2.5 gives an overview of these smart materials, which are introduced in the following section.

TABLE 2.5

Overview of Non-PU Thermoplastic SMPs

Polymer	Switching Phase/ T_{switch} (°C)	Fixing Phase	Ref./Pub. Time
PEO–PET copolymer	PEO M_n = 2,000–10,000 T_m = 44–55	PET crystalline	[61–63]/1997–1999
Poly(L-LA-*co*-GA)	T_g = 51–60	Poly(L-LA) crystalline	[64]/2007
Poly(L-LA-*co*-CL)	T_g = 34–59		
Poly(L-LA-*co*-dioxanone)	T_g = 26–58		
Poly(L-LA-*co*-GA-*co*-TMC)	T_g = 12–42	Chain entanglement	[65]/2007
trans-Poly(1,4-butadiene)-PS copolymer	Poly(1,4-butadiene) T_m = 68; T_{switch} = 80	PS phase	[66,67]/1994, 1997
PE-PS	T_{switch} = 70–90	PS phase and chain entanglement	[68]/2008
PE-PBS	T_{switch} = 70–80		
Polynorbornene	T_g = 57	Chain entanglement	[69–71]/1989, 1993, 2000
Polynorbornene-POSS	T_{switch} > 70		
Poly(methylene-1,3-cyclopentane) (PMCP)	T_m = 63–68	Chain entanglement and PE crystalline	[72]/2002
PMCP–PE	T_{switch} = 60–85		

2.3.2 Thermoplastic SMPs Based on Poly(Ethylene Oxide–Ethylene Terephthalate) Copolymers

Poly(ethylene oxide–ethylene terephthalate) (PEO–PET) segmented copolymers were originally studied in an attempt to simultaneously reduce the crystallinity of poly(ethylene terephthalate) (PET) and to increase the hydrophilicity to improve stainability with hydrophilic dyes. The PEO–PET segmented copolymers were synthesized in two steps: (1) transesterification reaction and (2) polycondensation reaction under vacuum (Scheme 2.3).

The shape-memory effect of a series of PEO–PET segmented copolymers with long soft segments has systematically been studied.[61–63] All the samples except for the sample with poly(ethylene oxide) (PEO) with M_w of 2000 g mol^{-1} exhibited a shape-memory effect. The switching temperature was in the range of 45°C–55°C, and increased when higher molecular weight PEO was used. The shape recovery ratio was in the range of 84%–99%, and improved when a higher molecular weight of PEO was used or when the hard-segment content was increased.

The crystalline structural characteristics of the copolymers during the memory process were investigated by polarizing microscopy and wide-angle XRD.[63] PEO crystals in stretched PEO–PET copolymer preferentially oriented along the fiber axis or the stretch direction. During stretching, the structure of the copolymer underwent a transformation from a spherulite-like to

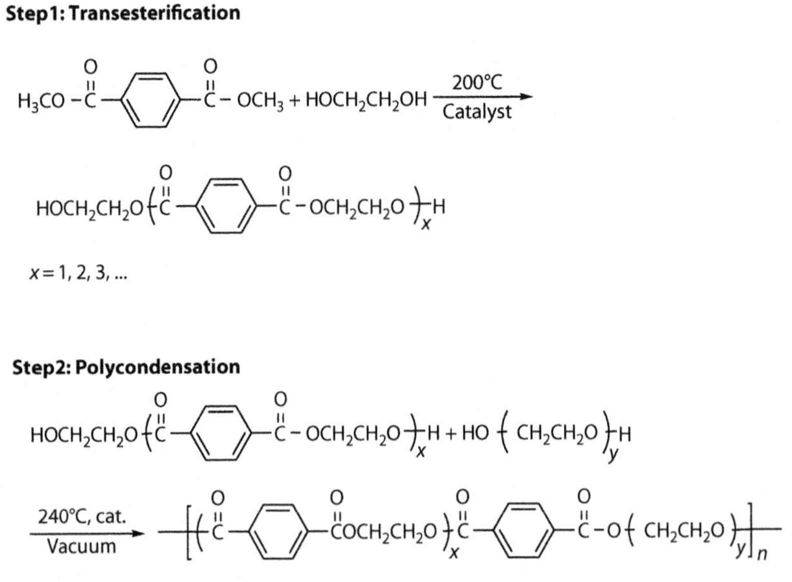

Step1: Transesterification

$$H_3CO-\overset{O}{\overset{\|}{C}}-C_6H_4-\overset{O}{\overset{\|}{C}}-OCH_3 + HOCH_2CH_2OH \xrightarrow[\text{Catalyst}]{200°C}$$

$$HOCH_2CH_2O\left(\overset{O}{\overset{\|}{C}}-C_6H_4-\overset{O}{\overset{\|}{C}}-OCH_2CH_2O\right)_x H$$

$x = 1, 2, 3, \ldots$

Step2: Polycondensation

$$HOCH_2CH_2O\left(\overset{O}{\overset{\|}{C}}-C_6H_4-\overset{O}{\overset{\|}{C}}-OCH_2CH_2O\right)_x H + HO\left(CH_2CH_2O\right)_y H$$

$$\xrightarrow[\text{Vacuum}]{240°C,\ cat.} \left[\left(\overset{O}{\overset{\|}{C}}-C_6H_4-\overset{O}{\overset{\|}{C}}OCH_2CH_2O\right)_x \overset{O}{\overset{\|}{C}}-C_6H_4-\overset{O}{\overset{\|}{C}}-O\left(CH_2CH_2O\right)_y\right]_n$$

SCHEME 2.3
Two step synthesis of EO-ET segmented copolymers.

a fiber-like structure. The driving forces for the contraction could thus be attributed to the orientation of the chains, and only such oriented or extended chains can contribute to the recovery of deformation; these extended chains involve both crystalline and amorphous segments. The study illustrates clearly how the shape-memory behaviors are the result of entropy-driven deorientation of oriented polymer chains.

2.3.3 Thermoplastic SMPs Based on Aliphatic Polyesters

Recently, some L-lactide-based copolymers were found to show a shape-memory effect, that include poly(L-lactide-*co*-glycolide) (PLGA), poly(L-lactide-*co*-*p*-dioxanone) (PLDO) and poly(L-lactide-*co*-ε-caprolactone) (PLC).[64] These biodegradable polymers are synthesized via a ring-opening polymerization of cyclic monomers shown in Figure 2.15. When the content of L-lactide exceeds 70%, most of the copolymers form a crystalline phase and an amorphous phase. The crystal part and chain entanglements of the copolymers build up the hard-segment phase and play the role of fixing the original shape of the copolymers. The amorphous part of the copolymers forms the soft-segment phase and plays the role of deforming and keeping the deformed shape. Under the condition of 100% elongation at (T_{trans}+5)°C all PLGA, PLDO and PLC copolymers showed remarkable shape-memory properties.

The shape-recovery ratio of the three copolymers was measured by recovering a strip of film from a spiral shape to a straight shape at a temperature of (T_{trans}+5)°C. All tested polymers recovered the original shape within 4–18 s, while the shape-recovery ratio increased with an increasing lactide content.

A series of polyester-based, single-phase thermoplastic SMPs is reported in a work on amorphous poly(L-lactide-*co*-glycolide-*co*-trimethylene carbonate) copolymers.[65] Similar to the poly(norbornene) reported in Section 2.3.5, these materials possess neither chemical cross-links, nor have a hard phase with a high thermal transition temperature. The physical cross-links, which are responsible for the shape-memory, are solely implied by chain entanglements in the high molecular weight materials. The T_g of the copolymers

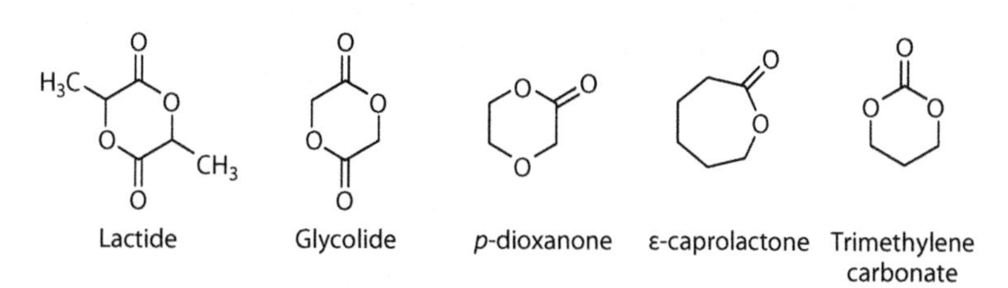

FIGURE 2.15
Cyclic monomers for biodegradable polymers.

can be changed from 12°C to 42°C in a predictable manner by varying the composition. All investigated materials displayed shape-memory properties and, after 100% deformation, they recovered their permanent primary shape in a time frame of seconds.

2.3.4 Thermoplastic SMPs Based on Polystyrene

Some types of poly(styrene-*b*-butadiene) block copolymers with lower polystyrene (PS) content and higher *trans*-poly(1,4-butadiene) (TPB) were reported as showing a shape-memory effect (see Figure 2.16a).[66,67] The copolymers form a microphase separation morphology. The PS block is amorphous with a T_g of 90°C, and the continuous TPB phase is semi-crystalline with a T_g of −90°C and a T_m of 68°C. The glassy, PS-rich hard domains play the role of forming physical cross-links and remembering the primary shape. A reversible deformation can be fixed by crystallizing the TPB phase at 40°C and recovering the stress-free state upon heating to 80°C to melt the TPB phase. A shape-recovery ratio of 80% was observed upon application of a maximum strain of 100%. The appearance of the shape-memory effect was explained by the formation of oriented crystallites upon application of high elongations. These oriented crystallites form a fibril-like structure with a high regularity over long distances. The single fibrils are linked through amorphous poly(1,4-butadiene) chains that are anchored in several crystallites at the same time.

Similarly, poly(ethylene–styrene) copolymer (PE–PS, with 44 mol% PS, $T_g = 21°C$ by differential scanning calorimetry [DSC], Figure 2.16b)[68] showed a shape-recovery ratio of 60%–70% when being heated up to 70°C–90°C, while poly(ethylene-*tert*-butyl styrene) (PE–PBS, with 43 mol% PBS, $T_g = 61°C$ by DSC) showed a higher shape-recovery ratio of more than 90% in this temperature range. The improvement in the shape-memory properties was derived from the entanglement of the substituent *tert*-butyl groups.

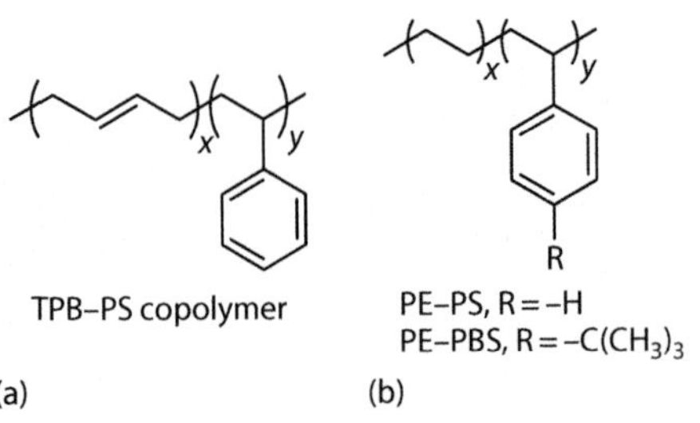

TPB–PS copolymer

(a)

PE–PS, R = –H
PE–PBS, R = $-C(CH_3)_3$

(b)

FIGURE 2.16
Chemical structures (a) poly(*trans*-butadiene-*co*-styrene) (TPB–PS) and (b) poly(ethylene-*co*-substituted styrene) elastomer PE–PS and PE–BPS.

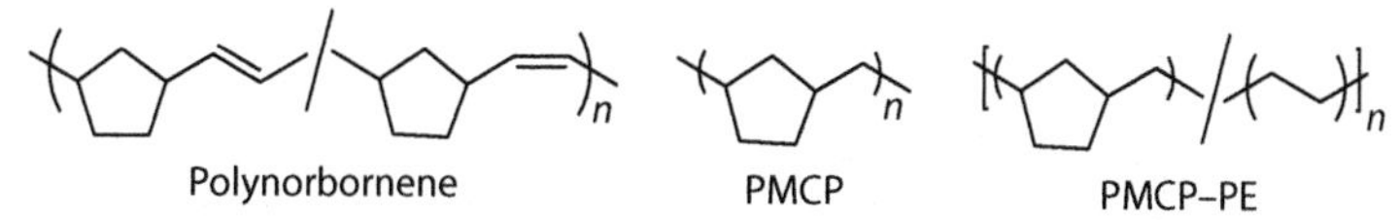

Polynorbornene PMCP PMCP–PE

FIGURE 2.17
Chemical structures of polynorbornene, PMCP and PMCP-PE.

2.3.5 Poly(Norbornyl)-Based SMPs

The shape-memory effect of polynorbornene (see Figure 2.17) is based on the formation of a physically cross-linked network as a result of entanglements of the high molecular weight linear chains, and on the transition from the glassy to the rubbery state. Stretching the polymer at 90°C and rapidly cooling it down to room temperature induced a crystallization of the polymer, showing a melting temperature at 85°C. In this sense, the crystallites could also act as physical cross-linking net points contributing to the shape-memory effect.[69,70]

The introduction of POSS, a polyhedral oligomeric silsesquioxane molecule, into polynorbornene, reduces the shape-recovery ratio.[71] The driving force for the shape-recovery is related to the relaxation of highly oriented chains during heating above the glass-transition temperature. In the case of POSS hybrid copolymers, the polynorbornene chain segments are relatively less oriented during tensile drawing, and the POSS–POSS interactions resist the strain recovery of the polynorbornene chain segments. However, the POSS–POSS interactions significantly enhance the thermal stability of the POSS–polynorbornene hybrid polymer.

With a similar chemical structure to polynorbornene, poly(methylene-1,3-cyclopentane) (PMCP, Figure 2.17), and its block copolymers with ethylene (PMCP-PE) were found to show a shape-memory effect.[72]

PMCP is synthesized by cyclopolymerization of 1,5-hexadiene with a new metallocene catalyst rac-(ethylenebis(1-indenyl))$Zr(N(CH_3)_2)_2$, and the PMCP-PE is produced by sequent polymerization of ethylene following PMCP synthesis. The PMCP has a T_g at 7.6°C and a small endothermic melting peak at 68.0°C, as analyzed through the DSC method. The PMCP–PE copolymers in addition, show a big endothermic melting peak at around 120°C–124°C resulting from the polyethylene segment. All these polymers show shape-memory effect. The shape-recovery ratio is enhanced by the copolymerization with ethylene, because crystalline polyethylene segments form strong physical cross-links which memorize the original shape.

2.3.6 Thermoplastic SMPs Based on Polymer Blends

The mixing of two or more different polymers makes it possible to achieve various property combinations within the final material in a cost-effective way. In the case of SMP blends, even if the two polymers do not show shape-memory effect on their own, they result in shape-memory behavior when being processed in a blend.

2.3.6.1 Polyethylene–Nylon-6 Blend

A blend of high-density polyethylene (PE) and nylon-6 produced by a reactive blending was reported as showing shape-memory properties.[73] Nylon-6, which has a high melting temperature of around 200°C, forms the hard segment in the matrix of semicrystalline PE. The switching temperature of the thermally induced shape-memory effect is given by the melting point of PE crystallites at 120°C. An excellent shape-fixity ratio of around 99% and a shape recovery ratio between 95% and 97% are determined for these blend materials.

2.3.6.2 Poly(Vinyl Acetate)-Based Polymer Blends

Two miscible blend systems based on poly(vinyl acetate) (PVAc) have been investigated: PVAc with PLLA, and PVAc with poly(vinylidene fluoride) (PVDF).[74,75]

PVAc is an amorphous polymer with a T_g at 35°C, while the PLLA is a semicrystalline material ($T_g = 56°C$, $T_m = 165°C$). The PLLA/PVAc blends showed complete miscibility at blending ratios of 50/50, 40/60, 30/70, 20/80, suggested by a single glass-transition temperature in the range of 40°C–50°C and one melting temperature in the range of 140°C–160°C. Here, the PLLA formed the hard-segment domains and the T_g of the blend served as the switching temperature of the shape-memory effect.

Similar to the PLLA, the PVDF showed semicrystalline features and had a degree of crystallinity of about 50%. The blends of the PVDF/PVAc showed a switching temperature in the range of 20°C–35°C. The samples recovered their original shape within 10 s.

2.3.6.3 Poly(Vinyl Chloride)/Thermoplastic Polyurethane Blends

The miscibility of poly(vinyl chloride) (PVC) and the segmented thermoplastic PUs formed from HDI, DHBP and the PCL diols was investigated, and the shape-memory properties of the resulting blends were examined.[76] The PVC was miscible with the PCL segment in PUs, and the mixed amorphous phase served as the switching segment, and no melting peak of the PCL segments was observed by the DSC analysis.

2.3.7 Summary

While the shape-memory effect has been reported for a wide variety of polymer structures, the discovery of the shape-memory effect in PUs in the 1980s has aroused great interest in studying shape-memory phenomenon in polymers. Among them, PU SMPs are the most systematically studied. By the introduction of different building blocks, a two-phase heterogeneous structure consisting of a hard segment and a soft segment is typical for PU shape-memory materials. The hard segments with a high thermal-transition temperature form the "net points" and determine the primary structure, while the soft/switching segments show a thermal-transition at the temperature of the shape-transition, and work as a reversible molecular switch.

The shape-memory effect was also found in several other types of thermoplastic polymers and blends. In some of these systems, molecular chain entanglements act as physical cross-links, and determine the primary shape. Due to the dynamic nature of entanglements, the materials show strong creep effects (and thus low shape-recovery ratios) when the polymer is stretched for a large deformation or when keeping the deformation at a high temperature for a longer period of time. The majority of thermoplastic SMPs, however, exhibit a microphase-segregated structure that is responsible for the formation of primary and secondary cross-links.

From the recent systematic studies, promising concepts for the future design of thermoplastic SMPs with superior properties can be expected, including enhanced creep-resistance and high thermal stability.

2.4 Remarks and Outlook

The commercial application of cross-linked polyethylene as heat-shrinkable tubes, and the discovery of the shape-memory effect in PUs in the 1980s have aroused much interest in studying and developing SMPs, leading to a variety of structures of these smart materials. Several reviews have been published reporting on different aspects of SMPs.[1,3] In this chapter we focused on the versatile structures, and the relationship between the structures and the properties of these smart polymers.

SMPs can be classified into thermoset and thermoplastic types. In each type of SMPs, either the crystalline or amorphous phase can serve as the switching segments (domains). The SMPs with a crystalline switching-segment have a relatively sharp thermal-transition in most cases, and offer relatively easy prediction and fine tuning of recovery temperature by the temperature of the soft-segment-forming macrodiol. On the other hand, SMPs with amorphous switching segments usually have a broad switching temperature range and in most cases a mixed T_g is formed from the hard segment and soft segment forming materials. However, amorphous polymers offer better transparency when compared to their crystalline counterpart.

Relatively speaking, thermoset SMPs offer better shape-fixity and recovery (including recovery ratio, recovery time and force, etc.), and intrinsically better chemical and thermal stability. Thermoplastic SMPs have the advantage of the availability of different processing techniques and their properties can be more easily modified.

Compared to shape-memory alloys (SMAs), SMPs can undergo large shape changes and can be modified to exhibit different thermal and mechanical properties. However, SMPs often exhibit lower recovery stress than SMAs. Different methods can be utilized to improve the mechanical properties, especially in the range of the transition temperature. Chemical methods include (1) increasing the cross-link densities in the case of the thermoset, (2) increasing the stability of the physically cross-linked hard segment by introducing aromatic or diamine chain extenders, a rigid and symmetrical aromatic diisocyanate or by increasing the hard-segment content. Physical methods include the optimized processing procedure comprising the deformation temperature, deformation time and deformation ratio. For example, deformation at $T < T_{trans}$ was reported to decrease the switching temperature. Thus, the shape switching at a temperature $< T_{trans}$ will enhance the recovery force.

Although the shape-memory effect has been found in a variety of structures of polymers, the majority of these materials are not specially designed as shape-memory materials, and many of them have not been systematically investigated. Application-driven research with the involvement of companies will greatly speed up the development of SMPs.[77,78] Catalyzed by the present investigations on the basic structure-property-relationships, we expect that SMPs with specific characteristics based on molecular design concepts will be available to meet the challenges for specific products and systems in sophisticated biomedical and technical applications.

References

1. Reviews on SMPs: (a) Liu C., Qin H., and Mather P. T., *J. Mater. Chem.*, 2007, *17*, 1543–1558; (b) Ratna D. and Karger-Kocsis J., *J. Mater. Sci.*, 2008, *43*, 254–269; (c) Rousseau I., *Polym. Eng. Sci.*, 2008, *48*, 2075–2089; (d) Mondal S. and Hu J. L., *Des. Monom. Polym.*, 2006, *9*(6), 527–550.
2. Monkman G. J., *Mechatronics*, 2000, *10*, 489.
3. Lendlein A. and Kelch S., *Angew. Chem. Int. Ed.*, 2002, *41*, 2034–2057.
4. Lendlein A., Jiang, H. Y., Jünger O., and Langer R., *Nature*, 2005, *434*, 879–882.
5. Schmidt A. M., *Macromol. Rapid Commun.*, 2006, *27*, 1168–1172.
6. Mishra J. K. and Das C. K. *Elastom. Plas.*, 2001, *33*, 137–153.
7. Hoffmann J. W., *IEEE Electr. Insul. Mag.*, 1991, *7*, 16–23.
8. Ota S., *Radiat. Phys. Chem.*, 1981, *18*, 81–87.
9. Morshedian J., Khonakdar H. A., Mhrabzadeh M., and Eslami H., *Adv. Polym. Technol.*, 2003, *22*, 112–119.
10. Hoffman J. W., *IEEE Electr. Insul. Mag.*, 1991, *7*, 33.

11. Gal O., Basic D., and Petric M., in *Integration of Fundamental Polymer Science and Technology*, Lemstra P. J. and Kleintjes L. A. (eds.), Elsevier, New York, 1989.
12. Liu C., Chun S. B., and Mather P. T., *Macromolecules*, 2002, *35*, 9868–9874.
13. Yang Z. H., Hu J. L., Liu Y. Q., and Yeung L. Y., *Mater. Chem.*, 2006, *98*, 368.
14. Chen W., Che C., and Gu X., *J. Appl. Polym. Sci.*, 2002, *84*, 1504–1512.
15. Zhang S., Feng Y., Zhang L., Sun J., Xu X., and Xu Y., *J. Polym. Sci. A*, 2007, *45*, 768–775.
16. Chun B. C., Chong M. W., and Chung Y. C., *J. Mater. Sci.*, 2007, *42*, 6542–6531.
17. Chun B. C., Chu T. K., Chung M. H., and Chung Y. C., *J. Mater. Sci.*, 2007, *42*, 9045–9056.
18. Feng Y., Zhang L., Li X., Xue Y., and Zhang S., *Polym. Mater. Sci. Eng.*, 2006, *95*, 585.
19. Lazdina B., Stima U., Tupureina V., and Sevastyanova I., *Mater. Chem.*, 2006, *13*, 26–33.
20. Alteheld A., Feng Y. K., Kelch S., and Lendlein A., *Angew. Chem.*, 2005, *117*, 2–6.
21. Hu J., Yang Z., Yeung L., Li F., and Liu Y., *Polym. Intern.*, 2005, *54*, 854–859.
22. Wilson T. S., Bearinger J. P., Herberg J. L., Marion J. E., Wright W. J., Evans C. L., and Maitland D. J., *J. Appl. Polym. Sci.*, 2007, *106*, 540–551.
23. Vernon L. and Harold M., U.S. Patent 2234993, 1941.
24. Lui C. and Mather P. T., *J. Appl. Med. Polym.*, 2002, *6*, 47–52.
25. Lendlein A., Schmidt A. M., Schroeter M., and Langer R. *J., Polym. Sci., A*, 2005, *43*, 1369–1381.
26. Kelch S., Steuer S., Schmidt A. M., and Lendlein A., *Biomacromolecules*, 2007, *8*, 1018–1027.
27. Lendlein A., Schmidt A. M., and Langer R., *PNAS*, 2001, *98*, 842–847.
28. Choi N. Y. and Lendlein A., *Soft Matter*, 2007, *3*, 901.
29. Kagami Y., Gong J. P., and Osada Y., *Macromol. Rapid Commun.*, 1996, *17*, 539–543.
30. Reintjens W., Prez F. E., and Goethals E. J., *Macromol. Rapid Commun.*, 1999, *20*, 251–255.
31. Li J., Viveros J. A., Wrue M. H., and Anthamatten M., *Adv. Mater.*, 2007, *19*, 2851–2855.
32. Bellin I., Kelch S., Langer R., and Lendlein A., *PNAS*, 2006, *103*, 18043–18047.
33. Cao Y., Guan Y., Du J., Luo J., Peng Y., Yip C. W., and Chan A. S., *J. Mater. Chem.*, 2002, *12*, 2957–2960.
34. Liu G., Ding X., Cao Y., Zheng Z., and Peng Y., *Macromol. Rapid Commun.*, 2005, *26*, 649–652.
35. Li F. and Larock R., *J. Appl. Polym. Sci.*, 2002, *84*, 1533–1543.
36. Zhu G., Xu S., Wang J., and Zhang L., *Rad. Phys. Chem.*, 2006, *75*, 443–448.
37. Rousseau I. A. and Mather P. T., *J. Am. Chem. Soc.*, 2003, *125*, 15300–15301.
38. Castonguay M., Koberstein J. T., Zhang Z., and Laroche G., Synthesis, physico-chemical and surface characteristics of polyurethanes, in *Biomedical Applications of Polyurethanes*, Vermette P., Griesser H. J., Laroche G., and Guidoin R. (eds.), Landes Bioscience, Austin, TX, 2000, pp. 1–21.
39. Lin J. R. and Chen L. W., *J. Appl. Polym. Sci.*, 1998, *69*, 1563–1574.
40. Lin J. R. and Chen L. W., *J. Appl. Polym. Sci.*, 1998, *69*, 1575–1586.
41. Wang W. S., Ping P., Chen X. S., and Jing X. B., *Polym. Int.*, 2007, *56*, 840–846.
42. Li F. K., Zhang X., Hou J. N. et al., *J. Appl. Polym. Sci.*, 1997, *64*, 1511–1516.

43. Li F. L., Hu J. L., Li T. C., and Wong Y. W., *Polymer*, 2007, *48*, 5133–5145.
44. Kim B. K., Lee S. Y., and Xu M., *Polymer*, 1996, *37*(26), 5781–5793.
45. Jeong H. M. Ahn B. K., Cho S. M., and Kim B. K., *J. Polym. Sci. B Polym. Phys.*, 2000, *38*(23), 3009–3017.
46. Jeong H. M., Ahn B. K., and Kim B. K., *Polym. Int.*, 2000, *49*, 1714–1721.
47. Chun B. C., Cho T. K., and Chung Y. C., *Eur. Polym. J.*, 2006, *42*, 3367–3373.
48. Takahashi T., Hayashi N., and Hayashi S., *J. Appl. Polym. Sci.*, 1996, *60*, 1061–1069.
49. Zhu Y., Hu J. L., Yeung L. Y., Liu Y., Ji F. L., and Yeung K. W., *Smart Mater. Struct.*, 2006, *15*, 1547–1554.
50. Ji F. L., Zhu Y. Z., Hu J. L., Liu Y., Yeung L. Y., and Ye G. D., *Smart Mater. Struct.*, 2006, *15*, 1385–1394.
51. Kim B. K., Shin Y. J., Cho S. M., and Jeong H. M., *J. Polym. Sci. B: Polym. Phys.*, 2000, *38*, 2652–2657.
52. Wang W. S., Ping P., Chen X. S., and Jing X. B., *J. Appl. Polym. Sci.*, 2007, *104*, 4182–4187.
53. Jeong H. M., Kim B. K., and Choi Y. J., *Polymer*, 2000, *41*, 1849–1855.
54. Kim B. K., Lee S. Y., Lee J. S. et al., *Polymer*, 1998, *39*(13), 2803–2808.
55. Zhu Y. Z., Hu J. L., Yeung K. W. et al., *J. Appl. Polym. Sci.*, 2007, *103*, 545–556.
56. Gunatillake P., Mayadunne R., and Adhikari R., *Biotechnol. Annu. Rev.*, 2006, *12*, 301–347.
57. Tuominen J., Kylma J. Kapanen A. et al., *Biomacromolecules*, 2002, *3*, 445–455.
58. Lendlein A. and Langer R., *Science*, 2002, *296*, 1673–1676.
59. Grablowitz H. G., PhD thesis, 2002, RWTH Aachen, Germany.
60. Min C. C., Cui W. J., Bei J. Z., and Wang S. G., *Polym. Adv. Technol.*, 2005, *16*, 608–615.
61. Luo X. L., Zhang X. Y., Wang M. T., Ma D. Z., Xu M., and Li F. K., *J. Appl. Polym. Sci.*, 1997, *64*, 2433–2440.
62. Wang M. T., Luo X. L., Zhang X. Y., and Ma D. Z., *Polym. Adv. Technol.*, 1997, *8*, 136–139.
63. Ma D. Z., Wang M. T., Wang M. C., Zhang X. Y., and Luo X. L., *J. Appl. Polym. Sci.*, 1998, *69*, 947–955.
64. Min C. C., Cui W. J., Bei J. Z., and Wang S. G., *Polym. Adv. Technol.*, 2007, *18*, 299–305.
65. Zini E., Scandola M., Dobrzynski P., Kasperozyk J., and Bero M., *Biomacromolecules*, 2007, *8*, 3661–3667.
66. Sakurai K., Shirakawa Y., Kashiwagi T., and Takahashi T., *Polymer*, 1994, *35*, 4238–4239.
67. Sakurai K, Tanaka H., Ogawa N., and Takahashi T., *J. Macromol. Sci. Phys. B*, 1997, *36*, 703–716.
68. Naga, N., Wakita Y., and Murase S., *J. Appl. Polym. Sci.*, 2008, *110*(6), 3770–3777.
69. Sakurai K. and Takahashi T., *J. Appl. Polym. Sci.*, 1989, *38*, 1191–1194.
70. Sakurai K., Kashiwagi T., and Takahashi T., *J. Appl. Polym. Sci.*, 1993, *47*, 937–940.
71. Jeon H. G., Mather P. T., and Haddad T. S., *Polym. Int.*, 2000, *49*, 453–457.
72. Jeong H. M., Song J. H., Chi K. W., and Kim K. T., *Polym. Int.*, 2002, *51*, 275–280.
73. Li F., Chen Y., Zhu W., Zang X., and Xu M., *Polymer*, 1998, *39*(26), 6929.

74. Liu C. and Mather P. T., *Proceedings of the Annual Technical Conference—Society of Plastics Engineers, 61st* (vol. 2), Society of Plastics Engineers, Brookfield, CT, 2003, pp. 1962–1966.
75. Campo C. J. and Mather P. T., *Polym. Mater. Sci. Eng.*, 2005, *93*, 933.
76. Jeong H. M., Song J. H., Lee S. Y., and Kim B. K., *J. Mater. Sci.*, 2001, *36*, 5457–5463.
77. www.mnemoscience.com
78. www.ctd-materials.com/products/emc.htm

3

Thermomechanical Behavior and Modeling Approaches

Hang Jerry Qi and Martin L. Dunn

Department of Mechanical Engineering, University of Colorado, Boulder, Colorado

CONTENTS

3.1 Introduction

Shape-memory polymers (SMPs) are an exciting class of materials that have the ability to store a temporary shape, possibly for long periods of time, and then recover a predetermined permanent shape when subjected to an environmental stimulus. We review the state-of-the-art with regard to the behavior of SMPs. Even though numerous environmental stimuli have recently been shown to effect shape-memory behavior, we focus on the most common driver for shape-memory behavior, i.e., temperature. After describing the basic thermomechanical behavior of SMPs, both in terms of macroscopic behavior and molecular mechanisms, we describe recent efforts to develop three-dimensional, large deformation constitutive models for SMPs. We classify the modeling approaches that have been developed and illustrate their capabilities in comparison to the thermomechanical response of various polymers during shape-memory cycles. To enable a rational engineering design using SMPs, constitutive models will ultimately have to be implemented in a finite element code. We describe one such implementation and illustrate the capabilities of such a modeling approach by simulating the behavior of a polymer sheet that folds into a cubic box during a shape-memory cycle.

3.1.1 Brief History—Current Status

Shape-memory polymers (SMPs), both thermosets and thermoplastics, are an exciting class of polymers that have the ability to store a temporary shape, possibly for long periods of time, and then recover a predetermined permanent shape when subjected to an environmental stimulus. The most common stimulus is temperature; however, recent efforts have focused on other stimuli, including light, magnetic fields, humidity, and chemical concentration fields [1]. While SMPs have existed and been used in engineering applications for decades, (e.g., cross-linked polyethylene was employed for heat shrink as early as the 1960s [2]), modern applications in the biomedical, aerospace, and transportation industries, among others, are fueling interest in them from new perspectives. These include the potential for SMPs to assume multiple stable shapes that can be switched among each other [3]. Shape-memory polymers share some macroscopic behavioral features with the more common metallic shape-memory alloys, but they promise to impact different and more diverse applications. The application potential is especially attractive because it can take advantage of many existing polymer processing techniques, e.g., various extrusion and molding processes. This can make SMP processing less expensive than SMA processing. Furthermore, SMPs can be tailored via the polymer chemistry to yield a wide range of macroscopic mechanical and physical properties. The synthetic design can also endow them with multifunctionality. Perhaps the most striking is the ability to biodegrade, which renders them attractive for implantable biomedical applications. SMPs can also be used as a matrix in various composite architectures ranging from nanocomposites [4–6] to deployable structural composites [7,8].

3.1.2 Macroscopic Response of SMPs—Thermomechanical

We think the best way to describe the behavior of SMPs is through a discussion of the macroscopic response during a thermomechanical shape-memory cycle in concert with the molecular mechanisms that drive the response. To this end, we adopt Behl and Lendlein's [1] somewhat broad characterization of an SMP as a polymer network endowed with permanent "net points" and molecular "switches" as shown in Figure 3.1. The net points (cross-links formed by chemical or physical bonds, depending on the material system)

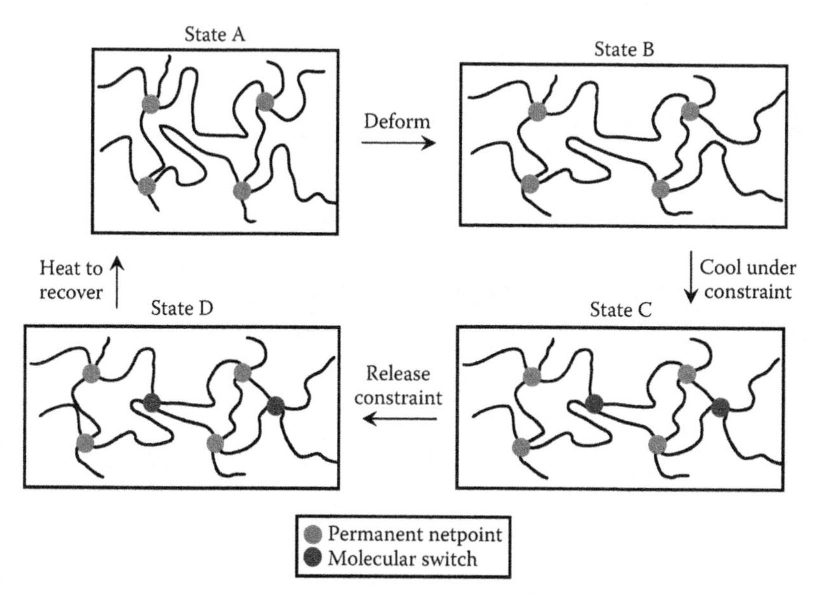

FIGURE 3.1
Idealization of a polymer network illustrating the molecular mechanisms that drive the shape-memory effect. (From Behl, M. and Lendlein, A., *Mater. Today*, 10(4), 20, 2007. With permission.)

between molecular chains are permanent and set a permanent shape to the polymer. The switches are essentially reversible cross-links and can also be formed by physical as well as chemical bonds.

A macroscopic shape-memory cycle can be considered as a subset of a more general thermomechanical response, but it is valuable to focus on it because it is perhaps the most fundamental characteristic that describes SMP behavior. Our approach will be to describe in detail the thermomechanical behavior during a shape-memory cycle; our discussion largely follows that of Liu et al. [9], Behl and Lendlein [1], and Qi et al. [10]. Using it as our reference, we will then describe the shape-memory response under other environmental loadings. Then, we will review the SMP materials' palette and explore the molecular mechanisms that lead to shape-memory behavior.

A three-dimensional strain-stress-temperature diagram of the shape-memory effect under uniaxial tensile loading is shown in Figure 3.2. A typical tensile predeformation and strain-recovery cycle for a polymer can be described by five states A→B→C→D→E with four processes between them: ①→②→③→④. The thermomechanical cycle begins with the material in its permanent shape at a temperature, T_h, that is above its glass-transition temperature, T_g. We call this state "A." Here, the material is rubbery; we take it to be stress and strain free. The first step of the cycle, process ①, involves deforming the sample, taking it from A→B. In state B, the deformation and stress are denoted by ε_h and σ_h. At both A and B we can consider the molecular switches to be "open."

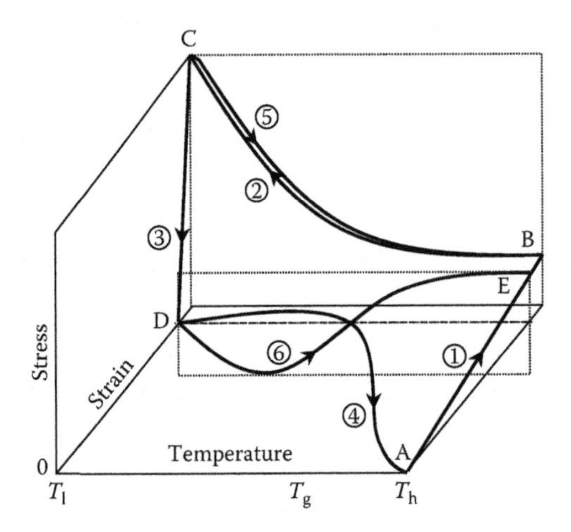

FIGURE 3.2
Macroscopic thermomechanical behavior during a shape-memory cycle. (From Liu, Y.P. et al., *Int. J. Plast.*, 22(2), 279, 2006. With permission.)

Process ② involves cooling the material from B→C. State C is at a temperature $T_l < T_{tr} < T_h$. T_{tr} is a transformation temperature, the nature of which depends on the specific material. For example, T_{tr} can be a glass-transition temperature, T_g, or a melting temperature, T_m. If T_{tr} represents a glass transition, the material is in a glassy state at C. During process ② the strain is held constant at ε_h and the molecular switches "close" at T_{tr}, helping to fix the shape of the material. The actual closing of the switches can occur over a temperature range centered at T_{tr} and the breadth of this range is material-dependent. Due to thermal contraction and potentially other molecular mechanisms, the tensile stress needed to maintain the predeformed shape increases to a value σ_l.

In process ③ the strain constraint at T_l, is removed, i.e., the material is unloaded from C→D. This typically results in some level of "springback," resulting in a final strain at T_l, of ε_l. A metric often used to describe the ability of an SMP to store a temporary strain is the strain fixity ratio:

$$R_f = \frac{\varepsilon_l}{\varepsilon_h} \tag{3.1}$$

An ideal material would have $R_f = 1$. In state D, the molecular switches are still closed and the temporary shape of the polymer is fixed. The material can be kept in "hibernation" at T_l for an extended period without the recovery of the strain ε_l.

Next, the sample is heated from T_l to T_h during which the molecular switches are reopened. As the material passes through T_{tr}; again, the switches actually open over a temperature range, the breadth of which is material-dependent.

This allows the SMP to "recover" its original permanent shape. The degree of recovery depends on the mechanical constraint applied during this process. Extreme cases are no constraint (process ④–free recovery) and rigidly clamped (process ⑤–constrained recovery), and these are shown in Figure 3.1. For free recovery ($D \rightarrow A$), the residual strain at T_h is denoted by ε_p. The strain-recovery ratio can be defined as:

$$R_r = \frac{\varepsilon_h - \varepsilon_p}{\varepsilon_h} \tag{3.2}$$

For an ideal material $R_r = 1$. This is nearly realized for many materials and in such cases, the final state of the material is essentially the starting state A. For constrained recovery, ($D \rightarrow B$) the strain is fixed at ε_l during heating. As the temperature is increased, the constraint prevents thermal expansion and induces a compressive stress on the sample. This compressive stress builds up and reaches a maximum in the glassy state in the range $T_l - T_{tr}$. As the temperature is further increased, the material softens as $T \rightarrow T_{tr}$ and the stress becomes tensile. At T_h (state E), both the stress and strain are typically close to those of the predeformation state B.

A thermomechanical SMP cycle can be repeated several times and the strain stored and recovered may vary as a function of the number of cycles. Furthermore, all of these properties, and the thermomechanical behavior in general, depend strongly on the temperature rate during the cycle [11]. These are important considerations for applications and are being actively studied by a number of research groups (e.g., Safranski and Gall [12]; Qi et al. [10]).

The discussion here has been cast in terms of a uniaxial tension deformation. Similar discussions in the context of the uniaxial compression can be found in Liu et al. [2] and Qi et al. [10]. Furthermore, the basic phenomena described here occur in a multiaxial loading scenario as well. Indeed, it is such a scenario that typically occurs in applications (see, for example, Liu et al. [2], Qi et al. [10], and Behl and Lendlein [1]), and understanding the shape-memory behavior in such cases is of utmost importance for the rational design of components with these novel materials.

3.1.3 Macroscopic Response of SMPs—Environmental

One of the most exciting and promising recent advances in shape-memory polymers over the last few years is the activation of the shape-memory effect via other environmental stimuli. Shape-memory activation by chemical, optical, electrical, and magnetic fields has recently been demonstrated [13–23]. One way these mechanisms work is to indirectly activate the thermal shape-memory effect, either by producing a temperature change that spans the appropriate transition temperature of the polymer, or by changing the transition temperature relative to the environment. For example, Behl and Lendlein [1] demonstrated a thermoplastic SMP with embedded iron (III) oxide core/silica shell nanoparticles. These materials change shape by converting energy from a remotely applied alternating magnetic field into heat. In general, details regarding thermomechanical behavior depend on molecular architecture and microstructural details, but these studies demonstrate performance comparable to that obtained by simply changing the environmental temperature. Recently, light-activated shape-memory polymers have been developed that exhibit the shape-memory effect when irradiated by light at specified wavelengths. The underlying mechanisms that drive light-activated polymer deformation differ significantly from those that drive thermally induced shape-memory. Light, as a stimulus, is particularly intriguing because it allows actuation of polymeric materials and devices without physical contact and offers the potential to control the location where the stimulus is applied. Lendlein et al. [24,25] and Scott et al. [26,27] have developed light activated polymers that exhibit the shape-memory effect. In these materials, photoactivation is achieved through a photochemically induced network adaptation/rearrangement, which leads to changes of the mechanical states of the material. Lendlein and co-workers [24,25] developed an amorphous, covalently cross-linked polymer with phototunable molecular cross-links, while Scott et al. [26,27] recently developed a photomechanical shape-memory polymer that is capable of photo-induced stress relaxation at particular wavelengths. In both classes of materials, the respective light-activated molecular mechanisms were exploited to demonstrate photo-induced shape-memory. A theoretical framework for modeling light activated SMPs has recently been developed by Long et al. [28] and it is in excellent agreement with the photomechanical response measured by Scott et al. [26,27].

3.1.4 Description of Mechanisms for Modeling

Before describing approaches to model the thermomechanical behavior of shape-memory polymers, we briefly describe mechanisms of the shape-memory effect in polymers. Lendlein et al. [29], Liu et al. [9], and Qi et al. [10] gave in-depth discussions of the underlying physical mechanisms of thermally induced shape-memory effects in various materials. Here, we briefly discuss the behavior broadly from the perspective of a thermoset with fixed and switchable cross-links that are controlled by the passage of the materials through its glass transition. As many authors have noted, a similar mechanistic description can be made for thermoplastics where crystallization, for example, is the operant mechanism. So, for the purpose of modeling, we consider that the shape-memory effect is caused by the transition of a cross-linked polymer from a state dominated by entropy (the rubbery state) to a state dominated by internal energy (the glassy state) as the temperature decreases. At temperatures above the glass transition temperature T_g, individual macromolecular chains undergo large random conformational changes.

Deforming the material reduces the number of possible configurations, and hence, the configurational entropy of the macromolecular chains, leading to the well-known entropic behavior of elastomers. After the removal of the external load at a temperature above T_g, the tendency of the material to increase its entropy will cause it to recover the undeformed (processed) shape defined by the spatial arrangement of cross-linking sites. However, this shape-recovery can be interrupted by lowering the temperature to below T_g. There, the mobility of the macromolecular chains is significantly reduced by the reduction in the free volume, and the conformational change of individual macromolecules becomes increasingly difficult. Instead, the cooperative conformational change of neighboring chains becomes dominant and the deformation thus requires a much higher energy. Therefore, the removal of the mechanical load at temperatures below T_g induces only a small amount of shape–recovery, and most of the deformation incurred at the temperature above T_g is retained (stored, or frozen). The shape-memory effect is invoked as the temperature increases above T_g, where the individual macromolecular chains become active again and the shape-recovery mechanism described above is permitted. In this sense, the shape-memory effect can be viewed simply as a temperature-delayed recovery.

3.2 Modeling of Thermomechanical Behavior of SMPs

3.2.1 Backdrop for Modeling

Almost all reported applications of SMPs involve their use in situations where they undergo large, 3D deformations. However, the understanding of the thermomechanical behavior of SMPs has until recently been limited to cases of small, uniaxial deformations. The development of components using SMPs will benefit tremendously from the constitutive models that describe the complex 3D, finite deformation thermomechanical response of SMPs. Ultimately, these need to be implemented in a finite element code to enable rational engineering analysis and design.

Based on the macroscopic and mechanistic understanding described above, several models have been developed in recent years to describe the complex thermomechanical behavior of SMPs. Broadly, these are considered constitutive models and can be categorized in two groups: models based on phase transition and models based on viscoelasticity. In the first group, the polymer is considered as a mixture of glassy and rubbery phases. Constitutive models developed by Liu et al. [9], Chen and Lagoudas [30,31], Qi et al. [10], Barot and Rao [32], and Barot et al. [33] belong to this category. For amorphous polymers, such an assumption is phenomenological. However, as discussed later, the phase transition model enjoys a great flexibility in extending existing polymer constitutive models for both rubbery and glassy behaviors to model shape-memory effects. In the models developed by Liu et al. [9] and Chen and Lagoudas [30,31], in addition to the phase transition, a concept of storage deformation is used. Whilst in recent models developed by Qi et al. [10], Barot and Rao [32], and Barot et al. [33], storage deformation is not used. As such, we further categorize the former models as models based on storage deformation and the latter models as models based on phase transition. In the second group, constitutive models for SMPs can be viewed as an extension of viscoelastic models for thermomechanical behavior through the glass transition by requiring viscosity to be a nonlinear function of temperature. Models in this category include those developed by Tobushi et al. [34,35] and Nguyen et al. [36].

In our brief review, we focus only on models developed and used specifically to model the shape-memory behavior of polymers. We omit a discussion on the broader efforts directed at constitutive modeling of the thermomechanical behavior of polymers through the glass transition, such as those of Angell et al. [37] and Adolf and Chambers [38]. Indeed, with proper interpretation, these models can be applied to describe shape-memory behavior of polymers.

3.2.2 Models Based on Storage Deformation

Recently, Liu et al. [9] presented a comprehensive study of the thermomechanical behavior of SMPs and developed a constitutive model for small deformation behavior. Diani et al. [39] extended this model to the case of finite deformation. A key concept in their model is that the polymer is assumed to be a mixture to two types of extreme phases: the active phase and the frozen phase. For the active phase, the conformational change of macromolecular chains is allowed and the deformation is entropic. For the frozen phase, the conformational change corresponding to high temperature entropic deformation is completely stored, while internal energetic changes can occur. Although the phase transition is phenomenological for SMPs based on amorphous polymers, it retains a physical significance as the mechanical behavior of SMPs undergo a transition from a state dominated by entropy (rubbery behavior) at high temperatures to a state dominated by internal energy (glassy behavior) at low temperatures, as discussed in detail by Qi et al. [10]. This transition can be captured by considering the change in Helmholtz free energy as:

$$H_{\text{total}} = f_{\text{r}}(T)H_{\text{r}} + f_{\text{g}}(T)H_{\text{g}}, \tag{3.3}$$

where

H_{total} is the total Helmholtz energy at a given temperature T
H_{r} is the Helmholtz energy of the rubbery material at $T \gg T_{\text{g}}$
H_{g} is the Helmholtz energy of the glassy material at $T \ll T_{\text{g}}$
f_{r} and f_{g} are functions of temperature
$f_{\text{r}} + f_{\text{g}} = 1$

At $T \gg T_g$, $f_r \approx 1$ and $f_g \approx 0$ so that $H_{total} \approx H_r$; at $T \ll T_g$, $f_r \approx 0$ and $f_g \approx 1$ so that $H_{total} \approx H_g$. Therefore, evolving f_r and f_g will capture the transition of Helmholtz energy. This energy consideration resembles a description of a phase-transition process. Therefore, for the sake of a model description, it is possible to phenomenologically assume that there exists a rubbery phase and a glassy phase in the material, and the transition between these two phases is realized through the change of the volume fraction of each phase.

Following the phase argument, the total stress and strain can be expressed as [9]:

$$\boldsymbol{\sigma} = f_f \boldsymbol{\sigma}_f + (1 - f_f)\boldsymbol{\sigma}_a, \quad \boldsymbol{\varepsilon} = f_f \boldsymbol{\varepsilon}_f + (1 - f_f)\boldsymbol{\varepsilon}_a, \tag{3.4}$$

where

- $\boldsymbol{\sigma}$ and $\boldsymbol{\varepsilon}$ are the total stress and the total strain
- $\boldsymbol{\sigma}_f$ and $\boldsymbol{\varepsilon}_f$ are the stress and the strain of the frozen phase
- $\boldsymbol{\sigma}_a$ and $\boldsymbol{\varepsilon}_a$ are the stress and the strain of the active phase
- f_f is the volume fraction of the frozen phase, and its evolution rule can be described as

$$f_f = 1 - \frac{1}{1 + c_f (T_h - T)^n}, \tag{3.5}$$

where

- T_h is a reference temperature
- c_f and n are the fitting parameters that can be determined via appropriate experiments

Another key concept in Liu et al. [9] is the introduction of the storage deformation. In Liu et al. [9], the deformation in the frozen phase arises from three parts:

$$\boldsymbol{\varepsilon}_f = \int_0^{f_f} \boldsymbol{\varepsilon}_e^f(x)\,\mathrm{d}f + \boldsymbol{\varepsilon}_f^i + \boldsymbol{\varepsilon}_f^T, \tag{3.6}$$

where

- $\boldsymbol{\varepsilon}_f^i$ is the internal energetic strain
- $\boldsymbol{\varepsilon}_f^T$ is the thermal strain

The first term on the right hand side of Equation 3.4 is called the "stored" (frozen) entropic strain,

$$\boldsymbol{\varepsilon}_s = \int_0^{f_f} \boldsymbol{\varepsilon}_f^e(\mathbf{x})\,\mathrm{d}f, \tag{3.7}$$

which stores the active strain as the material is cooled down. For the active phase,

$$\boldsymbol{\varepsilon}_{\mathrm{a}} = \boldsymbol{\varepsilon}_{\mathrm{a}}^{\mathrm{e}} + \boldsymbol{\varepsilon}_{\mathrm{a}}^{\mathrm{T}}. \tag{3.8}$$

The total strain at any moment during the thermomechanical history is

$$\boldsymbol{\varepsilon} = \boldsymbol{\varepsilon}_{\mathrm{s}} + \left[f_{\mathrm{f}}\boldsymbol{\varepsilon}_{\mathrm{f}}^{\mathrm{i}} + (1-f_{\mathrm{f}})\boldsymbol{\varepsilon}_{\mathrm{a}}^{\mathrm{e}}\right] + \left[f_{\mathrm{f}}\boldsymbol{\varepsilon}_{\mathrm{f}}^{\mathrm{T}} + (1-f_{\mathrm{f}})\boldsymbol{\varepsilon}_{\mathrm{a}}^{\mathrm{T}}\right]. \tag{3.9}$$

In Liu et al. [9], $\boldsymbol{\varepsilon}_{\mathrm{s}}$ is treated as an internal variable with the following evolution rule,

$$\frac{\mathrm{d}\boldsymbol{\varepsilon}_{\mathrm{s}}}{\mathrm{d}T} = \boldsymbol{\varepsilon}_{\mathrm{f}}^{\mathrm{e}}(\mathbf{x})\frac{\mathrm{d}f_{\mathrm{f}}}{\mathrm{d}T}. \tag{3.10}$$

In this model, the storage deformation concept plays the key role in capturing the shape-memory effect. Following Equation 3.7, the total strain consists of a storage strain, a mechanical strain, and a thermal strain, i.e.,

$$\boldsymbol{\varepsilon} = \boldsymbol{\varepsilon}_{\mathrm{s}} + \boldsymbol{\varepsilon}_{\mathrm{m}} + \boldsymbol{\varepsilon}_{\mathrm{T}}. \tag{3.11}$$

During the shape-fixing process, a total strain is maintained while the temperature is decreased. The evolution rule of Equation 3.8 states that the storage strain $\boldsymbol{\varepsilon}_{\mathrm{s}}$ increases with decreasing temperature. Without the thermal strain, Equation 3.9 results in a decrease in the mechanical strain. Eventually, the mechanical strain at a high temperature will be completely stored by the storage strain. When the thermal expansion is considered, a new mechanical deformation may be developed, but the mechanism of using a storage deformation to "memorize" the deformation remains the same.

Under the small deformation linear elastic condition, this model compared well with experimental results. Figure 3.3 shows a comparison between the experimental results and the model predictions for the stress evolution during cooling under different prestrain conditions (Figure 3.3A) and for free strain recovery during heating for polymers predeformed at different levels (Figure 3.3B).

Recently, a more comprehensive development based on the Liu et al. [9] model was carried out by Chen and Lagoudas for both, large deformation [30] and linearized small deformation cases [31]. Their model is in excellent agreement with the experimental results of Liu et al. [9].

3.2.3 Models Based on Phase Transition

Recently, Qi et al. [10] developed a 3D finite deformation constitutive model using the concept of phase-transition. A major difference of this model

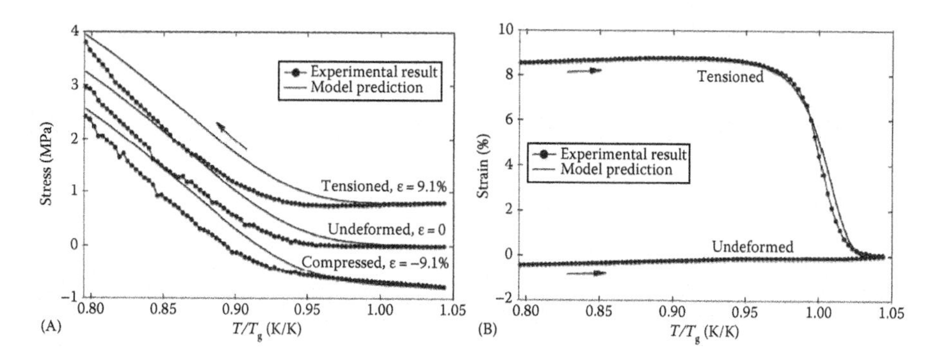

FIGURE 3.3
Comparison between experimental results and model predictions from Liu et al. [9] for (A) stress evolution during cooling under different prestrain conditions; (B) free strain recovery during heating for polymers predeformed at different levels. (From Liu, Y.P. et al., *Int. J. Plast.*, 22(2), 279, 2006. With permission.)

from the model by Liu et al. [9] is that it does not use the concept of storage deformation. Instead, it is assumed that the deformation storage process can be captured by requiring that the glassy phase formed during cooling be of a different stress-free configuration. Therefore, the glassy phase can be divided into two phases: the initial glassy phase and the frozen glassy phase. Therefore, there are three phases in the model:

1. *Rubbery phase (RP)*: The RP dominates at temperatures well above the glassy transition temperature T_g. The RP volume fraction is f_r.
2. *Frozen glassy phase (FGP)*: The FGP refers to the newly formed glassy phase caused by a decrease in the temperature. The FGP volume fraction is f_T. As the temperature decreases, the volume fraction of this phase increases.
3. *Initial glassy phase (IGP)*: The IGP refers to the glassy phase in the initial configuration of the material. This phase of the material deforms when an external load is applied at the beginning of an analysis. The IGP volume fraction is f_{g0}.

The volume fractions of the phases satisfy $f_T + f_{g0} = f_g$ and $f_r + f_{g0} + f_T = 1.0$. Following the argument of dividing the glassy phase into an initial glassy phase and a frozen glassy phase, one can also distinguish the rubbery phase into an initial rubbery phase and a melting rubbery phase. Such a distinction for the rubbery phase is only necessary when one increases the temperature after the initial mechanical loading. However, such a case is not a typical thermo-mechanical loading in SMP applications.

With the three phases defined above, the total stress is

$$\mathbf{T} = f_r \mathbf{T}_r + f_{g0} \mathbf{T}_{g0} + f_T \mathbf{T}_T. \tag{3.12}$$

As the temperature changes, the volume fractions of individual phases will change. In Qi et al. [10], the volume fractions of individual phases as functions of the temperature are defined as

$$f_r = \frac{1}{1+\exp\left[-(T-T_r)/A\right]}, \quad f_g = 1 - \frac{1}{1+\exp\left[-(T-T_r)/A\right]} \tag{3.13}$$

where

A is a parameter that characterizes the width of the phase transition zone

T_r is a reference temperature and is close to T_g

These equations are similar to the VTF functions [40–42], which are typically used to characterize the viscosity of a liquid as a function of temperature.

For the glassy phases, the IGP and the FGP evolve differently during cooling and reheating. Prior to a thermal loading, the initial values of the IGP and the FGP are

$$f_{g0}\big|_{t=0} = f_g\big|_{t=0}, \quad f_T\big|_{t=0} = 0.0. \tag{3.14}$$

During cooling, a fraction of the rubbery phase will transform into the glassy phase; this fraction is given by Δf_g. From the above discussion, the entire Δf_g becomes the FGP, i.e.,

$$f_{g0} = f_g\big|_{t=0}, \quad f_T\big|_{t=t2} = f_T\big|_{t=t1} + \Delta f_g \tag{3.15}$$

where $t_2 > t_1$, t_2 is the time immediately after t_1.

During reheating, the volume fraction Δf_g of the glassy phase transforms into the rubbery phase. Here, both the initial glassy phase and the frozen glassy phase will transform into the rubbery phase. Therefore, the partition of these two phases depends on their relative volume fraction ratio, i.e.,

$$\Delta f_{g0} = \frac{f_{g0}}{f_{g0}+f_T}\Delta f_g, \quad \Delta f_T = \frac{f_T}{f_{g0}+f_T}\Delta f_g \tag{3.16}$$

where

Δf_{g0} is the volume fraction from the IGP

Δf_T is the volume fraction from the FGP

Therefore,

$$f_{g0}\big|_{t=t4} = f_{g0}\big|_{t=t3} - \Delta f_{g0}, \quad f_T\big|_{t=t4} = f_T\big|_{t=t3} - \Delta f_T \tag{3.17}$$

where $t_4 > t_3$, t_4 is the time immediately after t_3.

As an SMP response at the temperature above T_g shows a rubber-like hyperelastic behavior, it is reasonable to model this behavior using a hyperelastic model. In Qi et al. [10], the Arruda–Boyce eight-chain model [43] is used, although other models could be used as well. The Cauchy stress tensor is defined as

$$\mathbf{T}_r = \frac{\mu_r}{3J}\frac{\sqrt{N}}{\lambda_{\text{chain}}}\mathcal{L}^{-1}\left(\frac{\lambda_{\text{chain}}}{\sqrt{N_r}}\right)\bar{\mathbf{B}}' + k\left[J - 1 - 3\alpha_1\left(T - T_0\right)\right]\mathbf{I} \tag{3.18}$$

where

μ_r is the initial shear modulus

N_r is the number of "rigid links" between the two cross-link sites (and/or strong physical entanglements)

k is the bulk modulus

α_1 is the thermal expansion coefficient at the temperature above T_g

T_0 is the initial temperature in an analysis

The first term on the left of Equation 3.18 describes the isochoric response of the material and the volumetric strain is taken out through $\bar{\mathbf{F}} = (\det[\mathbf{F}])^{-1/3}\mathbf{F}$. $\bar{\mathbf{B}}$ is the isochoric left Cauchy–Green tensor, $\bar{\mathbf{B}} = \bar{\mathbf{F}}\bar{\mathbf{F}}^{\mathrm{T}}$, and $\bar{\mathbf{B}}' = \bar{\mathbf{B}} - \mathrm{tr}(\bar{\mathbf{B}})\mathbf{I}/3$ are the deviatoric parts of $\bar{\mathbf{B}}$. $\lambda_{\text{chain}} = \sqrt{\bar{I}_1/3}$ is the stretch on each chain in the eight-chain network, and $\bar{I}_1 = \mathrm{tr}(\bar{\mathbf{B}})$ is the first invariant of $\bar{\mathbf{B}}$. $\mathcal{L}$ is the Langevin function defined as

$$\mathcal{L}(\beta) = \coth\beta - \frac{1}{\beta}. \tag{3.19}$$

The deformation of the initial glassy phase is represented by the total deformation gradient, $\mathbf{F}$. The viscoelastoplastic behavior of the SMPs at low temperatures can be captured by many models [44–57]. In Qi et al. [10], the method by Boyce and coworkers is used. The viscoelastoplastic behaviors of a polymer can be modeled by decomposing the stress response into an equilibrium time-independent behavior and a nonequilibrium time-dependent behavior. The total stress is

$$\mathbf{T}_{g0} = \mathbf{T}_{g0}^{\mathrm{r}} + \mathbf{T}_{g0}^{\mathrm{ve}}. \tag{3.20}$$

The hyperelastic spring (equilibrium response) can be modeled using the Arruda–Boyce eight chain model, as outlined in the previous section, but with different material parameters.

For the viscoelastoplastic deformation, the elastic deformation gradient is determined from the multiplicative decomposition of **F** into elastic and viscoplastic contributions,

$$\mathbf{F} = \mathbf{F}^{e}\mathbf{F}^{v} \tag{3.21}$$

where $\mathbf{F}^v$ is a relaxed configuration obtained by elastically unloading $\mathbf{F}^e$. The stress due to the viscoelastoplastic deformation can be calculated using $\mathbf{F}^e$, i.e.,

$$\mathbf{T}_{g0}^{ve} = \frac{1}{J^e}\left[\mathbf{L}^e : \mathbf{E}^e - \frac{\alpha_2 E}{1-2\nu}(T - T_0)\mathbf{I}\right] \tag{3.22}$$

where
- $J^e = \det(\mathbf{F}^e)$
- $\mathbf{E}^e = \ln \mathbf{V}^e$
- $\mathbf{V}^e = \mathbf{F}^e\,\mathbf{R}^e$
- α_2 is the thermal expansion coefficient at the temperature below T_g
- $\mathbf{L}^e$ is the fourth order isotropic elasticity tensor and

$$\mathbf{L}^e = 2\mu_g\,\mathcal{I} + \lambda_g\,\mathbf{I}\otimes\mathbf{I} \tag{3.23}$$

where
- μ_g and λ_g are Lame's constants
- $\mathcal{I}$ is the fourth-order identity tensor
- I is the second-order identity tensor

To determine $\mathbf{F}^e$ in Equation 3.21, the decomposition of the spatial velocity gradient is used

$$l = \dot{\mathbf{F}}\mathbf{F}^{-1} = \dot{\mathbf{F}}^{e}\mathbf{F}^{e-1} + \mathbf{F}^{e}l^{v}\mathbf{F}^{e-1} \tag{3.24}$$

where
- $\dot{\mathbf{F}}$ is the material-velocity gradient
- $l^v = \dot{\mathbf{F}}^v\,\mathbf{F}^{v-1}$ is the spatial-velocity gradient
- l^v can be further decomposed into a stretching and a spin:

$$l^v = \dot{\mathbf{F}}^v\mathbf{F}^{v-1} = \mathbf{D}^v + \mathbf{W}^v, \tag{3.25}$$

where $\mathbf{D}^v$ and $\mathbf{W}^v$ are the rate of the stretching and the spin, respectively. Boyce et al. [54] argued that it is reasonable to assume $\mathbf{W}^v = \mathbf{0}$ without losing generality in the isotropic case. The viscoplastic stretch rate $\mathbf{D}^v$ is constitutively prescribed to be

$$\mathbf{D}^{\mathrm{v}} = \frac{\dot{\gamma}^{\mathrm{v}}}{\sqrt{2}\bar{\tau}} \bar{\mathbf{T}}_{g0}', \tag{3.26}$$

where $\bar{\mathbf{T}}_{g0}$ is the stress acting on the viscoelasticplastic component convected to its relaxed configuration ($\bar{\mathbf{T}}_{g0} = \mathbf{R}^{\mathrm{eT}}\mathbf{T}_{g0}\mathbf{R}^{\mathrm{e}}$); the prime denotes the deviator; $\bar{\tau}$ is the equivalent shear stress that drives the viscous flow and is defined as

$$\bar{\tau} = \left[\frac{1}{2}\bar{\mathbf{T}}_{g0}' \cdot \bar{\mathbf{T}}_{g0}'\right]^{1/2}. \tag{3.27}$$

$\dot{\gamma}^{\mathrm{v}}$ denotes the viscoplastic shear strain rate and is constitutively prescribed to take the form

$$\dot{\gamma}^{\mathrm{v}} = \dot{\gamma}_0 \exp\left[-\frac{\Delta G}{kT}\left\{1 - \left(\frac{\bar{\tau}}{s}\right)\right\}\right], \tag{3.28}$$

where
- $\dot{\gamma}_0$ is the preexponential factor
- ΔG is the zero stress level activation energy
- s is the athermal shear strength, which represents the resistance to the viscoplastic shear deformation in the material

To further consider the softening effects observed in experiments, the following evolution rule for s can be used:

$$\dot{s} = h_0\left(1 - s/s_s\right)\dot{\gamma}^{v}, \tag{3.29}$$

with the initial condition

$$s = s_0 \text{ when } \gamma^{v} = 0, \tag{3.30}$$

where
- s_0 is the initial value of the athermal shear strength
- s_s is the saturation value
- h_0 is a parameter to be determined from experiments.

One key component in the model developed by Qi et al. [10] is the introduction of a frozen glassy phase. The difference between the frozen glassy phase and the original glassy phase is their mechanical deformation state. As the phase formation is a continuous process, the new glassy phase will be formed ("frozen" from the rubbery phase) during cooling. However, the newly formed glassy phase does not inherit the deformation of the rubbery phase and will behave as an undeformed material. This effectively freezes the previous deformation in the rubbery phase. The initial deformation of the frozen glassy phase upon formation is zero. However, due to the 3D nature of the deformation, the vanishing deformation of the transforming rubbery phase will cause a redistribution of deformation in the material, which in turn will cause a new deformation in the frozen glassy phase. Since this new deformation is due to the redistribution of the overall deformation, it can be obtained as an incremental deformation gradient $\Delta\mathbf{F}_T$:

$$\Delta\mathbf{F}_T = \begin{cases} \dot{\mathbf{F}}\,dt & \text{if } \Delta T \neq 0 \\ \mathbf{I} & \text{if } \Delta T = 0 \end{cases} \tag{3.31}$$

Here, a monotonic thermal loading condition is considered, i.e., the cooling or reheating processes are not interrupted by the isothermal mechanical loading. This implies that during the cooling or reheating processes, the only deformation is due to the temperature change and the mechanical constraints. Such a monotonic thermal loading is typical of most SMP applications. Note that the concept of using the instantaneous configuration as the reference configuration for the newly formed phase is also used in [32,33] where the new configuration is called the natural configuration.

Strictly speaking, the FGP phases formed at different times during the cooling process have a different deformation history. Computationally, tracking the phases formed at different times is prohibitively expensive. Such a challenge has recently been solved by Long et al. [58]. In Qi et al., this process is simplified by assuming a temporal and spatial average, i.e., all the FGPs have the same deformation gradient and therefore will not distinguish the FGPs formed at different times. The total deformation gradient acting on the FGP is

$$\mathbf{F}_T = \int_{t_T}^{t} \Delta\mathbf{F}_T \, dt, \tag{3.32}$$

where t_T is the time when a thermal loading starts. It is noted that in typical SMP applications, the programming is achieved by holding the sample in a deformed shape, and then lowering the temperature. In this process, the new deformation in the frozen glassy phase should be small when compared to the deformation imposed on the material at high temperatures. Therefore, Equation 3.32 provides a reasonable approximation of the deformation gradient in the FGP. The stress in the FGP can then be calculated using Equations 3.11 through 3.22, with $\mathbf{F}_T$ instead of $\mathbf{F}$ in Equations 3.20 through 3.30.

One important feature of this model is that it considers material behaviors at temperatures above and below T_g, separately. This allows one to use various models for the rubbery and the glassy behavior, and to capture the shape-memory effects. This feature also simplifies the process of material parameter identification. For example, using the experimental stress-strain curve at temperatures well above T_g, the material parameters associated with the rubbery phase can be identified. Using the experimental stress-strain curve at the temperature well below T_g, the material parameters associated with the glassy phase can be identified.

Figure 3.4A shows the comparison between the model prediction and the experimental results for isothermal stress-strain behavior at different temperatures. The constitutive model developed by Qi et al. has been implemented in the commercial finite element software ABAQUS. Figure 3.4B shows the free recovery of the strain during heating and Figure 3.4C shows snapshot images of a deformed SMP sample during the free recovery

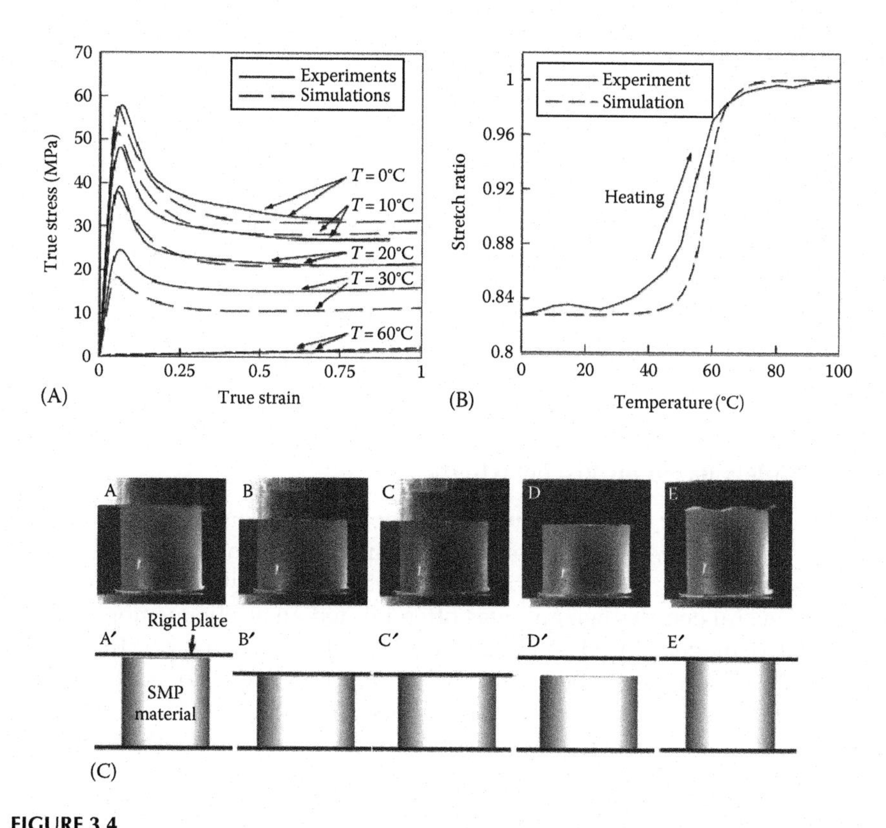

FIGURE 3.4
Comparison between model prediction and experimental results (A) isothermal stress-strain behavior at different temperatures; (B) recovery of the strain under free condition due to heating; (C) snapshot images of SMP deformations during the free-recovery experiment and the simulation.

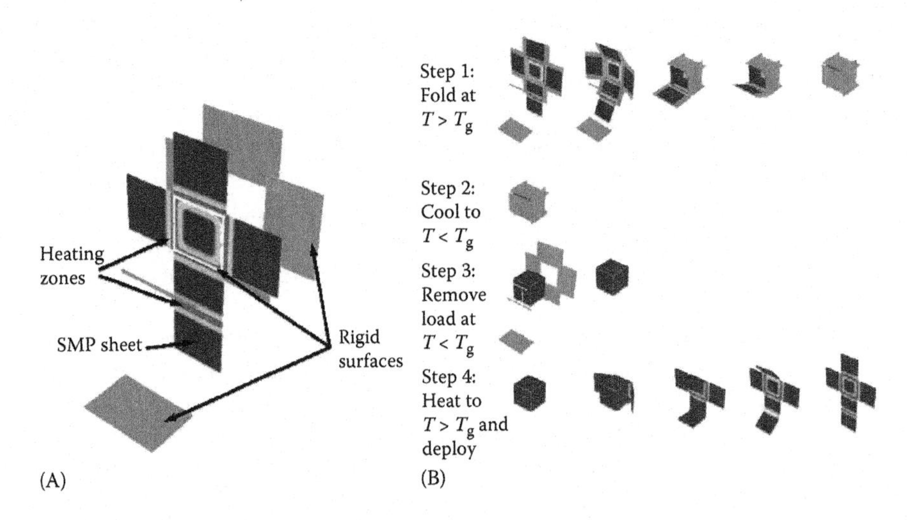

FIGURE 3.5
Finite element simulations of an SMP deployable box with heating zones: (A) Model; (B) The FE simulation of the process of programming and recovering. (From Castro, F. et al., *J. Mech. Time Depend. Mater.*, 2009. With permission.)

experiment and simulation. In addition, Figure 3.5 shows the finite element simulations of an SMP deployable box with heating zones where the SMP sheet contains five heating zones which can be controlled individually [59]. In Figure 3.5, an SMP sheet was folded into a box by five rigid surfaces and five rigid blockers. 7980 nodes and 5616 3D 8 node brick elements were used in the simulation. The SMP retained the box shape after the temperature was lowered and recovered the box shape after the temperature was increased above T_g.

3.2.4 Models Based on Viscoelasticity

The shape-memory effect in amorphous polymers is due to the dramatic change of viscosity, or relaxation time of the polymer as the temperature traverses the glass transition. It was realized in the polymer physics community several decades ago and was often termed an elastic-memory effect. Constitutive models, based on viscoelasticity, and that can describe the effect, have recently been developed. Tobushi and coworkers developed a 1D constitutive model based on a modified viscoelastic model [34,35]. In Tobushi et al., the stress–strain relationship is described by modifying the standard linear viscoelastic (SLV),

$$\dot{\varepsilon} = \frac{\dot{\sigma}}{E} + \frac{\sigma}{\eta} - \frac{\varepsilon - \varepsilon_s}{\tau_r} + \alpha \dot{T}, \tag{3.33}$$

where

σ and ε are the stress and the strain

E, η, and τ_r are the elastic modulus, the viscosity, and the retardation time, respectively

α and T are the thermal expansion coefficient and the temperature

The difference between Equation 3.33 and the SLV is the introduction of ε_s, the irrecovery strain, which is defined as

$$\varepsilon_s = \begin{cases} 0 & \varepsilon_c < \varepsilon_l \\ C(\varepsilon_c - \varepsilon_l) & \varepsilon_c > \varepsilon_l \end{cases} \tag{3.34}$$

where

ε_c is a creep strain

ε_l is a critical strain

C is a proportional coefficient

Both C and ε_l are functions of temperatures,

$$C = C_g \exp\left[a_c\left(\frac{T_g}{T} - 1\right)\right], \quad \varepsilon_l = \varepsilon_g \exp\left[-a_\varepsilon\left(\frac{T_g}{T} - 1\right)\right] \tag{3.35}$$

where C_g, ε_g, a_c, and a_ε are constants. In addition, E, η, and τ_r are functions of the temperature following the similar functional form of C. Although the Tobushi model captures the shape-memory behavior, it is unclear what the contribution of ε_s is. Indeed, as shown by a model developed by Nguyen et al. [36], the shape-memory effect can be captured by varying E, η and τ_r.

More recently, a thermoviscoelastic model was developed by Nguyen et al. [36]. This model was motivated by the physical studies of the glass-forming process. It is well known that the structure of an amorphous polymer below its glassy transition temperature (T_g) is in a nonequilibrium state [60–64]. This is due to the dramatic reduction of the macromolecular chain mobility and the free volume as the temperature falls below T_g. During cooling, as the material passes the T_g, due to the reduced free volume, the adjustment of the material structure toward the equilibrium becomes slow; this becomes pronounced as the temperature decreases to far below T_g: the material structure adjustment becomes so sluggish that the structure of the polymer virtually cannot reach equilibrium. Macroscopically, this translates into a nonlinear thermal expansion and a dramatic increase in viscosity. Following the Hodge-Scherer equation for the relaxation time and the Erying type viscous flow rule, the effective viscosity can be formulated as

$$\eta = \eta_0 \frac{Q}{T}\frac{s}{s_y} \exp\left[\frac{C_1}{0.433}\left(\frac{C_2\left(T - T_f\right) + T\left(T_f - T_g^{ref}\right)}{T\left(C_2 + T_f - T_g^{ref}\right)}\right)\right]\left[\sinh\left(\frac{Q}{T}\frac{s}{s_y}\right)\right]^{-1} \quad (3.36)$$

where

η_0 is the viscosity at a reference temperature (typically the glassy-transition temperature) T_g^{ref}

T_f is the fictitious temperature used in Tool [65]

Q is an activation parameter

s is the flow stress

s_y is the athermal shear resistance and characterizes the yielding behavior of polymers at low temperatures

C_1 and C_2 are two parameters in the WLF equation

The viscous flow rule can then be obtained as

$$\dot{\gamma}^{v} = \frac{s_y}{\sqrt{2}\eta_0}\frac{T}{Q} \exp\left[\frac{C_1}{0.433}\left(\frac{C_2\left(T - T_f\right) + T\left(T_f - T_g^{ref}\right)}{T\left(C_2 + T_f - T_g^{ref}\right)}\right)\right]\left[\sinh\left(\frac{Q}{T}\frac{s}{s_y}\right)\right] \quad (3.37)$$

Following the same decomposition of the total deformation gradient into an elastic part and a viscous part, the stress from the nonequilibrium branch can be calculated by using Equations 3.21 through 3.27. Furthermore, the total stress is the summation of the nonequilibrium and equilibrium contributions, where the latter can be calculated using a hyperelastic model, such as the Arruda–Boyce eight chain model, as described above.

Equation 3.37 described viscosity as a function of time and of temperature and plays a key role in capturing the shape-memory effect. To illustrate, Equation 3.28 under small stress conditions can be considered and simplified as

$$\eta = \eta_0 \exp\left[\frac{C_1}{0.433}\left(\frac{C_2\left(T - T_f\right) + T\left(T_f - T_g^{ref}\right)}{T\left(C_2 + T_f - T_g^{ref}\right)}\right)\right]. \quad (3.38)$$

In Equations 3.36 and 3.38, the fictitious temperature T_f is a function of the time and the temperature and τ_R, the relaxation time; at the temperature below T_g, T_f approaches T_g and at the temperature above T_g, T_f approaches T. The time and the temperature dependence of T_f renders a temperature rate dependence, hence, a hysteresis effect in the variation of the viscosity. To simplify the discussion, one can take $T_f = T_g$ for the temperature below T_g and $T_f = T$ for the temperature above T_g. Figure 3.6 shows the variation of

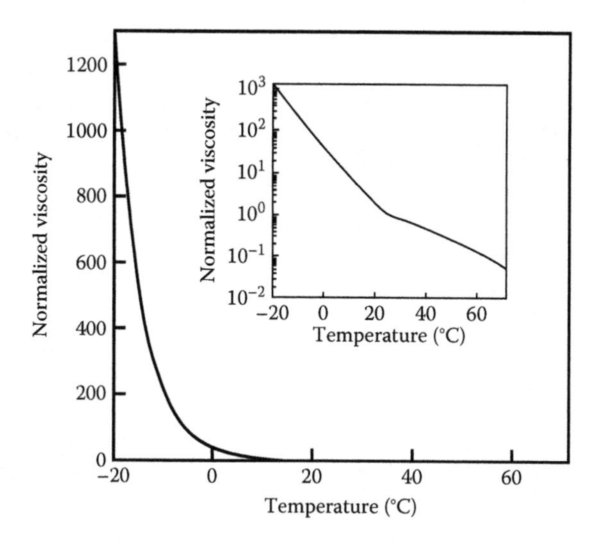

FIGURE 3.6
Variation of viscosity as a function of temperature. Note that the dependence of viscosity on temperature typically shows a hysteresis effect due to the time and the temperature dependency of T_f. Here, to simplify the discussion, the hysteresis effect is eliminated by taking $T_f = T_g$ for the temperature below T_g and $T_f = T$ for the temperature above T_g.

the normalized viscosity η/η_0 as a function of the temperature. It can be seen that the viscosity increases dramatically as the temperature decreases. At $T = T_g^{ref} = 25°C$, $\eta/\eta_0 = 1$; at $T = T_g^{ref} - 40°C$, $\eta/\eta_0 = 1292$; at $T = T_g^{ref} + 40°C$, $\eta/\eta_0 = 0.06$.

The dramatic change in viscosity plays a key role in capturing the shape-memory effect [66]. This can be illustrated by a SLV model shown in Figure 3.7. The left element represents the long-term equilibrium response of the material and the right Maxwell element represents the nonequilibrium response. Typically, the spring in the Maxwell element has a much higher stiffness than the equilibrium spring. At high temperatures, the dashpot has a very low viscosity. Mechanical loading (step 1 in Figure 3.7) at high temperatures therefore, will not activate the nonequilibrium spring as the dashpot will be highly elongated which accommodates most deformations. As the temperature is lowered to below T_g, the viscosity in the dashpot increases tremendously and results in an extremely long relaxation time. When the external load is removed, the equilibrium can be reached by the force balance between the equilibrium spring and the nonequilibrium spring. However, due to the high modulus of the nonequilibrium spring, only a very small recovery can be observed. Finally, when the temperature increases, the viscosity in the dashpot becomes very low again, resulting in the release of the deformation in the dashpot and the shape recovery of the SMP.

Figure 3.8 shows the comparison of the stress-strain behavior under isothermal conditions between the experiment and the model prediction.

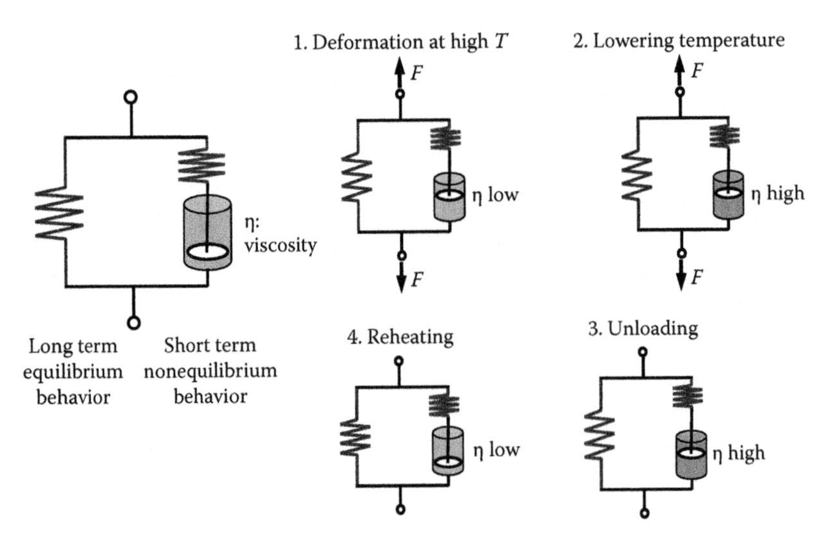

FIGURE 3.7
Schematic of the shape-memory effect due to the dramatic change of viscosity in shape-memory polymers. (From Castro, F. et al., *J. Mech. Time Depend. Mater.*, 2009. With permission.)

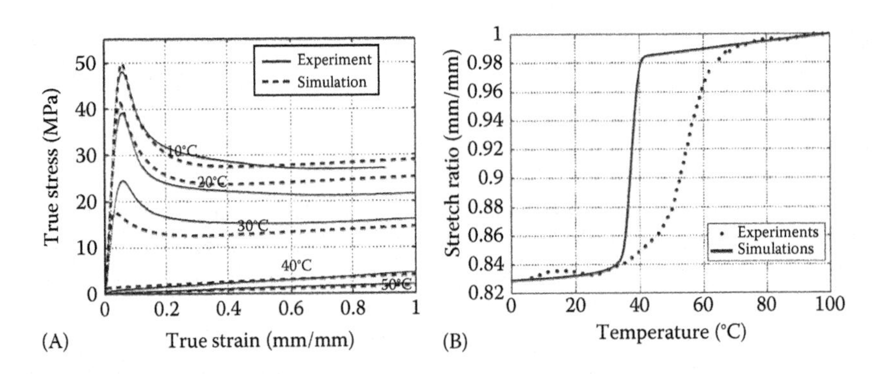

FIGURE 3.8
Comparison between model prediction and experimental results (A) isothermal stress-strain behavior at different temperatures; (B) recovery of the strain under free condition due to heating. (From Nguyen, T.D. et al., *J. Mech. Phys. Solids*, 56(9), 2792, 2008. With permission.)

Figure 3.8B shows the variation of the stretch ratio as a function of temperature during the free recovery. It can be seen that the model captures the stress-strain behavior to a fairly large extent but to a much lesser extent the strain recovery during the free recovery. In addition, Figure 3.8B confirms that the dramatic viscosity change in the polymer as the temperature traverses the glassy temperature is indeed the key mechanism in determining shape-memory effects. Also, the discrepancy between the model prediction and the experiment demands further studies.

3.3 Conclusions

We briefly reviewed the recent progress in modeling the 3D, finite thermo-mechanical deformation of shape-memory polymers. Considerable progress has been made to this end over the past few years and this is reflected in experimentally verified constitutive models now being implemented in finite element codes to facilitate engineering design with this exciting class of active materials. Nevertheless, considerable effort is still required to fully understand the myriad macroscopic physical and mechanical properties of SMPs and their complex connection to underlying molecular mechanisms.

References

1. Behl, M. and A. Lendlein, Shape-memory polymers. *Materials Today*, 2007. 10(4): 20–28.
2. Liu, C., H. Qin, and P.T. Mather, Review of progress in shape-memory polymers. *Journal of Materials Chemistry*, 2007. 17(16): 1543–1558.
3. Bellin, I., S. Kelch, and A. Lendlein, Dual-shape properties of triple-shape polymer networks with crystallizable network segments and grafted side chains. *Journal of Materials Chemistry*, 2007. 17(28): 2885–2891.
4. Liu, Y.P. et al., Thermomechanics of shape memory polymer nanocomposites. *Mechanics of Materials*, 2004. 36(10): 929–940.
5. Gall, K. et al., Shape memory polymer nanocomposites. *Acta Materialia*, 2002. 50(20): 5115–5126.
6. Gall, K. et al., Internal stress storage in shape memory polymer nanocomposites. *Applied Physics Letters*, 2004. 85(2): 290–292.
7. Gall, K. et al., Carbon fiber reinforced shape memory polymer composites. *Journal of Intelligent Material Systems and Structures*, 2000. 11(11): 877–886.
8. Lan, X. et al., Fiber reinforced shape-memory polymer composite and its application in a deployable hinge. *Smart Materials & Structures*, 2009. 18(2): 024002.
9. Liu, Y.P. et al., Thermomechanics of shape memory polymers: Uniaxial experiments and constitutive modeling. *International Journal of Plasticity*, 2006. 22(2): 279–313.
10. Qi, H.J. et al., Finite deformation thermo-mechanical behavior of thermally induced shape memory polymers. *Journal of the Mechanics and Physics of Solids*, 2008. 56(5): 1730–1751.
11. Liu, Y.P. et al., Thermomechanical recovery couplings of shape memory polymers in flexure. *Smart Materials & Structures*, 2003. 12(6): 947–954.
12. Safranski, D.L. and K. Gall, Effect of chemical structure and crosslinking density on the thermo-mechanical properties and toughness of (meth) acrylate shape memory polymer networks. *Polymer*, 2008. 49(20): 4446–4455.
13. Buckley, P.R. et al., Inductively heated shape memory polymer for the magnetic actuation of medical devices. *IEEE Transactions on Biomedical Engineering*, 2006. 53(10): 2075.

14. Huang, W.M. et al., Water-driven programmable polyurethane shape memory polymer: Demonstration and mechanism. *Applied Physics Letters*, 2005. 86(11): 114105.
15. Lendlein, A. and S. Kelch, Shape-memory polymers. *Angewandte Chemie—International Edition*, 2002. 41(12): 2035.
16. Monkman, G.J., Advances in shape memory polymer actuation. *Mechatronics*, 2000. 10(4–5): 489.
17. Schmidt, A.M., Electromagnetic activation of shape memory polymer networks containing magnetic nanoparticles. *Macromolecular Rapid Communications*, 2006. 27(14): 1168.
18. Leng, J.S. et al., Comment on water-driven programable polyurethane shape memory polymer: Demonstration and mechanism [*Applied Physics Letters* 86, 114105 (2005)]. *Applied Physics Letters*, 2008. 92(20): 206105.
19. Lv, H.B. et al., Shape-memory polymer in response to solution. *Advanced Engineering Materials*, 2008. 10(6): 592–595.
20. Leng, J.S. et al., Electroactivate shape-memory polymer filled with nanocarbon particles and short carbon fibers. *Applied Physics Letters*, 2007. 91(14): 144105.
21. Leng, J.S. et al., Significantly reducing electrical resistivity by forming conductive Ni chains in a polyurethane shape-memory polymer/carbon-black composite. *Applied Physics Letters*, 2008. 92(20): 204101.
22. Leng, J.S. et al., Electrical conductivity of thermoresponsive shape-memory polymer with embedded micron sized Ni powder chains. *Applied Physics Letters*, 2008. 92(1): 014104.
23. Liu, Y.L. et al., Electro-activate shape memory composites. *Composite Science and Technology*, 2009. doi:10.1016/j.compscitech.2008.08.016.
24. Lendlein, A. et al., Light-induced shape-memory polymers. *Nature*, 2005. 434(7035): 879–882.
25. Jiang, H.Y., S. Kelch, and A. Lendlein, Polymers move in response to light. *Advanced Materials*, 2006. 18(11): 1471–1475.
26. Scott, T.F., R.B. Draughon, and C.N. Bowman, Actuation in crosslinked polymers via photoinduced stress relaxation. *Advanced Materials*, 2006. 18(16): 2128–2132.
27. Scott, T.F. et al., Photoinduced plasticity in cross-linked polymers. *Science*, 2005. 308(5728): 1615–1617.
28. Long, K.N. et al., Photomechanics of light-activated polymers. *Journal of the Mechanics and Physics of Solids*, 2009. 57(7): 1103–1121.
29. Lendlein, A. et al., Shape memory polymers, in *Encyclopedia of Materials: Science and Technology*. 2005.
30. Chen, Y.C. and D.C. Lagoudas, A constitutive theory for shape memory polymers. Part I—Large deformations. *Journal of the Mechanics and Physics of Solids*, 2008. 56(5): 1752–1765.
31. Chen, Y.C. and D.C. Lagoudas, A constitutive theory for shape memory polymers. Part II—A linearized model for small deformations. *Journal of the Mechanics and Physics of Solids*, 2008. 56(5): 1766–1778.
32. Barot, G. and I.J. Rao, Constitutive modeling of the mechanics associated with crystallizable shape memory polymers. *Zeitschrift Fur Angewandte Mathematik Und Physik*, 2006. 57(4): 652–681.
33. Barot, G., I.J. Rao, and K.R. Rajagopal, A thermodynamic framework for the modeling of crystallizable shape memory polymers. *International Journal of Engineering Science*, 2008. 46(4): 325–351.

34. Tobushi, H. et al., Thermomechanical constitutive modeling in shape memory polymer of polyurethane series. *Journal of Intelligent Material Systems and Structures*, 1997. 8(8): 711–718.
35. Tobushi, H. et al., Thermomechanical constitutive model of shape memory polymer. *Mechanics of Materials*, 2001. 33(10): 545–554.
36. Nguyen, T.D. et al., A thermoviscoelastic model for amorphous shape memory polymers: Incorporating structural and stress relaxation. *Journal of the Mechanics and Physics of Solids*, 2008. 56(9): 2792–2814.
37. Angell, C.A. et al., Relaxation in glassforming liquids and amorphous solids. *Journal of Applied Physics*, 2000. 88(6): 3113–3157.
38. Adolf, D.B. and R.S. Chambers, A thermodynamically consistent, nonlinear viscoelastic approach for modeling thermosets during cure. *Journal of Rheology*, 2007. 51(1): 23–50.
39. Diani, J., Y.P. Liu, and K. Gall, Finite strain 3D thermoviscoelastic constitutive model for shape memory polymers. *Polymer Engineering and Science*, 2006. 46(4): 486–492.
40. Fulcher, G.S., Analysis of recent measurements of the viscosity of glasses. *Journal of American Ceramic Society*, 1925. 8: 789–794.
41. Vogel, H., The law of viscosity change with temperature. *Physik Z.*, 1921. 22: 645–646.
42. Tammann, G. and W. Hesse, Dependence of viscosity on temperature in supercooled liquids. *Zeitschriftfiir Anorganische Und Allgemeine Chemie*, 1926. 156: 245–257.
43. Arruda, E.M. and M.C. Boyce, A 3-dimensional constitutive model for the large stretch behavior of rubber elastic-materials. *Journal of the Mechanics and Physics of Solids*, 1993. 41(2): 389–412.
44. Simo, J.C., On a fully 3-dimensional finite-strain viscoelastic damage model—Formulation and computational aspects. *Computer Methods in Applied Mechanics and Engineering*, 1987. 60(2): 153–173.
45. Miehe, C. and J. Keck, Superimposed finite elastic-viscoelastic-plastoelastic stress response with damage in filled rubbery polymers. Experiments, modelling and algorithmic implementation. *Journal of the Mechanics and Physics of Solids*, 2000. 48(2): 323–365.
46. Bergstrom, J.S. and M.C. Boyce, Constitutive modeling of the large strain time-dependent behavior of elastomers. *Journal of the Mechanics and Physics of Solids*, 1998. 46(5): 931–954.
47. Boyce, M.C. et al., Deformation of thermoplastic vulcanizates. *Journal of the Mechanics and Physics of Solids*, 2001. 49(5): 1073–1098.
48. Qi, H.J. and M.C. Boyce, Stress-strain behavior of thermoplastic polyurethanes. *Mechanics of Materials*, 2005. 37(8): 817–839.
49. Reese, S. and S. Govindjee, A theory of finite viscoelasticity and numerical aspects. *International Journal of Solids and Structures*, 1998. 35(26–27): 3455–3482.
50. Lion, A., Constitutive modelling in finite thermoviscoplasticity: A physical approach based on nonlinear rheological models. *International Journal of Plasticity*, 2000. 16(5): 469–494.
51. Govindjee, S. and S. Reese, A presentation and comparison of two large deformation viscoelasticity models. *Journal of Engineering Materials and Technology-Transactions of the ASME*, 1997. 119(3): 251–255.

52. Govindjee, S. and J. Simo, A micro-mechanically based continuum damage model for carbon black-filled rubbers incorporating mullins effect. *Journal of the Mechanics and Physics of Solids*, 1991. 39(1): 87–112.
53. Boyce, M.C., G.G. Weber, and D.M. Parks, On the kinematics of finite strain plasticity. *Journal of the Mechanics and Physics of Solids*, 1989. 37(5): 647–665.
54. Boyce, M.C., D.M. Parks, and A.S. Argon, Large inelastic deformation of glassy-polymers.1. rate dependent constitutive model. *Mechanics of Materials*, 1988. 7(1): 15–33.
55. Boyce, M.C., D.M. Parks, and A.S. Argon, Large inelastic deformation of glassy-polymers.2. numerical-simulation of hydrostatic extrusion. *Mechanics of Materials*, 1988. 7(1): 35–47.
56. Boyce, M.C., D.M. Parks, and A.S. Argon, *Computational Modeling of Large Strain Plastic-Deformation in Glassy-Polymers*. Abstracts of Papers of the American Chemical Society, 1988. 196: 156-POLY.
57. Weber, G. and L. Anand, Finite deformation constitutive-equations and a time integration procedure for isotropic, hyperelastic viscoplastic solids. *Computer Methods in Applied Mechanics and Engineering*, 1990. 79(2): 173–202.
58. Long, K.N., M.L. Dunn, and H.J. Qi, *Mechanics of Soft Active Materials with Evolving Phases*. Submitted, 2009.
59. Qi, H.J., F. Castro, and K.N. Long. Finite element simulations of thermally induced shape memory polymers based applications. in *NSF CMMI Grantee Conference*. 2008. Knoxville, TN.
60. Kovacs, A.J., Transition vitreuse dans les polymeres amorphes. Etude phenomenologique, *Fortschr. Hochpolymer—Forsch*, 1963. 3: 394.
61. Kauzmann, W., The nature of the glassy state and the behavior of liquids at low temperatures. *Chemical Reviews*, 1948. 43(2): 219–256.
62. Kovacs, A.J. et al., Isobaric volume and enthalpy recovery of glasses.2. transparent multi-parameter theory. *Journal of Polymer Science Part B-Polymer Physics*, 1979. 17(7): 1097–1162.
63. McKenna, G.B., Glass formation and glassy behavior, in *Comprehensive Polymer Science*, Price C. and Booth C., (eds.). 1989, Pergamon, Oxford, U.K., pp. 311–362.
64. Hutchinson, J.M., Physical aging of polymers. *Progress in Polymer Science*, 1995. 20(4): 703–760.
65. Tool, A.Q., Relation between inelastic deformability and thermal expansion of glass in its annealing range. *Journal of the American Ceramic Society*, 1946. 29(9): 240–253.
66. Castro, F., Westbrook, K.K., Long, K.M., Shandas, R., and Qi, H.J., Effects of thermal rates on the thermomechanical behaviors of amorphous shape memory polymers, *Journal of Mechanics of Time Dependent Materials*, 2009, in press.

4

Thermomechanical Characterizations of Shape-Memory Polymers (Dual/Triple-Shape) and Modeling Approaches

Karl Kratz, Wolfgang Wagermaier, Matthias Heuchel, and Andreas Lendlein

Center for Biomaterial Development, Institute for Polymer Research, GKSS Research Center, Teltow, Germany

CONTENTS

4.1 Introduction to the Thermally Induced Shape-Memory Effect of Polymers

Shape-memory materials are able to change their shape by recovering a predetermined, memorized shape upon exposure to an external stimulus such as heat or light.

The shape-memory effect (SME) results from a combination of the material's molecular structure and a suitable processing and programming technology. In this context, it is important to differentiate between intrinsic material properties, which are given by nature, and the functionality of the polymer, resulting from a combination of the polymer's molecular architecture and a suitable process. The unexpected combination of different functionalities, such as the combination of biofunctionality, hydrolytic degradability, and shape-memory functionality, is referred to as multifunctionality. The development of such multifunctional materials is often motivated by the requirement of specific applications [1,2], such as intelligent textiles, packaging, (dis)assembly technologies, or biomedical applications.

For materials exhibiting a thermally induced dual-shape capability, the shape change occurring under stress-free recovery conditions is characterized by the switching temperature (T_{sw}), which is determined as the inflection point in the strain–temperature curve. Under constant strain conditions, the temperature related to the stress maximum ($T_{\sigma,max}$) or the temperature at the inflection point ($T_{\sigma,inf}$) of the stress–temperature curve is determined.

Shape-memory polymers (SMPs) [1,3–8] by far surpass the metallic alloys (SMAs), such as NiTi [9] in their shape-memory properties, e.g., in the maximum uniaxial deformation. SMPs also have significant economic advantages when compared with SMAs (e.g., lower materials, processing, and programming costs).

Thermoresponsive SMPs with dual-shape capability [4] can be classified with respect to the nature of their permanent netpoints and the thermal transition (T_{trans}) related to the switching domains into four different categories: (I) chemically cross-linked amorphous polymers ($T_{trans} = T_g$); (II) chemically cross-linked semicrystalline polymer networks ($T_{trans} = T_m$); (III) physically cross-linked thermoplastics with $T_{trans} = T_g$; and (IV) a physically cross-linked thermoplastics with $T_{trans} = T_m$ [6] (see Chapter 1). Furthermore, these four different types of SMPs are characterized by a specific dynamic thermomechanical behavior [6] as shown in Figure 4.1, where the tensile storage modulus vs. the temperature measured at the small oscillatory deformation at 1 Hz is plotted. Both the thermoplastic SMPs represented in Figure 4.1 as (III) and (IV) have crystalline hard domains characterized by a T_m at high temperatures.

Recently, thermosensitive multiphase polymer networks exhibiting a triple-shape effect have been introduced as a promising class of active polymers [10–13] enabling complex movements. Such triple-shape polymers consist of covalent cross-links (net points) and at least two phase segregated distinct domains acting as physical cross-links with individual thermal transitions assigned.

In the following sections, the characterization of the thermally induced dual-shape effect, as well as of the triple-shape effect by cyclic, thermomechanical tensile tests will be discussed, and a brief perspective on the suitability of different models for simulating the thermally induced SME of polymers will be given.

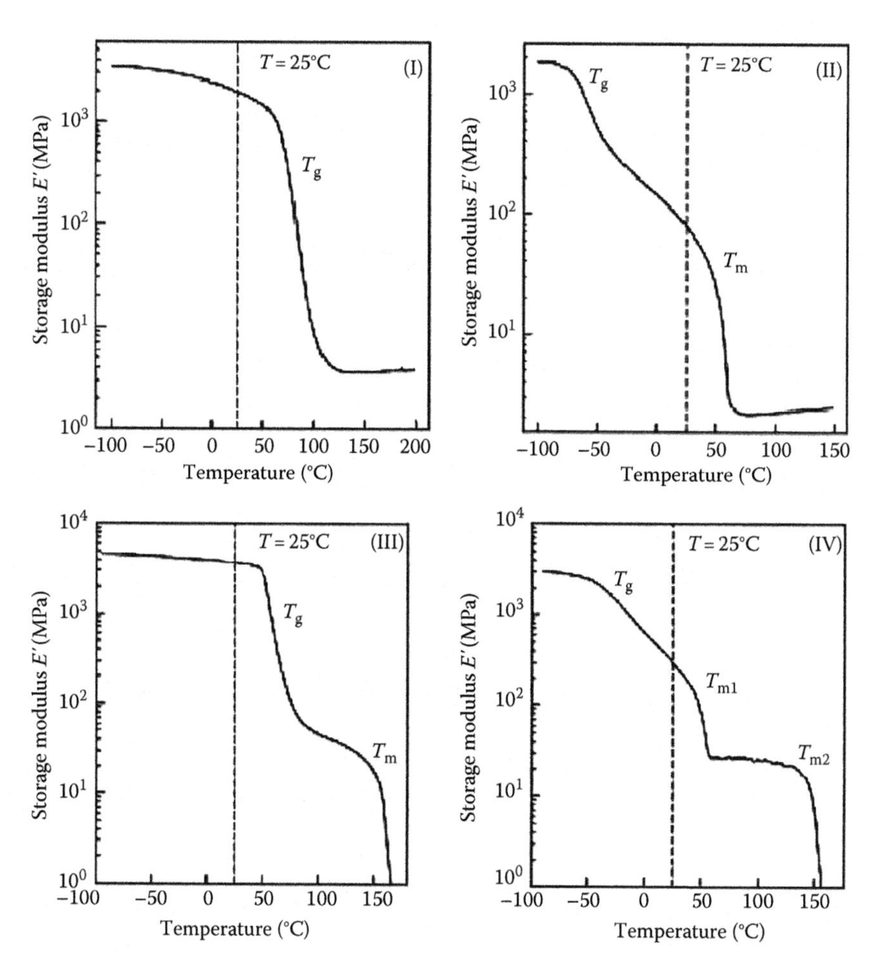

FIGURE 4.1
Four types of shape-memory polymers (dual-shape effect) depicted as a function of their dynamic thermomechanical behavior. Plotted is the tensile storage modulus vs. temperature as measured using a small oscillatory deformation at 1 Hz for (I) a chemically cross-linked amorphous polymer network ($T_{trans} = T_g$); (II) chemically cross-linked semicrystalline polymer network ($T_{trans} = T_m$); (III) physically cross-linked thermoplastic with $T_{trans} = T_g$; and (IV) physically cross-linked thermoplastic ($T_{trans} = T_m$). (Taken from Liu, C. et al., *J. Mater. Chem.*, 17, 1543, 2007. With permission.)

4.2 Investigation of the Dual-Shape Effect of Shape-Memory Polymers with Cyclic, Thermomechanical Tensile Tests

The quantitative analysis of shape-memory properties can be performed in cyclic, thermomechanical tests. Besides the determination of materials shrinkage [14–16] or the application of bending tests [17,18], the most common quantification of SME is presently described as the percentage of strain fixing and the extent of strain recovery determined in cyclic, thermomechanical tensile tests. Each single cycle consists of a programming part, where the temporary shape is created, and a recovery part, where the permanent shape is recovered [3]. Various test protocols are applied that differ in the programming procedure (cold drawing: deformation at $T_{deform} < T_{sw}$ or $T_{\sigma,max}$ or $T_{\sigma,inf}$ and the heating–stretching–cooling cycle: deformation at $T_{deform} > T_{sw}$ or $T_{\sigma,max}$ or $T_{\sigma,inf}$) [19,20]. Here cooling can be performed under stress- or strain-control [21]. While in strain-controlled tests the stress on the specimen $\sigma(t)$ is recorded at defined thermal conditions, the change in strain $\varepsilon(t)$ is measured in stress-controlled tests. Also, different recovery modes are employed: Under stress-free conditions (at $\sigma = 0$ MPa) the characteristic switching temperature T_{sw} is determined as the inflection point in the strain–temperature curve [22]. Under constant strain conditions (ε = const.) the maximum stress σ_{max} generated during recovery as well as the corresponding temperature, $T_{\sigma,max}$, are determined from the stress–temperature recovery curve [23–25] or the temperature at the inflection point ($T_{\sigma,inf}$). Besides the above mentioned variations of the principal test setup, the influence of further test parameters like the applied strain ε_m, strain rate $\dot{\varepsilon}$, cooling (β_c) and heating rates (β_h), as well as the applied temperatures for deformation (T_{deform}), fixation of the temporary shape (T_{low}), and recovery of the original permanent shape (T_{high}) on the shape-memory properties have been investigated. The possible combinations of the programming and recovery modules resulting in different cycle types are summarized in Table 4.1. As examples, two typical test protocols (cycle type A.3 and B.3) are described in the following.

A classical procedure with strain-controlled programming and stress-free recovery typically consists of four steps, which can be divided into a three-step strain-controlled programming procedure followed by a stress-free recovery module (cycle type A.3): (1) heating of the sample to the upper working temperature $T_{high} \geq (T_{trans} + 15\,K)$, stretching to a certain extension ε_m at a defined strain rate for a fixed time period (here: $T_{deform} = T_{high}$); (2) cooling to the lower working temperature $T_{low} \leq (T_{trans} - 15\,K)$ with a certain rate β_c while ε_m is kept constant for fixation of the temporary shape; (3) unloading of the specimen at T_{low}; (4) heating from T_{low} to T_{high} with a constant heating rate β_h under stress-free conditions allowing the restoration of the original permanent shape. After completing the first cycle, the subsequent cycles start again from step (1).

In the corresponding test (cycle type B.3) with stress-controlled programming and stress-free recovery, steps (1) and (2) are modified by keeping the stress constant at σ_m during cooling instead of keeping the sample at ε_m. Here, the deformation of the sample with respect to the distance of the clamps is monitored. In both cases, T_{low} as well as T_{high} are held constant at least for 10 min before the loading or the unloading of the specimen. For physically cross-linked polymers of category III and IV it is important not to exceed the highest thermal transition T_{perm}, which would cause the sample to melt.

TABLE 4.1

Definition of Cycle Types in Thermomechanical Tensile Tests for Characterization of the Dual-Shape Effect

	Programming Module			Recovery Module	
Cycle Type	$T_{deform} < T_{SW}$;	$T_{\sigma,max}$; $T_{\alpha,inf}$ [a]	$T_{deform} > T_{SW}$; $T_{\sigma,max}$; $T_{\alpha,inf}$ [b]	$T_{high} > T_{SW}$; $T_{\sigma,max}$; $T_{\alpha,inf}$	**Results**
A.1	Strain-controlled	Deformation to ε_m	Deformation to ε_m; cooling to T_{low} under $\varepsilon = \varepsilon_m$	Stress free: $\sigma = 0$ MPa	T_{SW}
A.2				Constant strain: $\varepsilon = \varepsilon_m$	σ_{max}; $T_{\sigma,max}$ or $T_{\alpha,inf}$
A.3				Stress free: $\sigma = 0$ MPa	T_{SW}
A.4				Constant strain: $\varepsilon = \varepsilon_m$	σ_{max}; $T_{\sigma,max}$ or $T_{\alpha,inf}$
B.1	Stress-controlled	Deformation to σ_m	Deformation to ε_m; determining σ_m cooling under $\sigma = \sigma_m =$ const.	Stress free: $\sigma = 0$ MPa	T_{SW}
B.2				Constant strain: $\varepsilon = \varepsilon_m$	σ_{max}; $T_{\sigma,max}$ or $T_{\alpha,inf}$
B.3				Stress free: $\sigma = 0$ MPa	T_{SW}
B.4				Constant strain: $\varepsilon = \varepsilon_1$	σ_{max}; $T_{\sigma,max}$ or $T_{\alpha,inf}$

Note: Each cycle consists of a programming and recovery part.

[a] Cold stretching: $T_{deform} = T_{low}$.

[b] Heating–cooling–heating: $T_{deform} = T_{high}$.

In cyclic, thermomechanical tensile tests, the cycle is typically repeated three to five times. While the first cycle is applied as a preconditioning procedure to erase the previous thermomechanical history of the sample, the subsequent cycles 2–5 are used for the determination of the shape-memory properties.

The results of such cyclic, thermomechanical measurements can be presented in a $\varepsilon - \sigma$ curve (Figure 4.2a; σ = tensile stress; ε = strain). Such a plot is called a "two-dimensional plot."

An example of a typical $\varepsilon - \sigma$ diagram is given in Figure 4.3a for a covalently cross-linked SMP with $T_{trans} = T_m$ prepared from crystallizable oligo(ε-caprolactone) dimethacrylate segments and *n*-butyl acrylate (cycle type A.3, $n = 5$). While a hysteresis in the $\varepsilon - \sigma$ diagram was obtained from the first to the second cycle, all subsequent cycles were identical [26]. A disadvantage of this type of graphical presentation is the fact that T_{sw} cannot be determined from the graph. Figure 4.2b represents a three-dimensional (3-D) diagram of a test procedure with stress-controlled programming and stress-free recovery. In this measurement, the sample is cooled under a constant tensile stress σ_m. When the stretched specimen is cooled (step ② in Figure 4.2a) different effects of the sample behavior have to be considered. Examples are

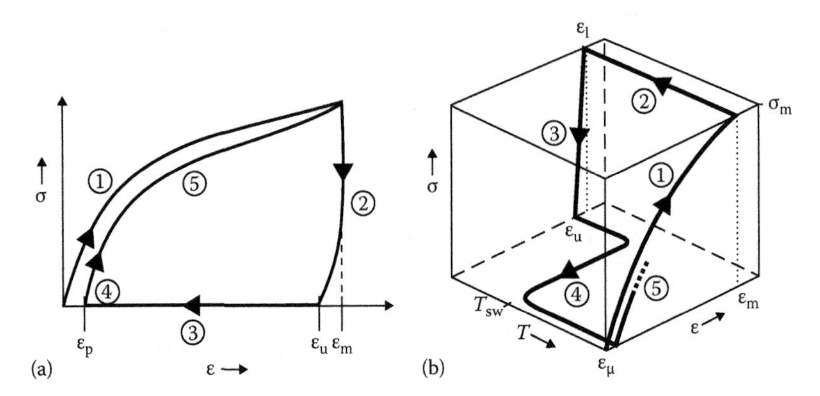

FIGURE 4.2
Schematic representation of the results of the cyclic thermomechanical tensile tests for two different cycle types: (a) strain-controlled programming with stress-free recovery ($\varepsilon-\sigma$ diagram): ①, stretching to ε_m at T_{high}; ②, cooling to T_{low} while ε_m is kept constant; ③, unloading to zero stress; ④, heating up to T_{high} while keeping $\sigma = 0\,\text{MPa}$; and ⑤, start of the second cycle. (b) Stress-controlled programming with stress-free recovery ($\varepsilon-T-\sigma$ diagram): ①, stretching to ε_m at T_{high}; ②, cooling down to T_{low} with cooling rate β_c while σ_m is kept constant; ③, clamp distance is reduced until the stress-free state $\sigma = 0\,\text{MPa}$ is reached; ④, heating up to T_{high} with a heating rate β_h at $\sigma = 0\,\text{MPa}$; and ⑤, start of the second cycle (From Lendlein, A. and Kelch, S., *Angew. Chem. Int. Ed.*, 41(12), 2034, 2002. With permission.)

the change of the expansion coefficient of the stretched specimen at temperatures above and below T_{trans}, as well as volume changes arising from crystallization in case of $T_{trans} = T_m$. The elastic modulus E (T_{high}) at T_{high} can be determined from the initial slope in the measurement range ① (Figure 4.2a). Additionally E of the stretched sample at T_{low} can also be determined from the slope of the curve at ② (see Figure 4.2a) [3]. A typical $\varepsilon-T-\sigma$ diagram (cycle type B.3 consisting of a stress-controlled programming module and a stress-free recovery module) for a phase-segregated multiblock copolymer (PDC) synthesized from the PPDO-diol ($M_n = 4200\,\text{g}\cdot\text{mol}^{-1}$), the hard segments, and the oligo(ε-caprolactone)-diol switching segments ($M_n = 2400\,\text{g}\cdot\text{mol}^{-1}$) is displayed in Figure 4.3b [27].

The shape-memory properties are quantified by the determination of the strain-fixity rate R_f and the strain-recovery rate R_r. Both can be determined from cyclic, thermomechanical measurements according to Equations 4.1 and 4.2 for the strain-controlled programming case:

$$R_f(N) = \frac{\varepsilon_u(N)}{\varepsilon_m} \tag{4.1}$$

$$R_f(N) = \frac{\varepsilon_m - \varepsilon_p(N)}{\varepsilon_m - \varepsilon_p(N-1)} \tag{4.2}$$

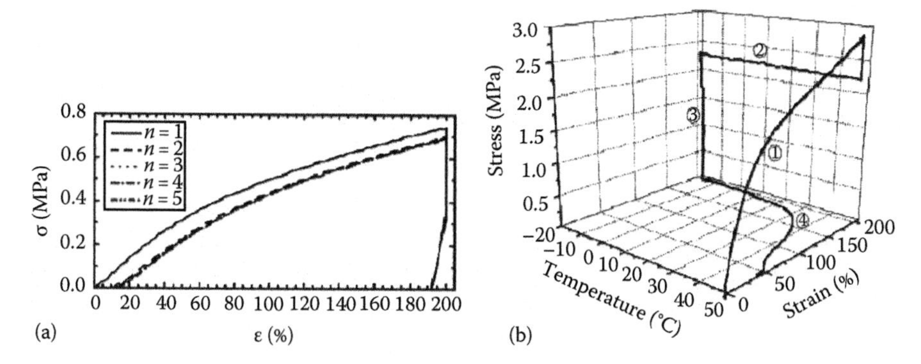

FIGURE 4.3
Examples for $\varepsilon-\sigma$ diagram and $\varepsilon-T-\sigma$ diagram of the SMP obtained in cyclic, thermomechanical tensile tests. (a) $\varepsilon-\sigma$ diagram of a thermomechanical tensile test (cycle type A.3 consisting of a strain-controlled programming module with $\varepsilon_m=200\%$, $T_{deform}=T_{high}=70°C$, $T_{low}=0°C$ and a stress-free recovery module) for a covalently cross-linked polymer network prepared from crystallizable oligo(ε-caprolactone) dimethacrylate (PCLDMA; $M_n=10{,}000\,g\cdot mol^{-1}$), and n-butyl acrylate (50 wt%). Cycle number $n=1$ (solid line); $n=2$ (dashed line); $n=3$ (dotted line); $n=4$ (dash-dotted line); and $n=5$ (dash-double dotted line). (Reprinted and modified from Lendlein, A. et al., *Proc. Natl. Acad. Sci. USA*, 98(3), 842, 2001.) (b) $\varepsilon-T-\sigma$ diagram of a thermomechanical tensile test (cycle type B.3 consisting of a stress-controlled programming module and a stress-free recovery module) for a phase-segregated multiblock copolymer (PDC) synthesized by cocondensation of PPDO-diol ($M_n=4200\,g\cdot mol^{-1}$) and oligo(ε-caprolactone)-diol precursor ($M_n=2400\,g\cdot mol^{-1}$) with TMDI. ①, stretching to $\varepsilon=200\%$ at $T_{high}=45°C$ and setting the stress to a constant level of σ_m; ②, cooling down to $T_{low}=-15°C$ with cooling while σ_m is kept constant; ③, releasing the stress to $\sigma=0\,MPa$; and ④, heating up to $T_{high}=50°C$ with a heating rate $\beta_h=2\,K\cdot min^{-1}$. (Reprinted and modified from Lendlein, A. and Langer, R., *Science*, 296, 1673, 2002; Behl, M. et al., *Adv. Funct. Mater.*, 18, 1, 2008. With permission.)

and Equations 4.3 and 4.4 for the stress-controlled test protocol:

$$R_f(N)=\frac{\varepsilon_u(N)}{\varepsilon_l(N)} \tag{4.3}$$

$$R_r(N)=\frac{\varepsilon_l(N)-\varepsilon_p(N)}{\varepsilon_l(N)-\varepsilon_p(N-1)} \tag{4.4}$$

In a strain-controlled protocol, R_f is given by the ratio of the strain in the stress-free state after the retraction of the tensile stress in the Nth cycle $\varepsilon_u(N)$ and the maximum strain ε_m (Equation 4.1). In such a protocol, R_r quantifies the ability of the material to memorize its permanent shape and is a measure of how far a strain, that was applied in the course of the programming, is recovered in the following shape-memory transition. For this purpose, the strain that occurs during the programming step in the Nth cycle $\varepsilon_m(N)-\varepsilon_p(N-1)$ is related to the change in strain that occurs during the present SME $\varepsilon_m(N) - \varepsilon_p(N)$ (Equation 4.2). The strain of the samples in two successively passed cycles in the stress-free state before application of yield stress is represented by $\varepsilon_p(N-1)$ and $\varepsilon_p(N)$. In case of a stress-controlled programming protocol, R_f is given by the ratio of the retraction of the tensile strain $\varepsilon_u(N)$ and the strain under stress $\sigma = \sigma_m$ in the strain-free state after cooling to T_{low} of the Nth cycle $\varepsilon_l(N)$ (Equation 4.3). In such a protocol, R_r quantifies the ability of the material to memorize its permanent shape and is a measure of how far a strain that was applied during programming is recovered in the following shape-memory transition. For this purpose the strain that occurs during the programming step in the Nth cycle $\varepsilon_l - \varepsilon_p(N-1)$, is compared to the change in the strain, that occurs with the shape-memory effect $\varepsilon_l - \varepsilon_p(N)$ (Equation 4.4).

R_r-values being reached for the SMP or the composites thereof are typically in the range of 80%–99%, while the shape recovery occurs within a temperature range ΔT_{rec} typically lower than 40 K [3,5]. ΔT_{rec} is defined as the difference between the temperature at which the recovery starts and the temperature at which the recovery is completed. An additional value for quantification of the recovery behavior, the recovery rate v_r, is given as the ratio of the strain-recovery rate over the recovery-temperature interval ΔT_{rec} [28,29].

4.3 Investigation of the Triple-Shape Effect of Shape-Memory Polymers with Cyclic, Thermomechanical Tensile Tests

The realization of the triple-shape capability requires suitable multiphase polymer networks and the application of a specific multistep thermomechanical programming procedure [10–13] (see Chapter 1). In the following text, the thermomechanical characterization of such triple-shape polymers will be discussed for recently introduced triple-shape polymers (TSP) that can change on demand from a first shape (A) to a second shape (B) and from there to a third shape (C), when stimulated by two subsequent temperature increases.

The structural concept for the TSP is based on multiphase polymer networks, which are able to form at least two segregated distinct domains. The original shape (C) is defined by covalent net points resulting from a cross-linking reaction. Shapes (A) and (B) are created by a two-step thermomechanical programming process. Shape (B) is determined by physical cross-links associated to the highest transition $T_{trans,B}$, and shape (A) relates to the domains associated to the second highest transition temperature $T_{trans,A}$. Both thermal transitions $T_{trans,B}$ and $T_{trans,A}$ can either be a T_m or a T_g.

For triple-shape functionalization, an individual two-step programming process for the creation of the temporary shapes (B) and (A) was developed, based on the thermal and mechanical characteristics of the polymer networks. At the beginning of the programming process, the polymer network is heated to $T_{high} \geq (T_{trans,B} + 15\,K)$ at which the material is in a rubber-elastic state. After equilibration at T_{high}, the sample is deformed to ε_B^0. Then the material is cooled to a temperature T_{mid} which is in-between $T_{trans,A} < T_{mid} < T_{trans,B}$ and the external stress is maintained until the physical cross-links are established (ε_B^{load}). After release, the external stress shape B (ε_B) is obtained. In the second step, shape A is created. The sample, which presently is in shape B, is further deformed at T_{mid} to ε_A^0. Cooling under external stress to $T_{low} \leq (T_{trans,A} - 15\,K)$ leads to a second set of physical net points related to $T_{trans,A}$, stabilizing ε_A^{load}. These new physical cross-links stabilize shape (A) (ε_A), which the material takes when the external stress is released. By reheating to T_{high}, the shapes (B) (ε_B^{rec}) and (C) (ε_C^{rec}) are sequentially recovered.

For characterization of the triple-shape effect, a specific cyclic thermomechanical tensile experiment was developed [10,11,13]. In each cycle, the two additional shapes (B and A) were created by a two-step uniaxial deformation, followed by recalling shape (B) and finally shape (C).

Quantification of the triple-shape effect in cyclic thermomechanical tests, include, determination of the strain-fixity ratios after creation of shape B $R_f(C \rightarrow B)$ or shape A $R_f(B \rightarrow A)$ and the shape-recovery ratio for the recovery of shape B $R_r(A \rightarrow B)$ as well as the total shape-recovery ratio $R_r(A \rightarrow C)$ after restoration of the initial shape C (see Equations 4.5 through 4.8).

$$R_f(C \rightarrow B) = \frac{\varepsilon_B - \varepsilon_C}{\varepsilon_B^{load} - \varepsilon_C} \tag{4.5}$$

$$R_r(A \rightarrow B) = \frac{\varepsilon_A - \varepsilon_B^{rec}}{\varepsilon_A - \varepsilon_B} \tag{4.6}$$

$$R_f(B \rightarrow A) = \frac{\varepsilon_A - \varepsilon_B}{\varepsilon_A^{load} - \varepsilon_B} \tag{4.7}$$

$$R_r(A \rightarrow C) = \frac{\varepsilon_A - \varepsilon_C^{rec}}{\varepsilon_A - \varepsilon_C} \tag{4.8}$$

Typical results obtained from such cyclic, thermomechanical tests for two different multiphase network systems with excellent triple-shape properties are displayed in Figure 4.4. Both multiphase networks exhibit a pronounced triple-shape effect (TSE) characterized by two distinct switching temperatures (T_{sw}), which can be determined as inflection points from the strain-temperature recovery curve under stress-free conditions.

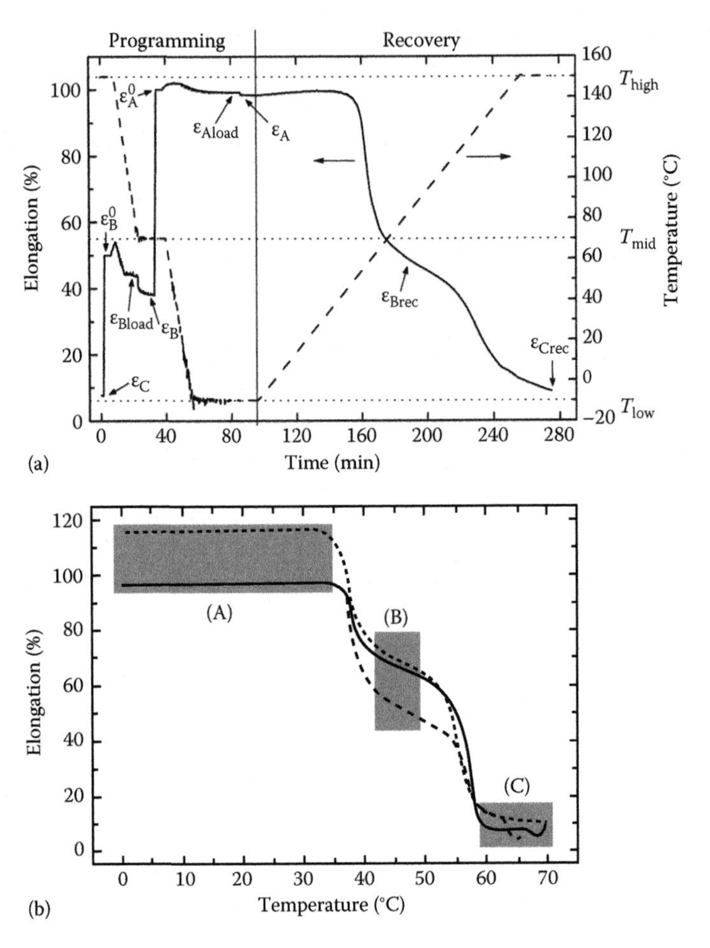

FIGURE 4.4
Example for cyclic, thermomechanical experiments for quantification of the triple-shape effect. (a) Strain and temperature as a function of time taken from the fifth cycle for the MACL(45) multiphase network composed of crystallizable PCL segments and amorphous poly(cyclohexyl methacrylate) segments with a 45 wt% PCL content ($T_{trans,A} = T_{m,PCL} = 50°C$ and $T_{trans,B} = T_g = T_{g,PCHMA} = 140°C$). The solid line indicates the strain; the dashed line indicates the temperature. In this triple-shape experiment, the sample is first stretched from ε_c to ε_B^0 at $T_{high} = 150°C$, then cooled to T_{mid} with a cooling rate of $\beta_c = 5\,K \cdot min^{-1}$ under stress-control resulting in ε_B^{load} and after unloading ε_B is fixed. Then the sample is further elongated at T_{mid} to ε_A^0 and subsequently cooled to T_{low} under stress-control with $\beta_c = 5\,K \cdot min^{-1}$ whereas the elongation decreases to ε_A^{load}. Shape (A), corresponding to ε_A, is obtained by unloading. The recovery process of the sample is monitored by reheating with a heating rate of $\beta_h = 1\,K \cdot min^{-1}$ from T_{low} to T_{high} while the stress is kept at 0 MPa and the sample contracts to the recovered shape (B) at ε_B^{rec} and finally shape (C) at ε_C^{rec} is recovered. (b) $T - \varepsilon$ diagram showing the recovery of shapes B and C in cyclic, thermomechanical experiments (third cycle) for multiphase network CLEG(40) composed of crystallizable PEG and PCL segments with 40 wt% PCL content ($T_{trans,A} = T_{m,PEG} = 38°C$ and $T_{trans,B} = T_{m,PCL} = 55°C$) for different combinations of ε_B^0 and ε_A^0. Solid line, $\varepsilon_B^0 = 50\%$ and $\varepsilon_A^0 = 100\%$; dashed line, $\varepsilon_B^0 = 30\%$ and $\varepsilon_A^0 = 100\%$ B; dotted line, $\varepsilon_B^0 = 50\%$ and $\varepsilon_A^0 = 120\%$. (Reprinted and modified from Bellin, I. et al., *Proc. Natl. Acad. Sci. USA*, 103(48), 18043, 2006.)

4.4 Thermomechanical Model Approaches for Simulation of the Shape-Memory Behavior of Polymers

As the thermally induced shape-memory effect can be understood as a polymer functionalization [1], which results from a combination of a polymer morphology and a specific processing and is not an intrinsic property of the polymer, particular constitutive theories for the description of the SME are required. Therefore, the research activities of the last decade are also focused on the development of constitutive theories that describe the thermomechanical properties of the SMP at a macroscopic level. In some earlier theoretical approaches, rheological models consisting of spring, dashpot, and frictional elements are applied, but they are limited to certain classes of constitutive relations and their predictions agree only qualitatively with experimental observations [30–32].

All the model approaches presented so far rely on covalently cross-linked polymers with either $T_{trans} = T_g$ (SMP class I) [33–38] or $T_{trans} = T_m$ (SMP class II) [20,39–42]. The models for SMPs with crystallizable switching segments follow a mechanical approach, in which the stress–strain behavior is described as a combination of spring or dashpot units [41] or, the observed expansion and contraction is predicted based on so-called constitutive equations for a rubbery phase, a semicrystalline phase, the crystallization, and the melting process which are adjusted to predict the conditions of a specific SMP [39].

A mechanical model developed by Morshedian et al. [41] allows a qualitative and quantitative prediction of the stress–strain–time behavior of heat-shrinkable polymers like cross-linked polyethylene during the heating–stretching–cooling cycle applying a strain up to 200%. This model consists of a combination of a Kelvin unit and a dashpot unit: (a) for simulation of different steps for generating a heat-shrinkable system; (b) for stretching by a fixed strain and cooling; (c) for stress relaxation followed by orientation freeze-up; and (d) for partial recovery after shrinkage as displayed in Figure 4.5a. This rather simple mechanical model is in exceptional congruence with the uniaxial deformation experiments of radiation cross-linked polyethylene, containing two damping units with different viscosities. If these viscosities are known or good estimations are available, shape-recovery rates can be predicted (see Figure 4.5b). Recently, this model was extended to describe the influence of different stretching temperatures during programming of such heat-shrinkable polymers [20].

A second approach for modeling the thermomechanical behavior of crystallizable SMPs within a thermodynamic framework for homogenous and inhomogeneous deformation in different geometries was developed by Barot and coworkers [39,40,42]. In this model four different polymer related processes during the SME have to be considered: the rubbery, totally amorphous state at $T > T_m$ of the switching domains, the semicrystalline phase at temperatures $T < T_m$, the crystallization process on cooling below T_m,

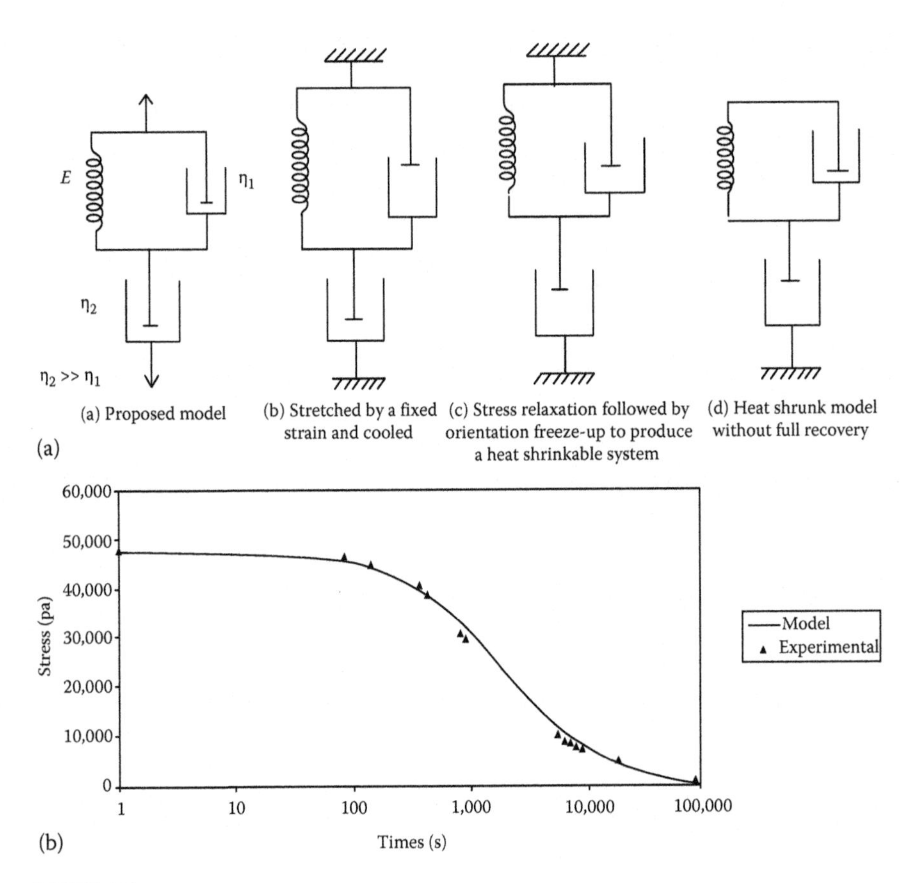

FIGURE 4.5
Example of a thermomechanical model to predict stress–strain–time behavior. (a) Proposed thermomechanical model for shape-memory induction and thermally induced recovery. (b) Comparison of experimental and model results for stress relaxation behavior. Stress relaxation of cross-linked polyethylene under 200% strains at 160°C. (Reprinted and modified from Morshedian, J. et al., *Macromol. Theory Simul.*, 14(7), 428, 2005. With permission.)

and the melting process while heating again above T_m. The homogenous deformation was studied in an uniaxial deformation experiment in which crystallization took place either under constant stress (stress controlled) or constant strain (strain controlled), while the inhomogeneous deformation was investigated in circular shear deformation experiments either under constant moment or constant shear [39].

For covalently cross-linked SMPs with $T_{trans} = T_g$, a 3-D thermomechanic constitutive model was reported by Liu et al. [33,34], assuming active and frozen phases, representing the multiphase character of thermoplastic SMPs. In this approach, two kinds of idealized C–C bonds that exist in the polymer, i.e., "frozen bonds" and "active bonds," are defined. The frozen bond represents the fraction of the C–C bonds that is fully disabled with

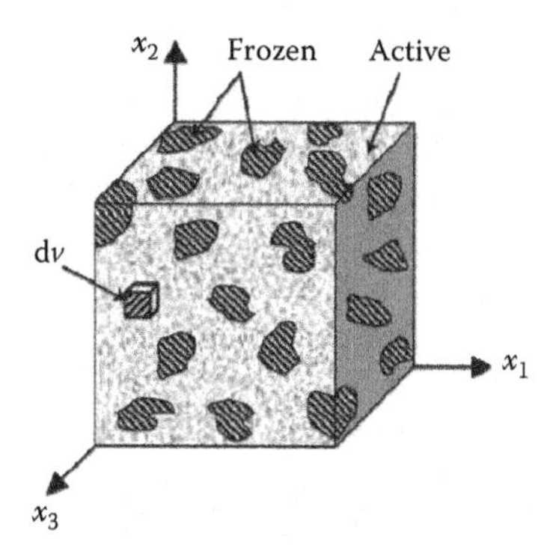

FIGURE 4.6
Schematic diagram of the micromechanics foundation of the 3D shape-memory polymer constitutive model. The diagram represents a polymer model with two extreme phases in the glass transition state consisting of a predominant active phase (gray areas) and a frozen phase (shaded areas). (Reprinted and modified from Liu, Y. et al., *Int. J. Plast.*, 22(2), 279, 2006. With permission.)

regard to the conformational motion in the glassy state at $T < T_g$, while the active bond represents the rest of the C–C bonds that can undergo localized free conformational motions in the rubbery state at $T > T_g$. In the glassy state, the major phase of a polymer is the frozen phase composed of frozen bonds, where conformational motions of the chains are locked. In contrast, the active phase consists of active bonds and the free conformational motion can potentially occur, and the polymer exists in the full rubbery state. Figure 4.6 shows a schematic picture of a simplified 3-D shape-memory polymer model with a "frozen phase" (dark-shaded region) and an "active phase" (light-shaded region).

From a thermodynamic point of view, the frozen volume fraction ϕ_f is an internal state variable of the system and it is assumed that ϕ_f depends only on temperature (T). Based on experimental results, an analytical phenomenological function $\phi_f = \phi_f(T)$ can be determined. This function captures the fraction of strain storage and release during the thermomechanical cycle as a function of temperature. The total strain ε can be represented as the sum $\varepsilon = \phi_f\varepsilon_f + (1 - \phi_f)\varepsilon_a$ of two strain contributions from the frozen phase and the active phase. The strain in the frozen phase ε_f can be determined from three contributions, i.e., the (stored) entropic strain, the internal energetic strain, and the thermal strain. In the active phase, the strain deformation ε_a consists of two parts: the external stress-induced entropic strain, and the thermal strain. For all single-strain contributions, analytical relations can be derived. For details see reference [33]. Finally, one obtains as model, a set of equations depending on five polymer specific coefficients. The major limitation of this model was the prediction of stress and strain dependent on temperature for only small unidirectional deformations of about 10%, which were systematically investigated for epoxy resins. Based on this theory, Chen and

Lagoudas [37,38] reported an approach to describe the thermomechanical properties of SMPs under large deformations, where the general constitutive functions of the neo-Hookean type for nonlinear thermoelastic materials are used for the active and frozen phases. The relation between the overall deformation and the stress is derived by integration of the constitutive equations of the coexisting phases.

In addition to the so far presented theories for the prediction of shape–memory properties, recently, a mesomechanical concept was applied to model the shape-memory phenomenon [35].

Further, a thermoviscoelastic constitutive model, based on the nonlinear Adam–Gibbs model of structural relaxation, and a modified Eyring model of viscous flow into a continuum finite–deformation thermoviscoelastic framework was developed by Nguyen et al. [36] for amorphous SMPs following the hypothesis that structural and stress relaxation are the primary molecular mechanisms of the shape-memory effect and its time-dependence.

4.5 Summary and Outlook

With respect to the emerging number of scientific papers on shape-memory polymers published per annum within the last few years, methods for quantification of dual-shape or triple-shape properties and the modeling approaches for simulating the thermomechanical behavior of such polymers are of increasing scientific and technological significance. A detailed understanding of the underlying mechanism is necessary for the realization of the numerous potential applications of shape-memory polymers. For the description of the complex themomechanical processes in shape-memory polymers, multiple parameters have to be considered: chemical composition and morphology of the material, the conducted programming procedure, which can be stress- or strain-controlled, together with the included thermo-treatment characterized by parameters such as T_{deform} and β_c, and finally, the recovery process, which is characterized by parameters such as T_{high} and β_h.

The shape-memory effect is quantified by the strain recovery rate R_r and the strain fixity rate R_f. While R_r quantifies the ability of the polymer to memorize the initial permanent shape, R_f is a measure for fixation of the temporary deformed shape. Further, the common characteristic is the switching temperature T_{sw} obtained as the inflection point in the strain temperature recovery curve under stress-free conditions. Under constant strain conditions, the stress maximum σ_{max}, as well as the corresponding temperature $T_{\sigma,max}$, are characteristics for the physically cross-linked SMP, while the temperature

$T_{\sigma,inf}$ at the inflection point of the stress–temperature curve is determined in the case of covalently cross-linked SMPs.

Besides the different methods described for quantification of the shape-memory properties, the development in the field of modeling approaches will be crucial for a successful translation of this technology into industrial applications. The recently reported 3-D thermomechanical constitutive model assuming active and frozen phases, representing the multiphase character of thermoplastic shape-memory polymers, can be an especially fruitful approach for the future development of finite element models for the prediction of thermomechanical behavior.

References

1. Behl, M. and Lendlein, A. 2007. Shape-memory polymers. *Materials Today* 10:20–28.
2. Kelch, S. 2005. *Degradable, Multifunctional Polymeric Biomaterials with Shape-Memory*, VIII edn., Tech Trans Publications, Zurich, Switzerland.
3. Lendlein, A. and Kelch, S. 2002. Shape-memory polymers. *Angewandte Chemie—International Edition* 41(12):2034–2057.
4. Behl, M. and Lendlein, A. 2007. Actively moving polymers. *Soft Matter* 3:58–67.
5. Beloshenko, V. A., Varyukhin, V. N., and Voznyak, Y. V. 2005. The shape memory effect in polymers. *Russian Chemical Reviews* 74:265–283.
6. Liu, C., Qin, H., and Mather, P. T. 2007. Review of progress in shape-memory polymers. *Journal of Materials Chemistry* 17:1543–1558.
7. Ratna, D. and Karger-Kocsis, J. 2008. Recent advances in shape memory polymers and composites: A review. *Journal of Materials Science* 43:254–269.
8. Otsuka, K., Saburi, T., Wayman, C. M., Tadaki, T., Maki, T., Suzuki, Y., Humbeeck, R. J. V., Stalmans, K. U., and Miyazaki, S. 1998. *Shape Memory Materials*, Cambridge University Press, Cambridge, U.K.
9. Hornbogen, E. 2006. Comparison of shape memory metals and polymers. *Advanced Engineering Materials* 8:101–106.
10. Bellin, I., Kelch, S., Langer, R., and Lendlein, A. 2006. Polymeric triple-shape materials. *Proceedings of the National Academy of Sciences of the United States of America* 103(48):18043–18047.
11. Bellin, I., Kelch, S., and Lendlein, A. 2007. Dual-shape properties of triple-shape polymer networks with crystallizable network segments and grafted side chains. *Journal of Materials Chemistry* 17:2885–2891.
12. Kolesov, I. S. and Radusch, H.-J. 2008. Multiple shape-memory behavior and thermal-mechanical properties of peroxide cross-linked blends of linear and short-chain branched polyethylenes. *eXPRESS Polymer Letters* 2:461–473.
13. Behl, M., Bellin, I., Kelch, S., Wagermaier, W., and Lendlein, A. 2008. One-step process for creating triple-shape capability of AB polymer networks. *Advanced Functional Materials* 18:1–7.

14. Capaccio, G. and Ward, I. M. 1982. Shrinkage, shrinkage force and the structure of ultra high modulus polyethylenes. *Colloid and Polymer Science* 260:46–55.
15. Chowdhury, S. R. and Das, C. K. 2000. Studies on blends of ethylene vinyl acetate and polyacrylic rubber with reference to their shrinkability. *Journal of Applied Polymer Science* 77:2088–2095.
16. Chowdhury, S. R., Mishra, J. K., and Das, C. K. 2000. Structure, shrinkability and thermal property correlations of ethylene vinyl acetate (EVA)/carboxylated nitrile rubber (XNBR) polymer blends. *Polymer Degradation and Stability* 70:199–204.
17. Lin, J. R. and Chen, L. W. 1998. Study on shape-memory behavior of polyether-based polyurethanes. I. Influence of the hard-segment content. *Journal of Applied Polymer Science* 69:1563–1574.
18. Lin, J. R. and Chen, L. W. 1998. Study on shape-memory behavior of polyether-based polyurethanes. II. Influence of soft-segment molecular weight. *Journal of Applied Polymer Science* 69:1575–1586.
19. Ping, P., Wang, W. S., Chen, X. S., and Jing, X. B. 2005. Poly(epsilon-caprolactone) polyurethane and its shape-memory property. *Biomacromolecules* 6:587–592.
20. Khonakdar, H. A., Jafari, S. H., Rasouli, S., Morshedian, J., and Abedini, H. 2007. Investigation and modeling of temperature dependence recovery behavior of shape-memory crosslinked polyethylene. *Macromolecular Theory and Simulations* 16:43–52.
21. Choi, N. Y. and Lendlein, A. 2007. Degradable shape-memory polymer networks from oligo[(L-lactide)-ran-glycolide]dimethacrylates. *Soft Matter* 3:901–909.
22. Kelch, S., Steuer, S., Schmidt, A. M., and Lendlein, A. 2007. Shape-memory polymer networks from oligo [(epsilon-hydroxycaproate)-co-glycolate] dimethacrylates and butyl acrylate with adjustable hydrolytic degradation rate. *Biomacromolecules* 8:1018–1027.
23. Miaudet, P., Derre, A., Maugey, M., Zakri, C., Piccione, P. M., Inoubli, R., and Poulin, P. 2007. Shape and temperature memory of nanocomposites with broadened glass transition. *Science* 318:1294–1296.
24. Liu, Y., Gall, K., Dunn, M. L., and McCluskey, P. 2003. Thermomechanical recovery couplings of shape memory polymers in flexure. *Smart Materials and Structures* 12:947–954.
25. Gall, K., Yakacki, C. M., Liu, Y. P., Shandas, R., Willett, N., and Anseth, K. S. 2005. Thermomechanics of the shape memory effect in polymers for biomedical applications. *Journal of Biomedical Materials Research, Part A* 73A:339–348.
26. Lendlein, A., Schmidt, A. M., and Langer, R. 2001. AB-polymer networks based on oligo(epsilon-caprolactone) segments showing shape-memory properties. *Proceedings of the National Academy of Sciences of the United States of America* 98:842–847.
27. Lendlein, A. and Langer, R. 2002. Biodegradable, elastic shape-memory polymers for potential biomedical applications. *Science* 296:1673–1676.
28. Lendlein, A., Kelch, S., Schulte, J., and Kratz, K. 2005. Shape-memory polymers, in *Encyclopedia of Materials: Science and Technology—Updates*, Buschow, K. H. J., Cahn, R. W., Flemings, M. C., Kramer, E. J., Mahajan, S., and Veyssiere, P. (Eds.), Elsevier Science Ltd., New York.
29. Choi, N. Y., Kelch, S., and Lendlein, A. 2006. Synthesis, shape-memory functionality and hydrolytical degradation studies on polymer networks from poly(rac-lactide)b-poly(propylene oxide)-b-poly(rac-lactide) dimethacrylates. *Advanced Engineering Materials* 8:439–445.

30. Tobushi, H., Hashimoto, T., Hayashi, S., and Yamada, E. 1997. Thermomechanical constitutive modeling in shape memory polymer of polyurethane series. *Journal of Intelligent Material Systems and Structures* 8:711–718.
31. Tobushi, H., Okumura, K., Hayashi, S., and Ito, N. 2001. Thermomechanical constitutive model of shape memory polymer. *Mechanics of Materials* 33:545–554.
32. Bhattacharyya, A. and Tobushi, H. 2000. Analysis of the isothermal mechanical response of a shape memory polymer rheological model. *Polymer Engineering and Science* 11:2498–2510.
33. Liu, Y., Gall, K., Dunn, M. L., Greenberg, A. R., and Diani, J. 2006. Thermomechanics of shape memory polymers: Uniaxial experiments and constitutive modeling. *International Journal of Plasticity* 22(2):279–313.
34. Diani, J., Liu, Y., and Gall, K. 2006. Finite strain 3D thermoviscoelastic constitutive model for shape memory polymers. *Polymer Engineering and Science* 46:486–492.
35. Kafka, V. 2008. Shape memory polymers: A mesoscale model of the internal mechanism leading to the SM phenomena. *International Journal of Plasticity* 24:1533–1548.
36. Nguyen, T. D., Qui, H. J., Castro, F., and Long, K. 2008. A thermoviscoelastic model for amorphous shape memory polymers: Incorporating structural and stress relaxation. *Journal of the Mechanics and Physics of Solids* 56:2792–2814.
37. Chen, Y. C. and Lagoudas D. C. 2008. A constitutive theory for shape memory polymers: Part I. *Journal of the Mechanics and Physics of Solids* 56:1752–1765.
38. Chen, Y. C. and Lagoudas D. C. 2008. A constitutive theory for shape memory polymers: Part II. *Journal of the Mechanics and Physics of Solids* 56:1766–1778.
39. Barot, G. and Rao, I. J. 2006. Constitutive modeling of the mechanics associated with crystallizable shape memory polymers. *Zeitschrift Fur Angewandte Mathematik Und Physik* 57:652–681.
40. Rao, I. J. 2002. Constitutive modeling of crystallizable shape memory polymers. *ANTEC Proceedings*, Dallas, TX, pp. 1936–1940.
41. Morshedian, J., Khonakdar, H. A., and Rasouli, S. 2005. Modeling of shape memory induction and recovery in heat-shrinkable polymers. *Macromolecular Theory and Simulations* 14(7):428–434.
42. Barot, G., Rao, I. J., and Rajagopal, K. R. 2008. A thermodynamic framework for the modeling of crystallizable shape memory polymers. *International Journal of Engineering Science* 46:325–351.

5

Electrical, Thermomechanical, and Shape-Memory Properties of the PU Shape-Memory Polymer Filled with Carbon Black

Wei Min Huang and Bin Yang
School of Mechanical and Aerospace Engineering, Nanyang Technological University, Singapore, Singapore

CONTENTS

This chapter presents a systematical study of the electrical and themomechanical properties of a thermo-moisture responsive polyurethane shape-memory polymer (SMP) filled with nano carbon powders (carbon black). It is revealed that the presence of a small amount of carbon powder dramatically improves the strength and electrical conductivity of the polymer, while preserving the good shape-memory. On the other hand, the influence of moisture on the thermomechanical properties of this type of polyurethane SMP is investigated. The so-called water-actuated recovery is characterized. This chapter aims at providing the necessary knowledge for engineers in applying this material for engineering applications.

5.1 Introduction

A thermo-responsive shape-memory polymer (SMP) is unique because of its ability to recover a significant deformation when heated to over its glass-transition temperature (T_g). Compared with other SMPs available at present, the thermoplastic polyurethane SMP has many advantages, namely, a wider range of shape-recovery temperature (from −30°C to 70°C), a higher shape recoverability (maximum recoverable strain >400%), a better processing ability and biocompatibility, etc. [1–4]. Till date, the structure, and the thermomechanical and shape-memory properties of many thermoplastic polyurethane SMPs have been developed, extensively investigated, and well documented in the literatures [5–9]. A few nonlinear constitutive models have been proposed to simulate the thermomechanical behavior (e.g., [10]).

So far, a few engineering applications of SMPs have been proposed and even realized (e.g., [11–13], and discussed in (Chapters 8 through 12) of this book). However, their full potential has not yet been explored. Compared with shape-memory metals and shape-memory ceramics, SMPs have a lower stiffness, which results in a lower recovery stress. Hence, they are more likely to be applicable in those applications in which the produced recovery stress is very low or virtually zero. Some researchers have tried to reinforce SMPs by doping them with various types of fillers with some success [14,15]. Another important issue, which significantly limits the application of SMPs, is that the shape recovery of the widely used thermal-responsive SMPs can only be triggered upon heating, e.g., by means of thermal heating using an external heater/laser or through radiation. It is a significant improvement to make SMPs electrically conductive, so that the actuation can be triggered by passing an electrical current directly (just like in shape-memory alloys), and stiffer by blending with conductive fillers, such as nano carbon powder (carbon black) or nano carbon tube, etc. [16,17].

Previously, we have investigated the influence of carbon powder and moisture on the T_g of a polyurethane SMP [18]. This polyurethane SMP and its composites filled with carbon powder are not only thermo responsive like other ordinary polyurethane SMPs, but also moisture responsive, i.e., shape-recovery can be triggered upon immersing the SMP into room temperature water. This unique moisture responsive feature can be further utilized for fabricating SMPs with a functionally gradient T_g for shape recovery in a programmable manner.

In this chapter, we present a systematical investigation on the electrical and thermomechanical properties of this polyurethane SMP and its nano carbon powder composites. In addition, the influence of moisture and moisture responsive feature are characterized.

This chapter aims at providing the necessary knowledge for engineers in applying this material for engineering applications. For other polyurethane SMPs and their composites filled with other electrically conductive fillers, readers may refer to, for instance, [17] or Chapters 7 and 12 in this book.

5.2 Experiment

5.2.1 Sample Preparation

The polymer matrix used in the course of this study is the thermoplastic SMP in the pellet form provided by Mitsubishi Heavy Industries (MHI), Japan. It is prepared from diphenylmethane-4,4′-diisocyanate, adipic acid, ethylene glycol, ethylene oxide, polypropylene oxide, 1,4-butanediol, and bisphenol A, with a glass transition temperature (T_g) of 55°C and a melting temperature of around 200°C. Carbon powders in an average diameter of 30 nm (from the JJ-Degussa Chemicals (s) Pte Ltd. Singapore.) were used as fillers.

Before the sample preparation, the polyurethane SMP and the carbon powders were dried in a vacuum oven at 80°C for 12 h to remove moisture. Then the pellets of the polyurethane SMP were melted at 200°C in the mixing head of a Haake Rheocord 90. Subsequently, carbon powders were added in and blended with the melted SMP for 15 min. Finally, the carbon powder–filled SMP was processed at 200°C into different shapes as samples for testing.

SMP composites with five different volume fractions of carbon powders, namely, 4%, 7%, 10%, 13% and 15%, were prepared. In this paper, CBXX denotes SMP filled with XX% volume fraction of carbon powder. Hence, CB0 denotes pure SMP. The volume fraction of the carbon powder (φ_f) was calculated by

$$\varphi_f = \frac{1}{\left(M_m/\rho_m\right) \times \left(\rho_f/M_f\right)} \times 100\% \tag{5.1}$$

where M_m, ρ_m, M_f, and ρ_f denote the mass of the SMP, the bulk density of the SMP (1.25 g/cm^3, from MHI), the mass of the carbon powder, and the bulk density of the carbon powder (1.85 g/cm^3, from the Degussa), respectively.

5.2.2 Testing

The electrical resistivity of the carbon powder–filled SMPs was derived from the measured resistance using a four-point resistivity probe system (SINGNTONE) with a temperature control unit and an upper limit of 10^{10} Ω m.

The resistance of SMP composites with a smaller volume fraction of the carbon powder is beyond this limit. Therefore, a digital high resistivity determiner (RP2680) was used instead.

In a series of previous tests using the same type of SMP from the same company but with a T_g of 35°C, we obtained the relationship of the electrical resistivity against φ_f as shown in Figure 5.1. We found that the percolation is at around 6% of φ_f. Figure 5.2 shows the scanning electronic microscope (SEM, Leica Cambridge S360) images of the distribution of carbon powder in the SMP at three different φ_f. It should be reasonable to take them as a reference for our current materials.

In real applications, the carbon powder–filled SMPs are normally deformed and then heated for recovery. Therefore, it is necessary to understand the effects of strain on the electrical resistivity of carbon powder–filled SMPs. Figure 5.3 illustrates the setup for this test. An Instron 5569 was used to stretch/compress the carbon powder–filled SMP samples at a constant crosshead speed, which gives a fixed strain rate of 10^{-3}/s. Carbon powder–filled SMP wires with a diameter of 2.0 mm and a gauge length of 30.0 mm were used for the extension test; while cylindrical samples with a diameter of 15.0 mm and gauge length of 20.0 were used for the compression test. Two copper electrodes were attached to the ends of samples. A multimeter was coupled with the Instron machine to measure the resistance of the sample (R'). The electrical volume resistivity (ρ) is calculated by,

$$\rho = R' \times S'/L \tag{5.2}$$

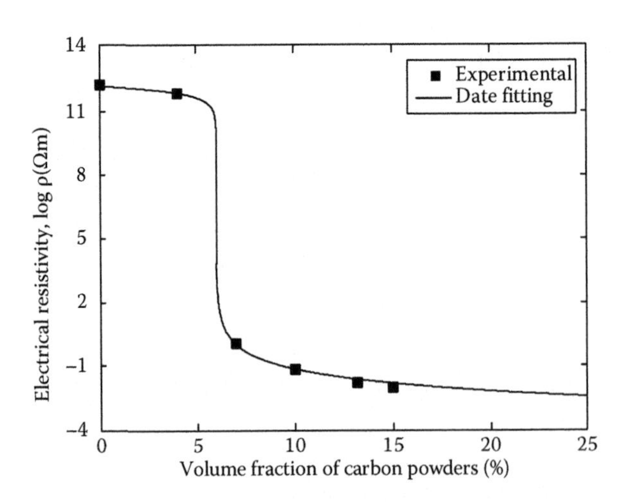

FIGURE 5.1
Electrical resistivity vs. volume fraction of carbon powders.

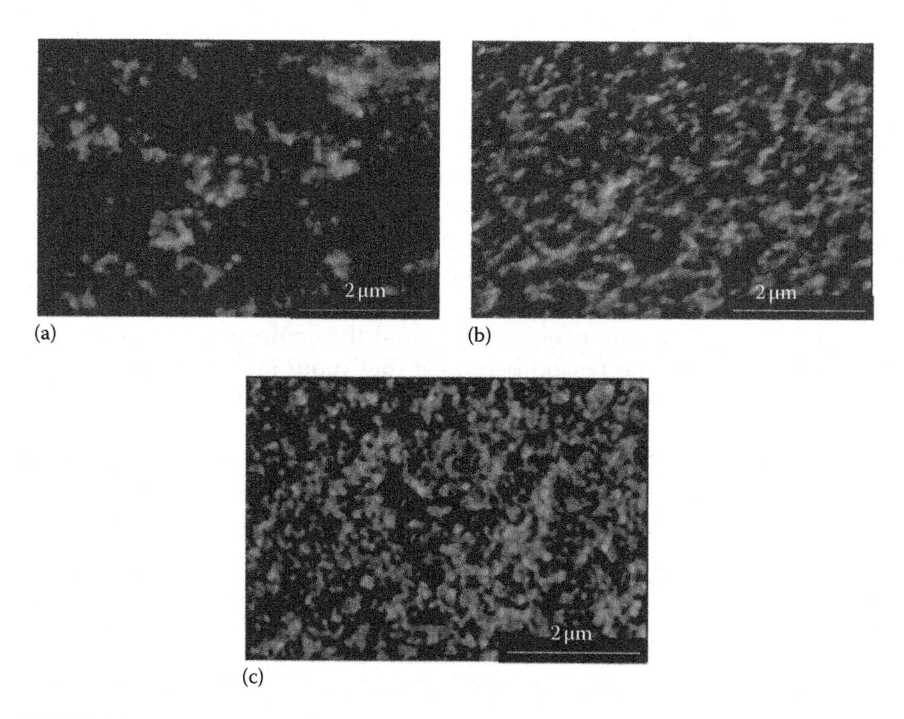

FIGURE 5.2
SEM images of cryofracture surfaces. (a) CB4; (b) CB7; and (c) CB13.

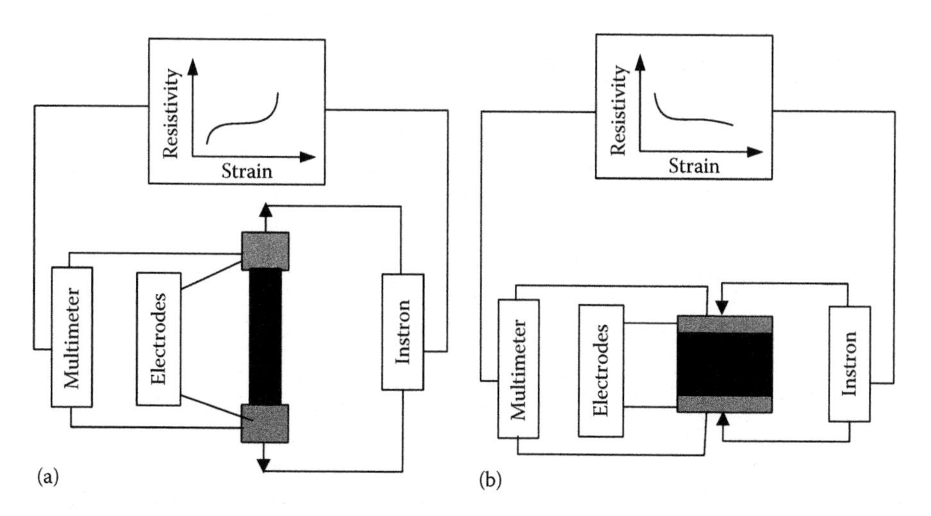

FIGURE 5.3
Experimental setup. (a) Tension test and (b) compression test.

where

S' is the cross-sectional area of the sample
L is the gauge length

The uniaxial tensile tests were carried out using an Instron microforce materials test system with a 1 kN load cell to investigate the uniaxial tensile properties of the SMP composites at room temperature (about 22°C). The samples with the dimensions as shown in Figure 5.4 were tested. The gauge length was set as 20 mm. In all tests, a constant strain rate of 5×10^{-3}/s was applied. We used the engineering strain and the engineering stress in this study. Simplicity is one reason behind it. But more importantly, from the engineering application point of view, the engineering strain and the engineering stress are more of the practical concerns.

Thermomechanical tests were carried out to investigate the shape-memory properties of SMP composites. The samples are with dimensions as shown in Figure 5.4. The testing procedure is illustrated in Figure 5.5. In step a, the sample is stretched uniaxially to a maximum strain (ε_m) of 20% at a constant strain rate of 5×10^{-3}/s by an Instron microforce materials test system with a 100 N load cell. The stretching is carried out inside a hot chamber at a constant temperature (T_h) of 65°C. Then, the sample is held at ε_m for four minutes and cooled to room temperature (T_r) (step b). During holding, the stress relaxes at the very beginning as a result of the viscous flow since the material is at the rubber state. In step c, the sample is unloaded to zero stress, which induces a small amount of strain-recovery. The fixed strain (ε_f) is defined as the strain right after unloading is finished.

It should be pointed out that normally, if a tensile test is carried out on a glass state SMP, the deformation is not uniform after the yielding point is reached. We can observe the necking and the subsequently propagation of the *plastic* zone, which is accompanied with a stress peak and then followed by a long stress plateau in the stress vs. strain curve. This is not ideal for predeforming an SMP. Thus, if possible, the tensile test should be performed on a rubber state SMP, in which the deformation is reasonably uniform.

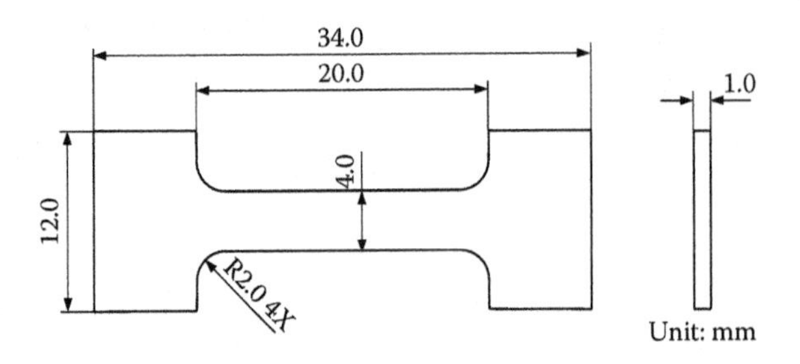

FIGURE 5.4
Sample dimensions.

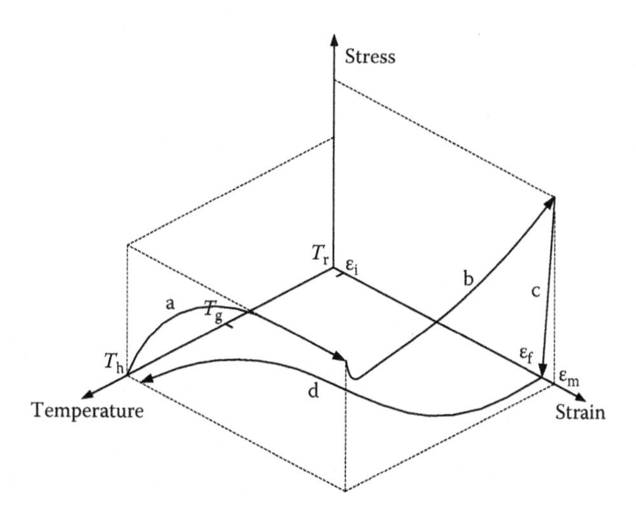

FIGURE 5.5
Procedure of the thermomechanical test.

From the thermomechanical point of view, the recovery stress and the recoverable strain are essential concerns in engineering applications of SMPs. Thus, after step c, two types of tests were performed, namely, the heat-induced free recovery test and the heat-induced constrained recovery test, in order to measure the recovery strain and the recovery stress, respectively. The samples were grouped into two for both tests. The samples in the first group were heated in the hot chamber at a constant rate of 2°C/min with one end free to move. The recovery strain was recorded against temperature. This is the so-called heat-induced free recovery test. The samples in the second group were tested under the same condition as in the first group except that the length of the samples was fixed during testing. The recovery stress was recorded against the temperature. This is the so-called heat-induced constrained recovery test.

The new feature of this SMP, i.e., water-actuated recovery, has been realized and reported in [19]. This new feature was found to be applicable in carbon powder–filled SMP composites [16]. For engineering applications, it is essential to know the recovery stress and recoverable strain in water-actuated recovery. Thus, we prepared two other groups of samples (CB0, CB4, and CB7) for the water-actuated free recovery test and the water-actuated constrained recovery test. A water tank was attached to the Instron machine and the prestretched samples were immersed completely into room temperature water. This time, no external heat was provided during testing. In the water-actuated free recovery test, the recovery strain was recorded against the immersion time, while in the water-actuated constrained recovery test, the recovery stress was recorded against the immersion time.

In order to evaluate the shape-memory properties of the SMP composites, the ratio of the fixable strain (R_f) and the ratio of the recoverable strain (R_r) are determined by

$$R_f = \varepsilon_f / \varepsilon_m \times 100\% \tag{5.3a}$$

and

$$R_r = (\varepsilon_m - \varepsilon_f + \varepsilon_r) / \varepsilon_m \times 100\% \tag{5.3b}$$

where ε_r is the total recovered strain at the end of the free recovery test.

In order to investigate the effects of moisture on the thermomechanical properties of the carbon powder–filled SMPs, the dynamic mechanical analysis (DMA) test [DMA 2980 (TA Instruments, New Castle, Delaware)] was carried out on the samples after different hours of immersion in room temperature water. The test was conducted in the film mode at a constant frequency of 1 Hz and a heating rate of 4°C/min. The samples for this test (34 × 4.0 × 1 mm) were cut from the samples as shown in Figure 5.4.

5.3 Results and Discussion

5.3.1 Electrical Properties

In general, the working temperature for this polyurethane SMP and its composites is from room temperature to about 100°C. Thus, each carbon powder–filled SMP sheet with a thickness of 1.0 mm was heated from room temperature to around 100°C, and the electrical resistivity of each composite was obtained against the temperature using the four-point resistivity probe system. Note that CB4 was not included as its electrical conductivity is too high to be used for electrical heating purposes. The experimental results are plotted in Figure 5.6. Figure 5.6 reveals that there is no apparent change in electrical resistivity in all carbon powder–filled SMP composites within this temperature range.

Figure 5.7a plots the tensile strain against the electrical resistivity. It shows that, generally speaking, with the increase of strain, the electrical resistivity increases, but in a nonlinear fashion. The increase of electrical resistivity is more remarkable in the lower-strain region, i.e., below 10%. With the increase of loading of carbon black, the increase of electrical resistivity in the lower-strain region becomes more significant.

When a conductive composite is subjected to a tensile strain, it is expected that two simultaneous processes will occur in the material: breakdown of existing conductive networks due to an increase in the gap between the

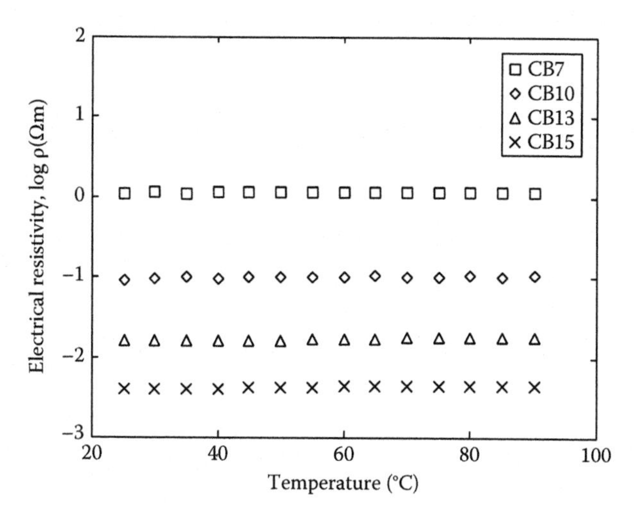

FIGURE 5.6
Electrical resistivity as a function of temperature.

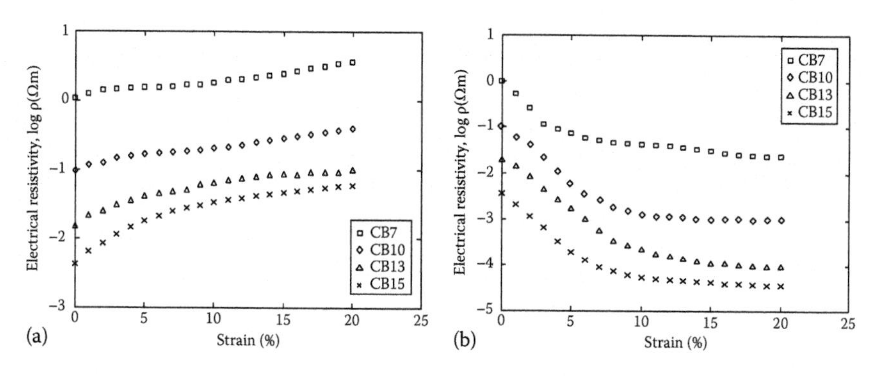

FIGURE 5.7
Electrical resistivity vs. strain relationship. (a) In tension and (b) in compression.

carbon powder aggregates, and the reformation of new conductive networks due to the reorientation of carbon aggregates [20,21]. At a lower-tensile strain, the breakdown process is more prominent than the formation process. Thus, the net result is a reduction in the number of conductive networks, which indicates an increase in resistivity. However, at a higher tensile strain, rather than formation of holes and destruction of the conductive networks in the material, the significant extension produces new conductive pathways and/or improves the existing pathways by the reorientation of carbon black. For high structured carbon powders, rotation, translation, and possible shape change in the asymmetric aggregates can preserve the number of contacts and conductive pathways. These processes somehow counterbalance the effect of a breakdown of the conductive networks, resulting in a slower rate of change in resistivity against the extension at a higher strain [22–24].

A compressive strain has a different effect on the electrical resistivity. The testing results are plotted in Figure 5.7b. It reveals that with the increase of the compressive strain, the electrical resistivity decreases monotonously. Similar to that in tension but opposite in direction, the most significant decrease occurs at a lower strain range, i.e., at a strain less than 10%. At a lower compressive strain, the gap between carbon aggregates in SMP composites is reduced so that the tunneling conduction becomes possible among aggregates and some aggregates may even directly contact each other. Thus, more conducive pathways are constructed, resulting in a decrease in resistivity. These effects become less significant in the high strain region because most of the possible conductive pathways upon compression have been formed in the low strain region.

5.3.2 Uniaxial Tension at Room Temperature

Figure 5.8 shows the stress–strain curves of all composites at room temperature. The relationships of Young's modulus and the elongation limit vs. φ_f are plotted in Figure 5.9. Note that Young's modulus was calculated from the early loading part of each stress–strain curve, since at room temperature and within a very small loading range, the composites are pretty rigid and the deformation is virtually linear elastic. As we can see in Figure 5.8, in general, SMP composites behave in more or less the same way upon loading, i.e., they experience elastic deformation, yielding, and cold drawing until failure. Figure 5.9 reveals that the addition of carbon powder improves the

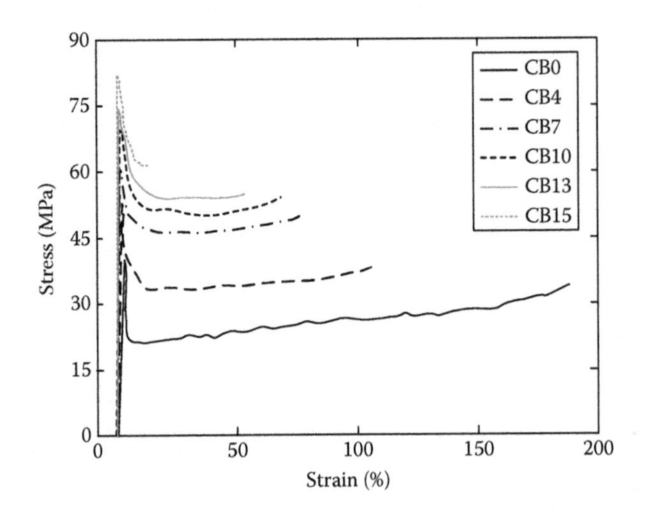

FIGURE 5.8
Stress–strain curves at room temperature.

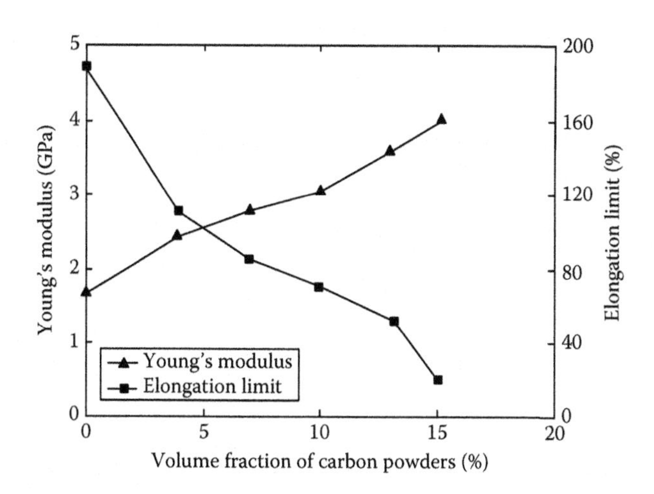

FIGURE 5.9
Relationship of Young's modulus and the elongation limit vs. φ_f.

stiffness of SMP composites so that Young's modulus increases almost linearly with the volume fraction of carbon powders. Young's modulus of CB15 is about two times of that of CB0. However, the presence of carbon powder reduces the elongation limit significantly. Even a small volume fraction of carbon powder, e.g., 4%, results in a dramatic decrease of the elongation limit by over 50%.

5.3.3 Shape Fixity

The stress–strain relationships obtained in the thermomechanical test (stretched at 65°C) are plotted in Figure 5.10. It shows that with the increase of the volume fraction of carbon powders the SMP composite requires a higher force for deformation. At the early holding/cooling stage, there is a significant drop in stress due to the relaxation induced by the viscous flow, as, at above T_g the material is in the rubber state. Upon further cooling to below T_g, the stress starts to increase. This occurs due to the thermal contraction upon cooling while Young's modulus increases as the material transits from the soft rubber state to the hard glass state. After the subsequent unloading, only a small amount of elastic strain (less than 3% in all tests) is recovered.

Figure 5.11 plots the ratio of the fixable strain against the volume fraction of carbon powders. It reveals that the presence of carbon powders deteriorates the shape fixity. With around 4% carbon powders, the ratio of the fixable strain is slightly lower than that of the pure SMP. A significant decrease occurs when the content of carbon powder reaches 7%. Further increase in carbon powder only marginally reduces the ratio of the fixable strain. We may conclude that in general, carbon powders do not dramatically reduce

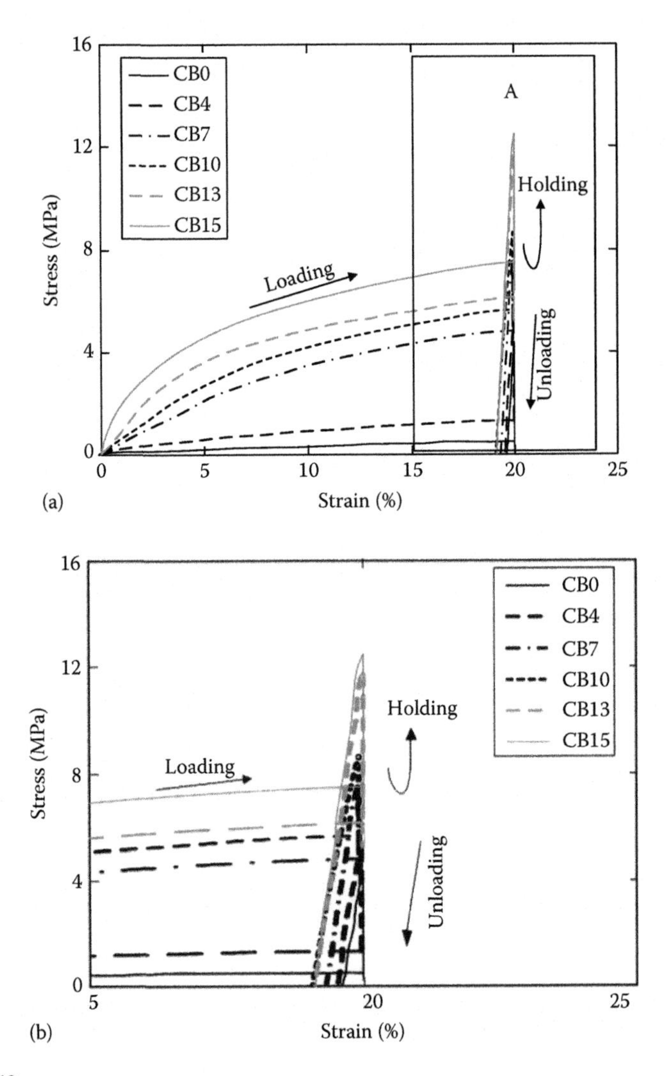

FIGURE 5.10
Stress–strain curves (loaded at 65°C). (a) Overall view and (b) zoom-in view of A in (a).

the shape fixity. In all tested SMP composites, over 95% of the prestrain is fixable after cooling and unloading.

5.3.4 Heating-Induced Recoverable Strain

The free recovery strain as a function of temperature in the free recovery test is plotted in Figure 5.12. It reveals that in all samples, a significant portion of the prestrain is recovered within the temperature range of 50°C–60°C.

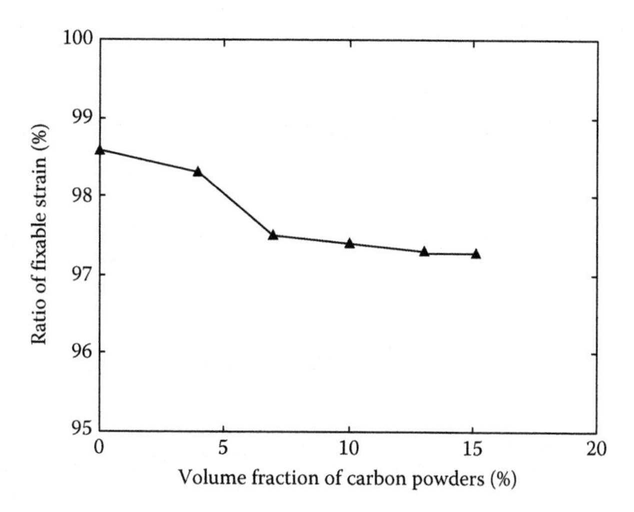

FIGURE 5.11
Ratio of the fixable strain vs. the volume fraction of carbon powders.

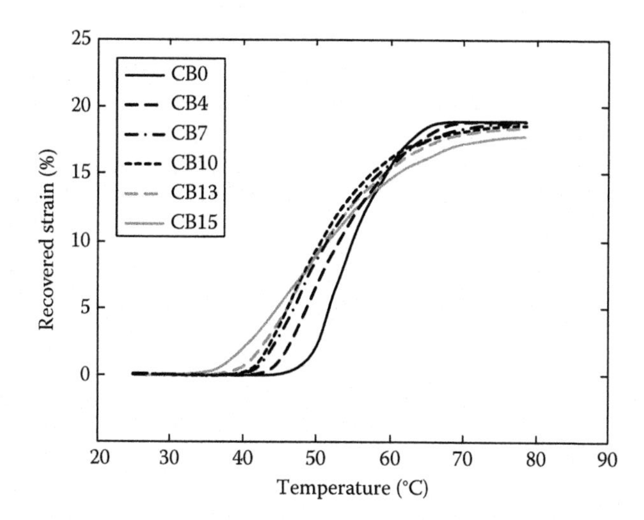

FIGURE 5.12
Recovered strain as a function of temperature.

With the increase of the carbon powders, the start temperature for the recovery becomes lower, while the finish temperature for the shape recovery turns out to be higher, i.e., the temperature range for the strain recovery is widened. Such effects may be attributed to the change of glass-transition kinetics due to the loading of carbon powders [15].

For a better view, Figure 5.13 plots the ratio of the recoverable strain against the volume fraction of carbon powders. It reveals that carbon powders have

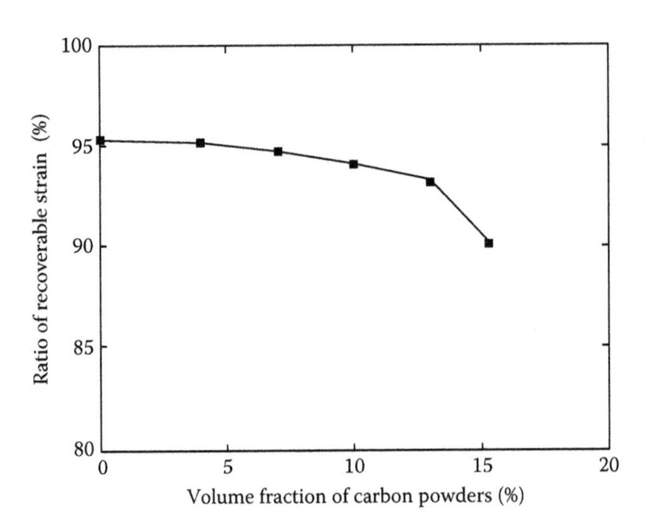

FIGURE 5.13
Ratio of the recoverable strain against the volume fraction of carbon powders.

a limited effect on the shape-recovery ability of the SMP at a low φ_f. Even when the volume fraction reaches 13%, the SMP composite is still able to recover over 92% of the prestrain, while that of the pure SMP is about 95%. However, with a further increase in carbon powders, the trend of the drop in the recoverable strain becomes remarkable. At a 15% carbon powder level, the recovery is less than 90%.

5.3.5 Heating-Induced Recovery Stress

Figure 5.14 plots the recovery stress against temperature in the constrained recovery test. Generally speaking, the change in the recovery stress is very small at below 35°C. A small amount of contraction due to shape recovery is largely canceled by the thermal expansion. With a further increase in temperature, the recovery stress increases progressively due to the shape-memory effect and then reaches a peak. After that, it decreases gradually and even approaches the zero stress in some of the tests conducted. The decrease of the recovery stress is due to the relaxation, which is attributed to the viscous flow of the polymer chains at a high temperature, and further softening upon heating.

Furthermore, the results reveal that the loading of carbon powders significantly increases the maximum recovery stress in SMP composites. With an increase in the amount of carbon powders, the decrease in the recovery stress upon further heating becomes more gradual after the maximum stress is reached. As carbon powders restrict the viscous flow in SMPs, a slower relaxation in the heavily carbon-loaded SMP composites is expected.

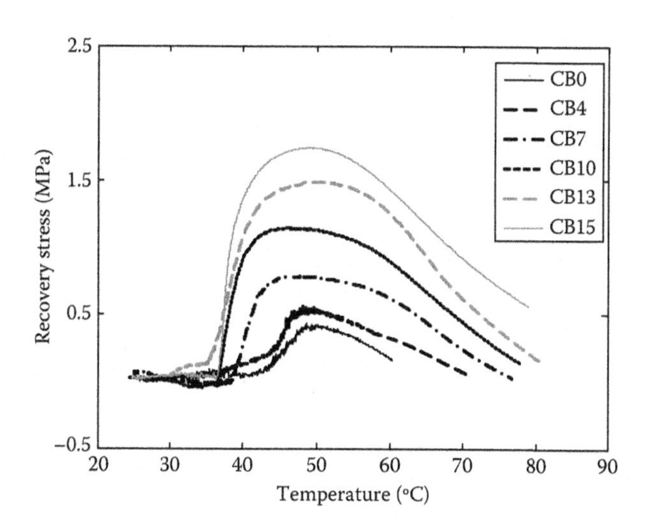

FIGURE 5.14
Recovery stress as a function of temperature.

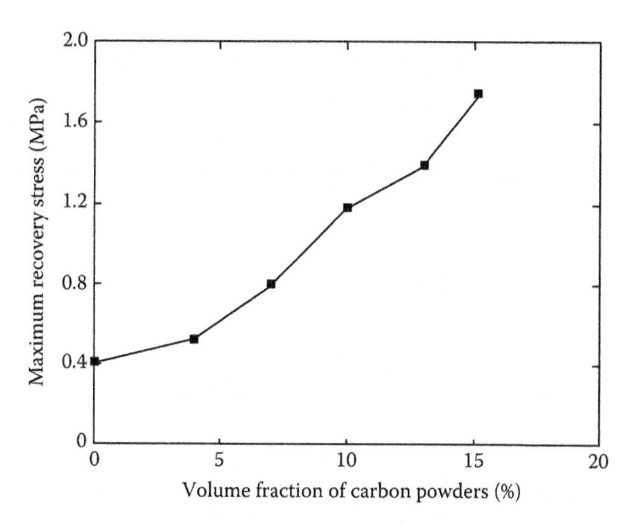

FIGURE 5.15
Maximum recovery stress against the volume fraction of carbon powders.

For a better view, the maximum recovery stress upon heating is plotted against the volume fraction of carbon powders in Figure 5.15. It shows that the maximum recovery stress is an exponential function of the loading of carbon powders. The maximum recovery stress in CB15 is about four times higher than that of the pure SMP. It demonstrates that the loading of carbon powders is a very efficient way to increase the recovery stress.

5.3.6 Water-Actuated Shape Recovery

In Figure 5.16, the recovered strain in the free recovery test is plotted against the immersion time. It shows that after immersion in room temperature water for 60 min, all SMPs start to recover. A CB0 can recover about 18% strain, which is about 90% of the prestrain. As a general trend, with an increase in the carbon powder content, the recovered strain decreases and the recovery becomes slower. A slower recovery speed can be partially attributed to the carbon powders, which delay the penetration speed of moisture into the polymer. The phenomenon of the less recovered strain should be the result of carbon powders as well. Upon predeformation, there is an unavoidable mismatch between the polymer and carbon powders as they have remarkably different mobility and deformability. The required driven force for a full recovery in the interface between them should be much higher than that for the polymer. As such, apart from the moisture, an additional driving force and/or a much longer time might be required for more recovery so that, we may expect both, a lesser recovery in the carbon powder–filled SMPs, and also a lower recovery speed. The recovery ratio is comparable to that in the thermally induced recovery (refer to Figure 5.12). The explanation provided here, on the influence of carbon powder, should be largely applicable for the thermally induced free recovery as well.

Figure 5.17 presents the relationship of the recovery stress vs. the immersion time in the constrained recovery test. It shows that, in general, the recovery stress begins to increase after around 60 min of immersion and then gradually reaches the maximum. After that, the recovery stress decreases

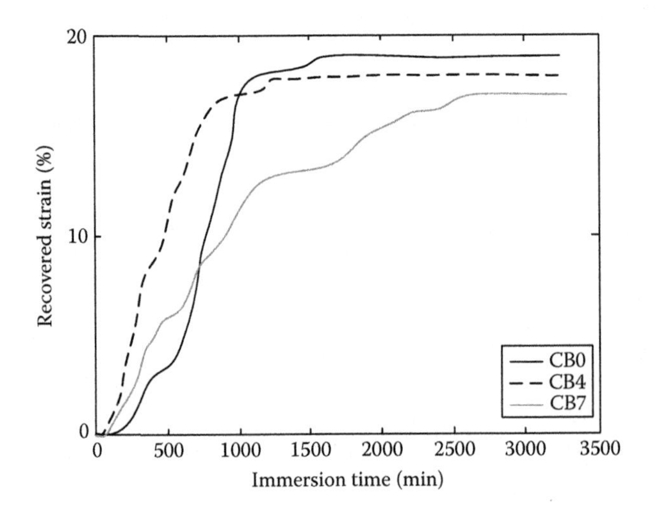

FIGURE 5.16
Recovered strain vs. immersion time.

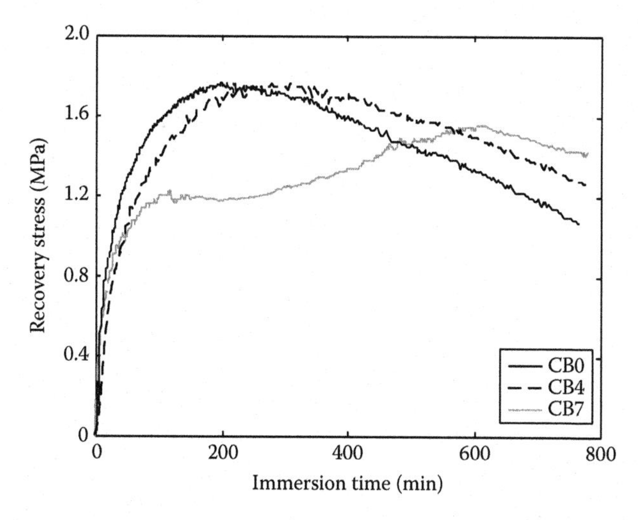

FIGURE 5.17
Recovery stress vs. immersion time.

slowly. A maximum stress about 1.8 MPa can be generated in the CB0 in the constrained recovery test. Except in CB7, the loading of carbon powders has no significant influence on the maximum recovery stress. This is remarkably different from that in the thermally induced constrained recovery test (Figure 5.14). With the increase of the carbon powder content, it takes a longer time for the SMP composites to reach the maximum recovery stress. After the peak is reached, the recovery stress decreases. It is noticed that the reduction in the recovery stress becomes remarkably less with the increase of the carbon powder content. Apparently, the carbon powder is, again, the major player behind the slower response as we have discussed in the above case of water-actuated free recovery.

5.3.7 Damping Capability

A typical result of the function of the tangent delta against the temperature in polymers after different immersion periods in water is plotted in Figure 5.18 (for the CB0). Note that the tangent delta is the ratio of the loss modulus to the storage modulus, which represents the damping capability of a material in shock or vibration. Generally speaking, the tangent delta reaches the maximum at around T_g. Figure 5.18 reveals that the maximum tangent delta of the dry CB0 is about 1.4, which is much higher than that of a typical high damping rubber. Therefore, the CB0 can be used as a superior damping material at around its T_g. Furthermore, the result reveals that the maximum tangent delta of the CB0 decreases significantly with the increase of the immersion time. The immersion time is directly related to the moisture content absorbed by the material. Therefore, the decrease

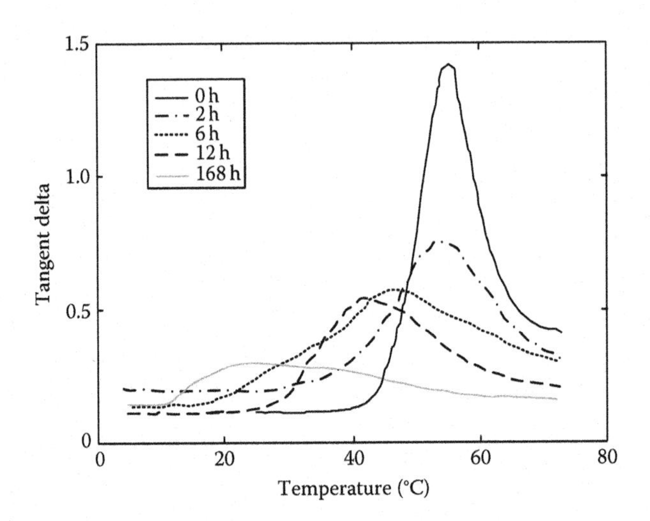

FIGURE 5.18
Tangent delta as a function of the temperature in the CB0.

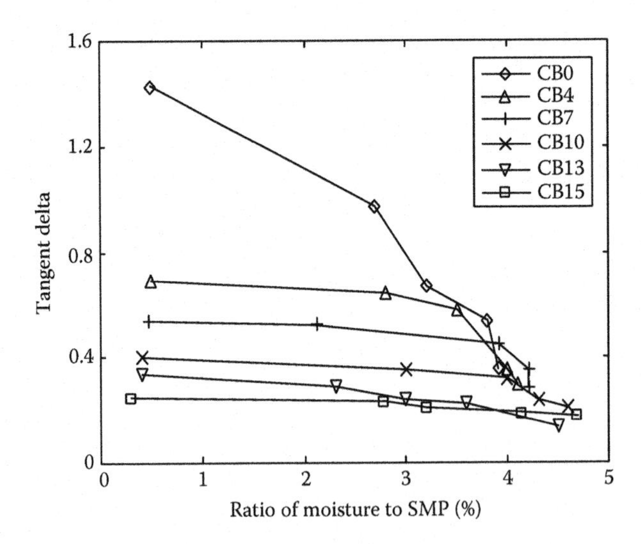

FIGURE 5.19
Change in the maximum tangent delta vs. the moisture content.

of the damping capability can be directly attributed to the increase of the moisture ratio in the material.

Figure 5.19 summarizes the relationships of the maximum tangent delta against the ratio of moisture for the SMP and its composites. Note that the ratio of moisture to the SMP is determined in the same way as that described in [16]. It reveals that, in general, the loading of the carbon powder remarkably lowers the damping capability. On the other hand, with a prolonged immersion time, the maximum tangent delta of the SMP and its composites decrease to about 0.2. The maximum tangent delta of the CB0 almost linearly decreases with the increase of the moisture content. However, with an increase in the loading of carbon powders, the decrease of the maximum tangent delta becomes less significant. This is even more remarkable in the CB15, in which there is only slight change in the maximum tangent delta.

In order to study whether all materials after being immersed in room temperature water can recover their original mechanical properties, two groups of the saturated samples were dried in a vacuum oven. According to the previous study in [19], 120°C is the critical point to divide the absorbed water into two parts, namely, the free water and the bound water. Thus, two temperatures, namely, 80°C and 120°C, were chosen for drying the samples in a vacuum oven. It was expected that upon drying at 80°C for 6h only some free water can evaporate, while upon drying at 120°C for 6h all the free water and some bound water can be removed. The typical results of a saturated sample (CB0) upon drying are plotted in Figure 5.20. In general, the saturated SMP and its composites only partially regain the damping capability after the removal of some of the free water. However, after drying at 120°C for 6h their damping capability is almost fully recovered.

5.3.8 Young's Modulus

Based on these DMA results, Young's modulus after immersion in water for various periods is obtained as a function of temperature. Figure 5.21 presents one typical result, which is the CB0. The dramatic change in Young's

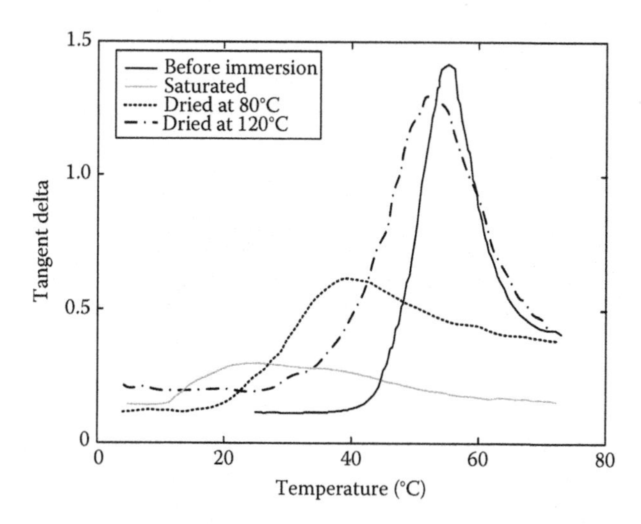

FIGURE 5.20
Tangent delta as a function of temperature in the saturated CB0 upon drying.

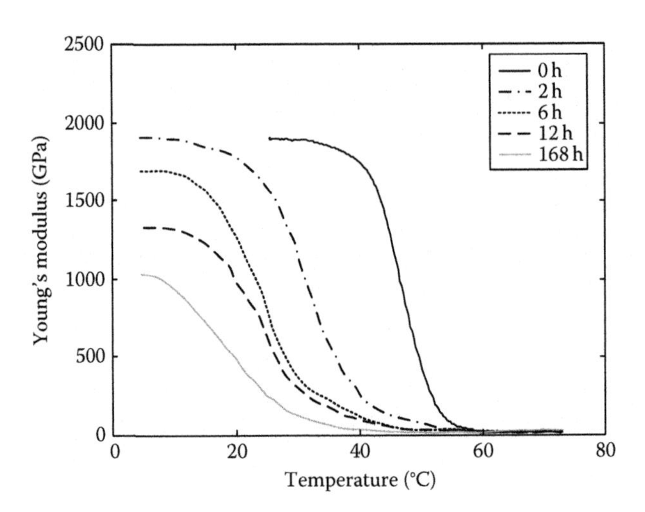

FIGURE 5.21
Young's modulus as a function of temperature in CB0.

modulus against the temperature after immersion in water is obviously the result of moisture.

As reported in, for instance, [10], Young's moduli of an SMP at the glass state and the rubber state are about constants. Normally, we can consider Young's modulus of an SMP at $T_g - 15°C$ as that of the glass state; while Young's modulus at $T_g + 15°C$ as that of the rubber state. For easy comparison, in this study, Young's moduli of all samples at $T_g - 15°C$ and $T_g + 15°C$ are summarized against the moisture content in Figure 5.22. Note that the T_g of an individual sample is determined as the temperature corresponding to the peak of the tangent delta in the previous DMA results. It reveals that Young's modulus of an SMP and its composites at the glass state decrease with the increase of moisture. The decrease in Young's modulus can be attributed to

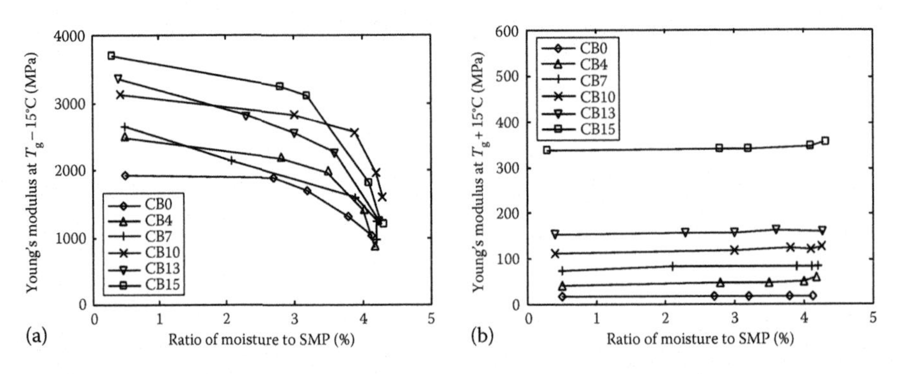

FIGURE 5.22
Evolution of Young's modulus at different moisture contents. (a) At T_g −15°C and (b) at T_g +15°C.

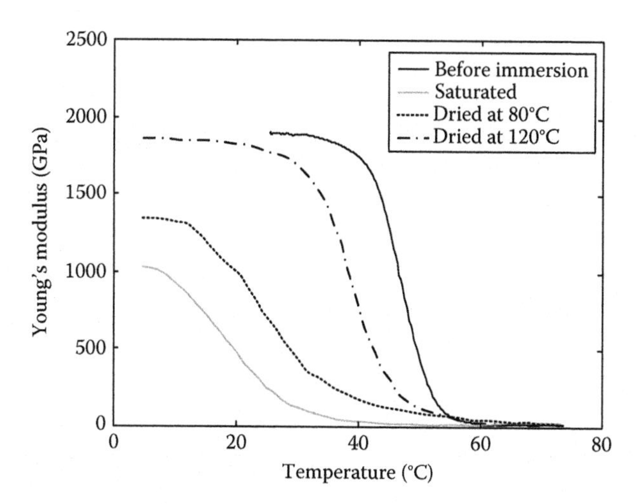

FIGURE 5.23
Young's modulus as a function of temperature in the saturated CB0 after drying.

the moisture, which interacts with the polymer chains and plasticizes the polymer. However, there is only some negligible change in Young's modulus for samples at the rubber state. It reveals that the polymer chains at the rubber state have very good mobility (evidenced by the significant drop in Young's modulus as compared with that in the glass state), so that the additional mobility due to moisture is ignorable.

Figure 5.23 presents one typical result of Young's modulus as a function of temperature in a saturated sample (CB0) after drying at 80°C and 120°C. It shows that the removal of free water in the sample results in a partial recovery of Young's modulus. After drying at 120°C for 6h its Young's modulus is largely recovered.

5.4 Conclusions

A series of experiments were carried out to study the electrical and thermomechanical properties of nano carbon powder–filled polyurethane SMPs. The influence of moisture on the thermomechanical properties of this type of polyurethane SMP is investigated. The so-called water-actuated recovery is characterized.

We find that within the temperature range of our concerns, the electrical resistivity of the carbon powder–filled SMP is temperature independent. However, a deformation by means of either uniaxial tension or compression can significantly alter the electrical resistivity.

The presence of carbon powder in the SMP improves the strength of the material. The carbon powder–filled SMPs, if not heavily loaded, also show good shape-memory properties. The shape recovery of carbon powder–filled SMPs can be actuated by water, in addition to the standard approach, i.e., upon heating (including Joule heating). The recovery of a predeformed carbon powder–filled SMP composite upon immersion into room temperature water is due to the decrease of the T_g, caused by the moisture. One may utilize this feature as an effective way to realize conductive SMPs with a functionally gradient T_g.

The recovery induced upon heating is quite different from that by moisture. The most important issue might be that the maximum recovery stress in the thermally induced constrained recovery increases with the powder content; while that in the water induced case is about a constant.

The damping capability of the nano carbon powder–filled SMP composites becomes worse after immersion in water. Young's modulus of the SMP composites at the glass state decreases with the increase of the water content. However, there is almost no change in Young's modulus at the rubber state upon immersion in water. Upon dehydration, the carbon powder–filled SMP composites remarkably regain the original damping capability and Young's modulus.

References

1. Ratna D and Anhalt M. *J Mater Sci* 2008; 43: 254–269.
2. Liang C, Rogers CA, and Malafeew E. *J Intell Mater Syst Struct* 1997; 8: 380–406.
3. Liu C, Qin H, and Mather PT. *J Mater Chem* 2007; 17: 1543–1558.
4. Lee BS, Chun BC, Chung YC, Sul KI, and Cho JW. *Macromolecules* 2001; 34: 6431–6437.
5. Tobushi H, Hayashi S, and Kojima S. *JSME Int J* 1992; 35: 296–302.
6. Tobushi H, Hara H, Yamada E, and Hayashi S. *Smart Mater Struct* 1996; 5: 483–491.
7. Yang JH, Chun BC, Chung YC, and Cho JH. *Polymer* 2003; 44: 3251–3258.
8. Poilane C, Delobelle P, Lexcellent C, Hayashi S, and Tobushi H. *Thin Solid Films* 2000; 379: 156–165.
9. Yang B, Huang WM, Li C, Lee CM, and Li L. *Smart Mater Struct* 2004; 13: 191–195.
10. Tobushi H, Okumura K, Hayashi S, and Ito N. *Mech Mater* 2001; 33: 545–554.
11. Chiodo JD, Billett EH, and Harrison DJ. *Proceedings of 7th IEEE International Symposium on Electronics and the Environment*, Danvers, MA, 1999; pp. 151–156.
12. Lee Pand Fitch JP. United States Patent 2000; 6,086,599, 2000.
13. Wache HM, Tarrtakowska DJ, Hentrich A, and Wagner MH. *J Mater Sci—Mater M* 2003; 14: 109–112.
14. Gall K, Mikulas M, Munshi NA, and Tupper M. *J Intell Mater Syst Struct* 2000; 11: 877–886.

15. Gall K, Dunn ML, Liu YP, Finch D, Lake M, and Munshi NA. *Acta Mater* 2002; 50: 5115–5126.
16. Yang B, Huang WM, Li C, and Chor JH. *Eur Polym J* 2005; 41: 1123–1128.
17. Meng Q and Hu J. *Composites A* 2008; 39: 314–321.
18. Yang B, Huang WM, Li C, Li L, and Chor JH. *Scripta Mater* 2005; 53: 105–107.
19. Huang WM, Yang B, An L, Li C, and Chan YS. *Appl Phys Lett* 2005; 86: 114105.
20. Aneli, JN, Zaikov GE, and Khananashvili LM. *J Appl Polym Sci* 1999; 74: 601–621.
21. Flandin L, Hiltner A, and Baer E. *Polymer* 2001; 42: 827–838.
22. Das NC, Chaki TK, and Khastgir D. *Polym Int* 2002; 51: 156–163.
23. Kost J, Narkis M, and Foux A. *J Appl Polym Sci* 1984; 29: 3937–3946.
24. Pramanik PK, Khastagir D, and Saha TN. *J Mater Sci* 1993; 28: 3539–3546.

6

Multifunctional Shape-Memory Polymers and Actuation Methods

Jinsong Leng, Haibao Lu, and Shanyi Du
Centre for Composite Materials and Structures, Harbin Institute of Technology, Harbin, P.R. China

CONTENTS

6.1 Introduction

Shape-memory polymer (SMP) is specially implied to thermal responsive SMP, of which the shape recovery actuation is always a thermal-induced process, though it can be triggered electrically [1], magnetically [2], or electromagnetically [3]. For some special type of SMPs that are incorporated with photo-responsive monomers can be induced by light with characteristic wavelength. For the SMP composites filled with functional filler or fillers in composites are activated by corresponding stimuli such as electric, magnetic, or electromagnetic field and thereby an inductive thermal effect is indirectly when the composites were heated above the transition temperature by inductive thermal heating, the recovery behavior occurs. Finkelmann et al. [4,5] reported a shape-memory liquid-crystalline (LC) elastomer that can be triggered by light through incorporating azo groups into the LC mesogens. Upon ultraviolet (UV) illumination, the azo groups isomerize to the configuration and sharply bend the mesogens, hampering the nematic ordering. Correspondingly, the material undergoes a photo-induced nematic-isotropic transition, accompanied with a large shape change. Recently, as shown in Figure 6.1, Lendlein et al. [6] reported a novel SMP that is in response to light by introducing molecular switches such as cinnamic acid [7] and cinnamylidene acetic acid [8]. The latter compounds have the capability of undergoing efficient photoreversible cycloaddition reactions when exposed to alternating wavelengths ($k > 260$ nm or $k < 260$ nm). The temporary shape of LC is kept resulting from the formation of new photo-responsive cross-links, while the photo-responsive cross-links can be reversibly cleaved by a UV light irradiation with a wavelength shorter than 260 nm, leading to regaining the original shape. The unique characteristics of the above-mentioned SMPs enable the shape recovery to be driven at ambient temperatures by remote activation rather than external heating; this method eliminates temperature constraints for biomedical and other applications.

Shape-memory effect of thermally induced SMPs is a mostly dual-shape feature in nature (always named as hard segment and soft segment); one is highly elastic to regain the original shape, and the other is able to remarkably reduce its stiffness in the presence of a particular stimulus. The latter can be either molecular switches or stimulus-sensitive domains. To enable the shape-memory feature, a special architecture is required, which consists of net points and molecular switches, which are sensitive to an external stimulus. Besides their dual-shape capability, these active materials are also biofunctional or biodegradable. Potential applications of these novel materials are attractively highlighted to be applied for medicine engineering. The field of actively moving polymers that have the capability of regaining their deformed shape by themselves is progressing rapidly [9]. Therefore, SMP materials are considered as one of the most important candidates for these applications. Fundamental shape-memory research is focusing on the implementation of stimuli other than direct heating to actuate SMPs, or to actuate them remotely, such as the light-induced stimulation (including light-responsive SMPs and indirectly light-heating thermal-responsive SMPs) of SMPs, electrical resistive heating, or the use of alternating magnetic fields for remote actuation. These novel activation approaches will largely explore the development and applications of SMPs. A highlight report of SMP is being used in active medical devices and implants for surgery medicine [10], and initial demonstrations have been presented [11,12]. This application requires many complex and special satisfactory such as bio-fitting, nontoxic natural tendency, adaptable and controllable transition temperature, so on. Therefore, a future look at the development of multifunctional materials can be summarized as tailored, complex shape change, and controllable [13]. Furthermore, there is a strong demand for actively moving materials able to perform complex movements. These requirements could be fulfilled by materials that are able to perform two or more predetermined shifts [14].

6.2 Thermal-Induced Shape-Memory Effect

In general, SMPs contain thermal SMPs, light-responsive SMPs, and shape-memory gels. Normally, SMPs are always used to indicate thermal SMP. And the shape-memory effect of thermal SMP is driven by thermal heating in traditional definition. There are several approaches for SMP materials division. First, SMPs are divided into two parts: physically cross-linked thermoplastic SMPs and chemically cross-linked SMPs. Moreover, SMPs can be divided into two components: block copolymer and covalently cross-linked polymer. The third approach is defined from glass transition temperature and melting transition temperature that act as the transition temperature of polymers.

Generally speaking, SMPs are always divided into block copolymer SMPs and covalently cross-linked SMPs. Investigation and application of these two types of SMPs have been greatly achieved in the last three decades. One of the most important types of physically cross-linked SMPs is linear block copolymers with melting transition temperature as transition temperature [15]. And polyurethanes and polyetheresters SMPs play a dominant role in development and application of physically cross-linked SMPs. In this polymer, the polymer chains are incorporated of two components. They are hard segment and soft segment. And hard segment is used to memory the originate shape of chain, while the soft segment works as a switching to control shape transition of chain. In polyesterurethanes, the urethane segments act as hard segments (with relative higher transition temperature), while the poly(ε-caprolactone) (T_m = 44°C–55°C) forms the switching segment [16–18]. Additionally, polyesterurethanes containing mesogenic moieties have been reported in recent research [19,20]. In the block copolymers incorporated with trans-(polyisoprene) switching segments and urethane hard segments, the microphase separation of the two segments has been studied and analyzed [21]. Here, the polyurethane segments act as physical cross-links and assemble into spherical domains. T_{trans} is determined by the melting transition of the *trans*-(polyisoprene) switching segments.

In other types of block copolymers, of which the transition temperature is determined by the glass transition temperature of soft segment ($T_{trans} = T_g$). And the transition temperature of polymer is always influenced by the thermal properties of hard segment. A typical example is polyurethanes, whose switching domains are, in most cases, mixed phases [22].

Another classic type of thermoplastic SMPs in which transition temperature is determined by glass transition temperature are polyesters. Taking for an example, in copolyesters made from poly(ε-caprolactone) and poly(butylene terephthalate), the poly(butylene terephthalate) segments work as switch role in the shape-memory effect [23]. The shape-memory property can also be introduced to a polymer with a polymer analogous reaction. A polymer analogous reaction is the approach of a standard organic interaction (like the reduction of a ketone to an alcohol) to a polymer owning several reactive groups. A promising example for the polymer analogous reduction is polyketone with NaBH4/THF, which results in a poly(ketone-*co*-alcohol) [24]. The polyketones are synthesized by late-transition-metal-catalyzed polymerization of propene, hex-1-ene, or a mixture of propene and hex-1-ene with CO. The T_g of the polymer is directly determined by the degree of reduction, and can be adjusted by the amount of NaBH4/THF. The most outstanding SMP material is a partly reduced poly(ethylene-*co*-propene-cocarbonoxide) that showed a phase-separated morphology with hard microcrystalline ethylene/CO-rich segments within a relative softer amorphous polyketone ethylene-propene/CO-rich matrix, in which the crystalline domains play a role of physical cross-linkers. This results in an elastic behavior above T_g, because the glass transition temperature ($T_{trans} = T_g$) is related to the switching phase. Partial reduction of the amount can be used to control T_g, which can be adjusted from below room temperature to 75°C.

Another important type of continuous network, of which the polymer network is cross-linked by radiation and chemical methods during synthesis or postpolymerization, is formed through covalently bonding. Thus, the dual-shape capability can be formed by adding to polyethylene [25] and its copolymers [26,27] through cross-linking interactions with ionizing radiation (γ-radiation, neutrons). Meanwhile, radiation cross-linking of poly(ε-caprolactone) leads to mainly molecular chain scission and loss of useful mechanical properties [28]. By mixing poly(ε-caprolactone) with polymethylvinylsiloxane, radiation cross-linking of the mixture can be achieved and shape-memory capability can be added to the material [29]. Alternative option is the chemical cross-linking of poly[ethylene-*co*-(vinylacetate)] where dicumylperoxide acts as a radical initiator [30]. The same radical initiator has also been used in semicrystalline polycyclooctene for cross-linking [31]. A covalently cross-linked polymer based on natural source has been made from cationic polymerization of soybean oil with styrene and divinylbenzene, where norbornadiene or dicyclopentadiene works as a cross-linker [32].

Through blending interpenetrating polymer network structures and post curing, as (co)condensation or poly-(co)condensation of one or several monomers, where at least one group has a trifunctional capability was introduced to stearyl acrylate, methacrylate and formed interpenetrating polymer network, or by copolymerization of monofunctional monomers, in which N, *N*-methylenebisacrylamide acted as the cross-linker. And crystalline domains of stearyl side chains worked as the switching function [33]. In multiphase copolymer networks formed by the radical polymerization or copolymerization of poly(octadecyl vinyl ether) diacrylates or dimethacrylates with butyl acrylate as the comonomer, the situation is similar with the above approach [34,35]. In both cases, the crystalline domains of octadecyl side chains again play a switching role in polymer. The radiation efficiency can be improved by blending special monomer before radiation crosslinking. And the switch temperature of the interpenetrating polymer network can be adjusted by controlling the crosslinking density.

An example of a cross-linked polymer network made from polyaddition of monofunctional monomers with low molecular weight or oligomeric cross-linkers has been carried out in polyurethanes by adding trimethylol to the reaction mixture [36].

Reaction of tetrafunctional silanes, which act as netpoints, with oligomeric silanes, as spacers, and to which two distinct benzoate-based mesogenic groups have been attached, results in the formation of a main-chain smectic-C elastomer [37]. These elastomers are blended with a polymer having a relative low transition temperature can generate polymer blends with shape-memory effect [38]. The cross-link process during synthesis determines the permanent

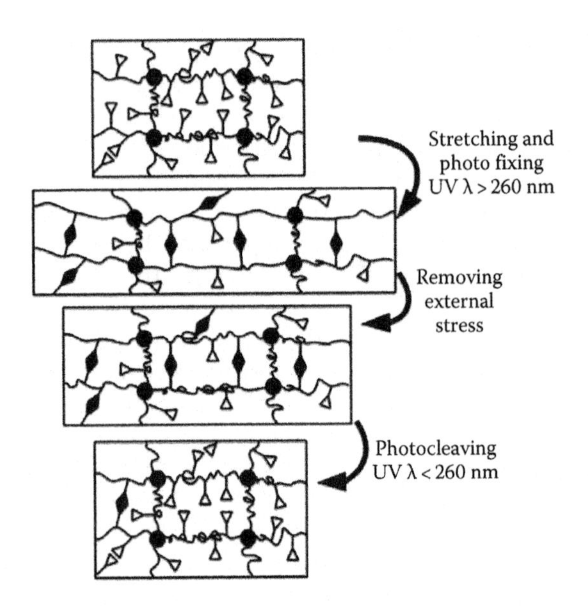

FIGURE 6.1
Molecular mechanism responsible for the light-induced shape-memory effect in a grafted polymer network. Chromophores (open triangles) are grafted onto the permanent polymer network (filled circles, permanent cross-links). Photoreversible cross-links (filled diamonds) are formed during fixation of the temporary shape by UV light irradiation of a suitable wavelength. Recovery is realized by UV irradiation at a second wavelength. (From Yu, Y.L. and Ikeda, T., *Macromol. Chem. Phys.*, 206, 1705, 2005. With permission.)

(or named original) shape. The shape-memory behavior is activated by the thermal transition of the liquid-crystalline domains. In the shape-memory recovery cycle, the polymer network is heated above melting transition temperature of the liquid-crystalline domains. And polymer will be deformed to a temporary shape, and then cooled down to the transition temperature of the smectic-C mesogens. The temporary shape is fixed. Upon reheating over this clearing transition, the permanent shape can be regained. In contrast to shape-changing liquid-crystalline elastomer systems, these polymers display shape-memory behavior due to the liquid-crystalline moieties that work as a switch to determine thermal transition of polymer. In shape-changing liquid-crystalline elastomers, the molecular movement of the single liquid crystal is converted into a macroscopic movement.

6.3 Light-Induced Shape-Memory Effect

6.3.1 Shape-Memory Polymer in Response to Light

Independency of shape recovery from any temperature effect has been achieved in light-responsive SMPs [6,39–41]. Here, light of different wavelengths is used for the fixation of the temporary shape and the shape recovery. This stimulation has no relation with any of the temperature effects. Therefore, the light-responsive SMPs are natural difference from thermal-responsive SMPs that always are directly and/or indirectly determined by temperature. For example, on the molecular level, this is realized by the incorporation of photosensitive molecular switches, such as CA or CAA. When irradiated by light of a suitable wavelength (λ), the photosensitive functional groups form covalent crosslinks with each other in a [2+2] cycloaddition reaction. In the course of programming, the polymer is strained and then irradiated by UV light of $\lambda > 260$ nm, so that new covalent bonds are created to fix the strained chain segments in their uncoiled conformation. These newly formed covalent bonds can be cleaved and the permanent shape recovered, when the sample is irradiated by light with $\lambda < 260$ nm.

There are two approaches having been used to achieve SMPs induced by light: a graft polymer and an interpenetrating polymer. For the graft polymer, CA molecules are grafted onto a permanent polymer network formed by the copolymerization of *n*-butylacrylate, hydroxyethyl methacrylate, ethylenegylcol-1-acrylate-2-CA, and poly(propylene glycol)-dimethacrylate ($M_n = 560$ g mol^{-1}) working as the cross-linker. Loading a permanent polymer network, made from butylacrylate and 3wt% poly(propylene glycol)-dimethylacrylate ($M_n = 1000$ g mol^{-1}) acting as a cross-linker, with 20wt% star-poly(ethylene glycol) end capped with terminal CAA groups yields photosensitive interpenetrating network to change shape in response to special wavelength light. In both cases, cross-links of an amorphous polymer network determine the permanent shape. In the course of programming, the polymer is strained and then irradiated by UV light of $\lambda > 260$ nm, so that new covalent bonds are created to fix the strained chain segments in their uncoiled conformation. These newly formed covalent bonds can be cleaved and the permanent shape recovered, as the sample is irradiated by light with $\lambda < 260$ nm.

6.3.2 Infrared Light-Induced Shape-Memory Effect

Indirect actuation of the shape-memory effect has been realized by two different strategies. One method involves indirect heating, e.g., by irradiation. The other possibility is to lower T_{trans} by diffusion of low molecular weight molecules into the polymer, which works as a plasticizer. This allows the triggering of the shape-memory effect while the sample temperature remains constant. Instead of increasing the environmental temperature, thermally induced SMPs can be heated by illumination with infrared light. This concept has been demonstrated in a laser-induced polyurethane medical device [41,42]. In such devices, heat transfer can be enhanced by incorporation of conductive fillers, such as conductive ceramics, carbon black, and carbon nanotubes (Figure 6.2) [43–45]. This incorporation of particles also

FIGURE 6.2
Strain recovery under infrared radiation (exposure from the left). Infrared absorption, non-radiative energy decay, and resulting local heating is constrained to the near-surface region of a stretched ribbon, resulting in strain recovery of the near-surface region and curling of the ribbon toward the infrared source within 5 s. (From Biercuk, M.J. et al., *Appl. Phys. Lett.*, 83, 2405, 2003; Li, F.K. and Larock, R.C., *J. Appl. Polym. Sci.*, 78, 1044, 2000; Liang, C. and Rogers, C.A., *J. Intell. Mater. Syst. Struct.*, 8, 285, 1997. With permission.)

influences the mechanical properties: incorporation of microscale particles results in increased stiffness and recoverable strain levels [46,47], and can be further enhanced by the incorporation of nanoscale particles [48,49]. To achieve an enhanced photothermal effect, the molecular structure of the particles has to be considered. Polyesterurethanes reinforced with carbon nanotubes or carbon black of similar size display increased strain and fixity. While carbon-black-reinforced materials show limited shape recovery of around 25%–30%, in carbon-nanotube-reinforced polymers, an R_r of almost 100% can be observed and has been attributed to a synergy between the anisotropic carbon nanotubes and the crystallizing polyurethane switching segments.

Infrared light possesses not only wide emission spectra but also unique heating effect, at the same time, the actuation for SMP with infrared light can be both non-contact and non-medium. Therefore, the infrared light-actuation method may dramatically expand the applications of SMP in smart structures. Jinsong Leng's group at Harbin Institute of Technology systematically investigated the relative performances of infrared light-induced shape-memory polymer filled with carbon black. In addition to the preparation of styrene-based SMP and its composite containing 10 wt% CB, shape-memory effect of these materials are actuated by infrared light. The factors that would influence infrared light-active shape-memory effect are also explored by results of infrared absorption characteristic tests and the shape recovery test. As shown in Figure 6.3, a bent sample fixed in an adiabatic clamp was placed in the quartz bottle. During the recovery process, time against recovery angles was recorded by digital camera.

As infrared light was absorbed in the form of molecule (or atom) resonance vibration, the capability of absorbing infrared light depends on

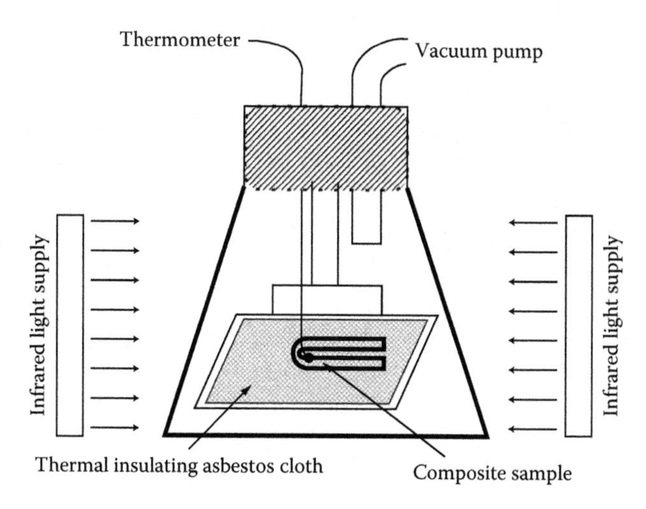

FIGURE 6.3
Illustration of actuation apparatus with infrared light in vacuum. (Reproduced from Leng, J.S. et al., *J. Appl. Polym. Sci.*, 114(4), 2455, 2009.)

molecular (or atom) constitution of materials. Therefore, absorption spectrum mirrors the resonance behavior of molecular (or atom) groups. Figure 6.4 shows the difference of infrared absorbing efficiency between SMP and SMP/CB composite from 400 to 4000 cm^{-1} in wave number. As shown in Figure 6.4, obvious absorption (above 3%) of SMP can be found in wave number ranges of 450 ~ 1960 cm^{-1} and 2760 ~ 3130 cm^{-1}. In these absorption ranges, there are three sharp peaks of 1060, 1830 and 2856 cm^{-1}, and at two of which infrared absorption strength of SMP reaches 6%. Moreover, a strong (above 6%) continuum absorption of SMP/CB composite can be found in the whole test wavenumber range. It is apparent that not only the absorption spectrum of the SMP/CB composite is more wide-ranged, but also the absorptivity of the SMP/CB composite is higher than that of neat SMP. Therefore, the existence of CB particles increases the capability to absorb infrared light for the SMP/CB composite remarkably.

When SMP and SMP/CB nanocomposite are irradiated by infrared light, most of the emitted energy is transmitted by the SMP, as it is transparent while most of the emitted infrared light is absorbed by the black and opaque nanocomposite, since it is black and opaque. In addition, SMP and the nanocomposite have approximately equal reflection, since their surfaces are similarly smooth. Therefore, the black CB absorbs infrared radiation much notably than that of pure SMP. The unique characteristic of infrared light is the notable heating effect. A previous study has proved that infrared radiation can be absorbed by CB remarkably and then produce notable heating effect during 500 ~ 3000 wavenumbers (cm^{-1}). In SMP/CB nanocomposite, heat is generated by the CB to be released into the matrix

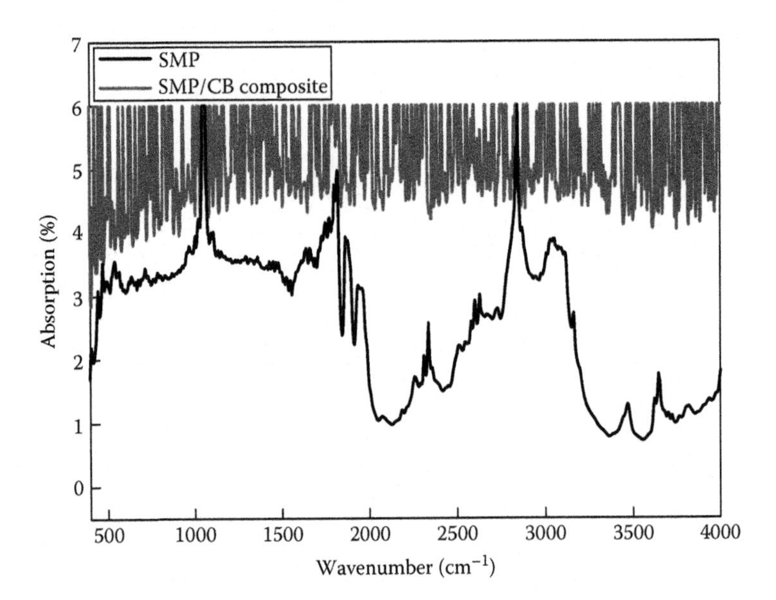

FIGURE 6.4
Spectra of infrared absorption in the range of 400–4000 (cm^{-1}). (Reproduced from Leng, J.S. et al., *J. Appl. Polym. Sci.*, 114(4), 2455–2460, 2009.)

(polymer) efficiently when light is absorbed. Then, the high capability to absorb infrared light of the CB will result in the more heat generated in the nanocomposite.

Based on the infrared absorption characteristic of the SMP materials, a setup is designed to accomplish the thermo-actuation method of infrared light in vacuum. The infrared actuation was carried out in vacuum condition to decrease the thermal inertia. Infrared radiation ranged from 2.5 to 8 μm is the strongly absorbing region for most of polymers, is adopted. Besides, the infrared-light actuation is non-contact as infrared heating can be realized in the form of electromagnetic radiation. Therefore, infrared light-actuation method for SMP materials is can be used for almost polymeric materials, making it possible to widen the application of SMP material in the future.

Shape recovery speed is an important performance for the application of SMP materials. Figure 6.5 shows the curves of shape recovery angle against time for SMP and SMP/CB composite at 90°C. To investigate the repeatability of shape-memory speed, each sample was tested five times. Results show that the additives of CB not only increase the recovery force but also enhance the capability to absorb infrared light for SMP/CB composite, and so the shape recovery ability and the shape recovery speed of the composite are much quicker than that of the SMP.

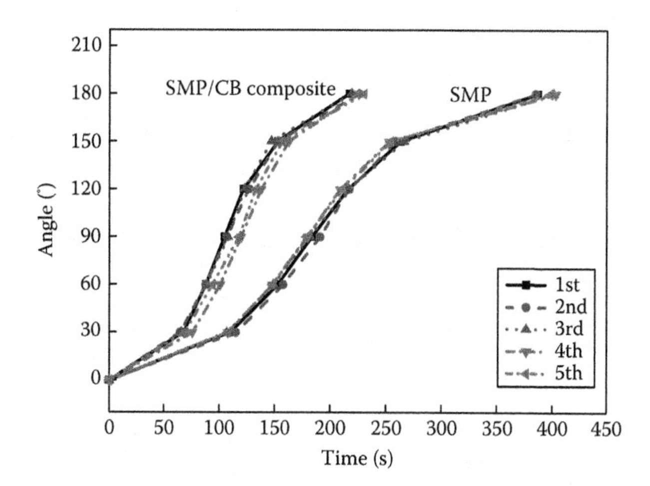

FIGURE 6.5
Recovery angle versus time relationship in five tests. (Reproduced from Leng, J.S. et al., *J. Appl. Polym. Sci.*, 114(4), 2455, 2009.)

6.3.3 Infrared Light-Induced Shape-Memory Polymer Embedded with an Optical Fiber

The thermally activated SMP possibly initiates an original shape by a non-contact and highly selective infrared laser stimulus [50–54]. For the light-induced SMP, the infrared laser stimulation of the SMP is of interest for sensor and actuator systems as well as medical applications. The infrared light-induced SMP embedded with optic fiber is proposed [55–57]. Figure 6.6 is the schematic illustration that shows the treated optical fiber and the SMP embedded with the treated fiber. The optical fiber was treated by the aqueous solution of sodium hydroxide and the cladding is removed. In this way, the light may be transmitted from the side of the fiber, which provides a better way for heating a large region. Furthermore, an optical fiber network may be embedded into the SMP for actuation of a fast response.

A styrene copolymer thermoset SMP was first synthesized. The experimental results are presented in Figure 6.7. It shows the storage modulus, loss modulus, and tan delta as functions of temperature. The tan delta curve exhibits a glass transition peak at approximately $T_g = 53.7°C$. The glass transition behavior of the polymer is basically identical during heating and cooling under the specified rate conditions.

Through IR spectra, it is found that in the SMP copolymer the strong absorption peak has been obtained from a Nicolet 60 SXR FI-IR Spectrometer. For the copolymer, the solid samples were ground into powder. Figure 6.8 indicates the strong absorption peak of approximately 3000 cm^{-1}, which indicates

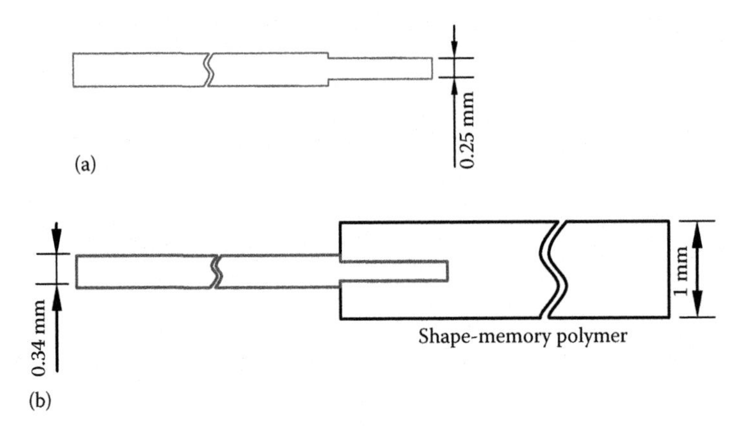

FIGURE 6.6
(a) Treated optical fiber; (b) the treated position of optical fiber is embedded in the shape-memory polymer. (Reproduced from Zhang, D. et al., Infrared laser-activated shape-memory polymer, 15th SPIE International Conference on Smart Structures/NDE, San Diego, CA, March 9–13, 2008, SPIE 6932.)

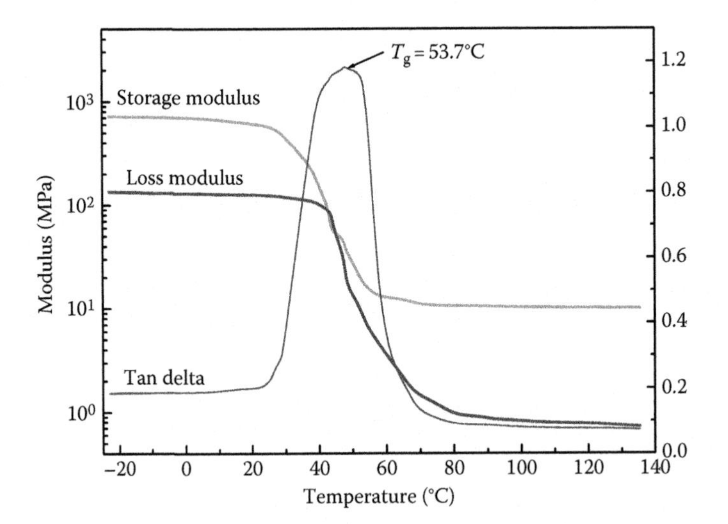

FIGURE 6.7
Storage modulus, loss modulus, and tan delta of the SMP. The DMA test is conducted in tensile using a dynamic scan analyzer a frequency of 1 Hz. (Reproduced from Zhang, D. et al., Infrared laser-activated shape-memory polymer, *15th SPIE International Conference on Smart Structures/NDE*, San Diego, CA, March 9–13, 2008, SPIE 6932. With permission.)

the characteristic of the absorbing peak of the C–H bond of benzene ring. So, the working length of the infrared laser is chosen at 3–4 μm.

An infrared light with a working wave band of 2–4 μm is coupled into the fiber in the SMP to study the realistic actuating effect of the infrared laser. In Figure 6.9a ($t = 2$ s), the only bright position is located at the end of the

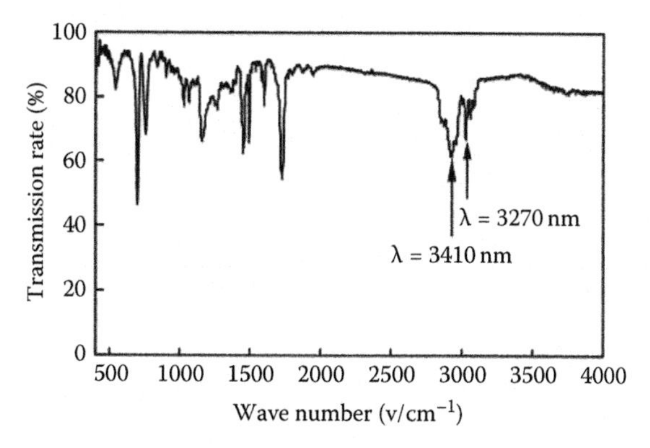

FIGURE 6.8
The IR spectra of the thermoset SMP. (Reproduced from Zhang, D. et al., Infrared laser-activated shape-memory polymer, *15th SPIE International Conference on Smart Structures/NDE*, San Diego, CA, March 9–13, 2008, SPIE 6932. With permission.)

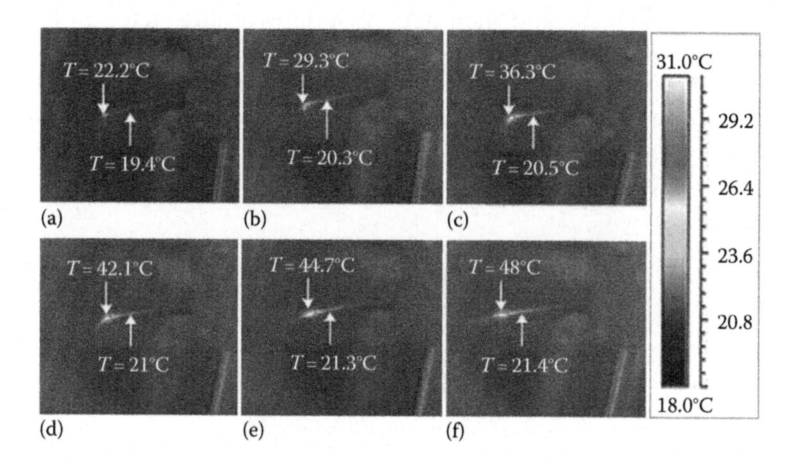

FIGURE 6.9
Temperature distribution snap shots of the SMP actuator during the shape recovery process. (a) $t = 2$ s; (b) $t = 4$ s; (c) $t = 6$ s; (d) $t = 8$ s; (e) $t = 10$ s; (f) $t = 12$ s.

fiber, which indicates that the infrared laser is mainly refracted at that position. In the subsequent snap shots, the actuator gets brighter globally. The temperature of the SMP around the treated optical fiber rises to 21.4°C with an increased temperature of 2.4°C when compared to the ambient temperature. In contrast, the temperature of the SMP at the end of the optical fiber increases to 48°C with an increased temperature of 29°C. It indicates that more light is refracted from the end than the side of the optical fiber. This experimental phenomenon can be explained, as the optical fiber is treated by chemical corrosion. Therefore, after removing the cladding, the surface of the core is uneven. The refractive index of the fiber core (n_c) is 2.8. In contrast, the refractive index of the polymer (n_p) is 1.4–1.46, where n_c is higher than n_p. Hence, according to the total reflection principle, partial laser is refracted back during the refraction of the infrared laser. The laser energy which is absorbed at the treated position of the optical fiber is a relatively small amount, and so the increase in temperature is slow. However, no laser is refracted back at the end of the fiber and a high temperature is engendered for more laser-energy absorption. Temperature at the treated position of the fiber is shown in Figure 6.9. It shows that the SMP actuator deploys promptly in about 12 s.

The SMP actuator, induced by the infrared laser transmitted through the embedded fiber in the SMP, may be applied to the flexible display. The schematic of the deployable flexible display actuated by SMP actuator is proposed as shown in Figure 6.10. The flexible system is the combination of the flexible display and the laser-activated SMP actuator. The optical fiber network is embedded in the SMP substrate (Figure 6.10a). Figure 6.10b shows the schematic of a working cycle, namely from a flat working state to a rolled storage state, and then deploys a flat working state again. First, the flexible display is in a flat working shape. When not in use, the infrared laser can be coupled with the optical fiber network resulting in the softening of the SMP substrate. In this way, the display can be rolled up into a storage shape. If heating SMP substrate in the same way, the flexible display can be deployed and the frozen shape can be fixed after cooling down.

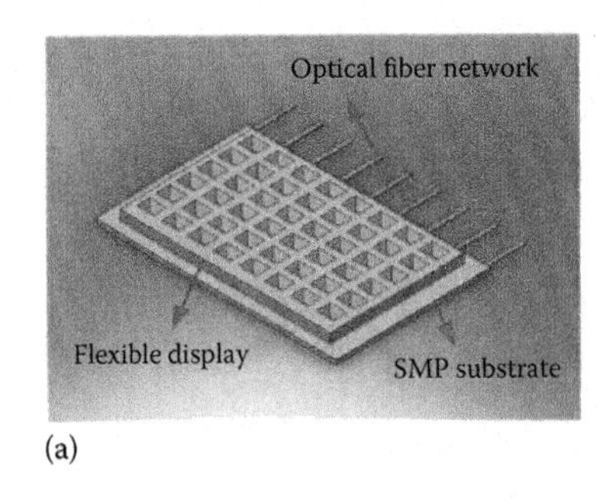

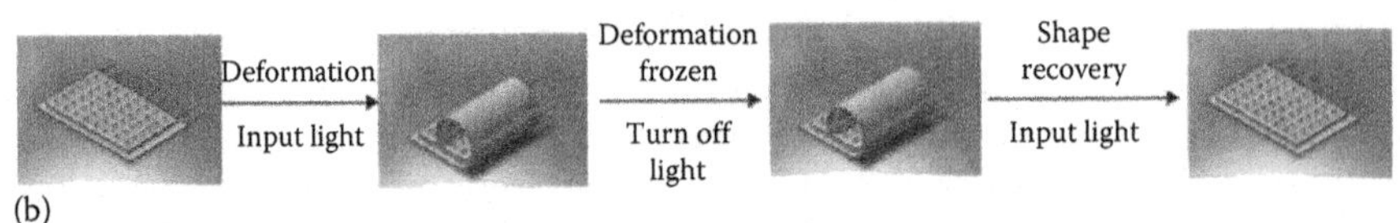

FIGURE 6.10
Infrared light-induced flexible display and its shape recovery process (a) light-induced flexible display; (b) shape recovery process of light-induced flexible display.

6.4 Electrically Induced Shape-Memory Effect

Many efforts have been made to achieve shape actuation of SMP composites filled with electrical conductive filler or fillers, by electrical resistive heating. These conductive fillers such as carbon nanotubes, carbon nanofibers, carbon particles, chopped carbon fibers, and continuous carbon fibers, and so on, have been incorporated into the SMP matrix to make electrical conductive SMP composites [58–60]. As it is known, most polymer materials have natural insulating tendencies. These inherent properties prevent pure SMPs from direct heating by electrical resistance effect. On this background, conductive fillers have been used to improve their conductive properties. As the electrical resistivity of SMP composites are reduced to a proper value, the Joule heating occurs into samples, resulting in recovery induced. Furthermore, these conductive fillers also could be used as reinforcement to improve mechanical properties of pure SMPs. Incorporation of 3.3 wt% nanofibers resulted in 200% increase in recovery stress of the SMP composites. The nanocomposites retain high strain recoverability (more than 90%) even after several cycles of training and are going to be a subject of major focus in the near future. Koerner et al. [61] investigated polyurethanes reinforced with carbon nanotubes (CNTs) or carbon black of similar size and reported that the nanocomposites display increased shape fixity. The CNT-reinforced materials show almost 100% shape recovery compared to the material (reinforced with carbon black), which exhibits only a limited shape recovery of around 30%. The difference in behavior of the two materials is attributed to the interactions between the anisotropic CNT and the crystallizing PU switching segments [1]. The nanocomposites offer electrically induced actuation, which can be controlled remotely, via Joule heating when current is passed through the conductive percolative network of the nanotubes within the composite system.

Meanwhile, SMP nanocomposites, in which shape-memory effect can be triggered by a magnetic field, have also been reported using a functional magnetic nanofiller such as Fe_3O_4 with a shell of oligo(e-caprolactone) [62] or Ni–Zn ferrite particles [63]. In these magnetic conductive composites, the sample temperature is increased by the inductive heating of the nanoparticles in an alternating magnetic field ($f = 258\,\text{kHz}$, $H = 7\text{–}30\,\text{kAm}^{-1}$). As electromagnetic high frequency field applied, this field energy is transformed to heat due to the relaxation process of functional particle. The effect has been demonstrated for a PU-based SMP and for a multiblock copolymer-based SMP with PCL-based crystalline segment.

6.4.1 Shape-Memory Polymer Filled with Carbon Nanotubes

A special amount of carbon nanotubes is added into SMPs, to prepare SMP nanocomposites, of which the store and release force are enhanced up to

50% more than that of the pure corresponding SMP sample. With similar low-volume addition of carbon nanotubes to resins, it could be produced that SMP nanocomposites can withstand a greater number of strain and release cycles. The conductive properties of pure SMP are significantly improved by adding carbon nanotubes; therefore, shape recovery of SMP nanocomposites could be remotely induced by infrared radiation. The radiation energy firstly is absorbed by the nanotubes. Then the nanotubes are heated, and the heat is transferred to the surrounding polymer matrix. The local temperature is increased above the transition temperature of SMP nanocomposites. Finally, recovery behavior of nanocomposites is achieved. Goo's group at the Konkuk University firstly added carbon nanotube into SMPs to fabricate conductive composites to induce shape recovery by electrical current [59]. In sequence, this method led to the SMPs being applied as electro-activated actuators, which is important in many practical applications such as smart actuators for controlling microaerial vehicles. To effectively obtain conducting SMPs, multiwalled carbon nanotubes (MWNTs) were incorporated before being chemically surface-modified in a mixed solvent of nitric acid and sulfuric acid, to improve the interfacial bonding between polymers with conductive fillers. These outcomes present in detail how to prepare electro-activated SMP nanocomposites, actuate shape recovery by electrical-resistive heating, and being used as an actuator.

The electrical resistivity of composite films, determined by the four-points probe method, was in the order of 10^{-3} S cm^{-1} for samples of 5 wt% modified-MWNT content. With an increment in modified-MWNT content, the electrical conductivity gradually increased, and the dependence of electrical conductivity of the composites on surface-modified MWNT was obvious. From the experimental results, it revealed that the electrical conductivity of the composites filled with surface-modified MWNT was lower than that of the composites filled with untreated MWNT, when both had the same amount of nanotube content. This is raised from the increased defects in the lattice structure of carbon–carbon bonds formed on the nanotube surface through the acid treatment. In practice, the severe modification of the nanotubes significantly lowered the electrical conductivity. As a result, both the mechanical and electrical properties had critical relation with the degree of surface modification of the MWNTs, and the acid treatment at 90°C gave desirable properties for SMPs.

Figure 6.11 shows the typical relation between the acid treatment temperature and the applied voltage for composites filled with 5 wt% surface-modified MWNT. The temperature of the samples was detected using digital multimeters (M-4660, DM-7241, and METEX) with a noncontact temperature measuring system. Typically, when voltage of 60 V was applied, the sample heated above 35°C in 8 s. It was, however, impossible to heat the sample above its transition temperature with a voltage lower than 40 V.

Though preparing electro-activate shape-memory composites and investigating their characteristics, the shape-memory effect could be shown to be

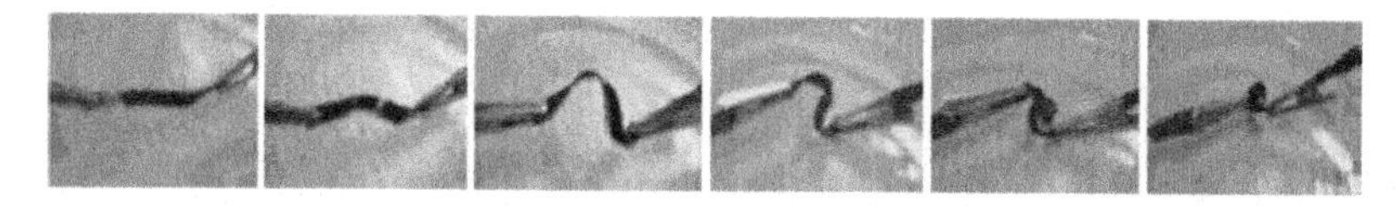

FIGURE 6.11
Electroactivate shape recovery behavior of PU-MWNT composites (MWNT content of 5 wt%). The sample undergoes the transition from temporary shape (linear, left) to permanent (helix, right) within 10 s when a constant voltage of 40 V is applied. (Reproduced from Cho, J.W. et al., *Macromol. Rapid Commun.*, 26(5), 412, 2005.)

dependent on the MWNT content and the degree of surface modification of the MWNTs. Mechanical properties of the composites filled with surface-modified MWNTs had been improved, and the modulus and stress at 100% elongation had been enhanced with increasing surface-modified MWNT content. Moreover, the electrical conductivity of the surface-modified MWNT composites was lower than that of the composites filled with unreacted MWNT. An order of 10^{-3} S cm^{-1} was obtained in the samples with 5 wt% modified-MWNT content. Consequently, the shape recovery of composites with surface-modified MWNTs could be driven with an energy conversion efficiency of 10.4% as well as a voltage of 60 V.

6.4.2 Shape-Memory Polymer Filled with Carbon Particles

To date, most of the studies regarding electroactive SMP composites are focused on thermoplastic SMP resins, such as polyurethane SMP. However, the relatively poor thermal and mechanical properties (e.g., strength, as well as temperature, moisture, and chemical resistance) of thermoplastic SMPs cannot fully meet the practical applications. In contrast, a new class of thermoset SMPs shows better above-mentioned properties than thermoplastic SMPs, and they can be widely used as functional materials or structural materials. Nevertheless, the study of electrical conductivity of thermoset SMP composites, which is a main factor in electroactive performance, received little attention because the synthesis technologies of thermoset SMPs were not mature enough until recently.

Jinsong Leng's group at Harbin Institute of Technology demonstrated that the thermoset styrene-based SMP presents better mechanical properties and moisture resistance than the thermoplastic polyurethane SMP. Naturally, it is proposed to incorporate thermoset styrene-based SMP with carbon powders to obtain a better functional material. A new system of thermoset styrene-based SMP filled with nanocarbon powders is presented [49]. In order to realize the electroactive stimuli of thermoset SMP, the styrene-based SMP composite filled with nanocarbon powders was fabricated. Then, its thermal and dynamic mechanical properties, distributions of fillers at micro/nano level, and electrical conductivity were investigated. Finally, the shape recovery demonstration of the SMP composite with 10 vol% nanocarbon powders was performed by passing an electrical current.

6.4.2.1 Differential Thermal Scanning Behavior

Differential scanning calorimetery (DSC) (modulated DSC) was used to examine the evaluation in T_g of SMP composites incorporated with various amounts of nanocarbon powders (CB0, CB2, CB4, CB6, and CB10). Figure 6.12 reveals that T_g decreases slightly with the increase of volume fraction of nanocarbon powders. For a better illustration, T_g (together with the onset and end point of glass transition) is summarized in Figure 6.13. The T_g of a pure SMP is about 69°C, accompanied by an onset point temperature of 66°C and an end point temperature of 73°C. According to the DSC results in Figures 6.12 and 6.13, with the increase in the content of nanocarbon powders, the glass transition shifts a little toward a lower temperature range, and the T_g decreases slightly from 69°C (CB0) to 56°C (CB10). It indicates the slight chemical interaction between nanocarbon powders and SMP.

6.4.2.2 Dynamic Mechanical Performances

A dynamic mechanical analysis (DMA) (TA Instrument, DMA 2980) was used to determine a storage modulus of SMP composite. Figure 6.14

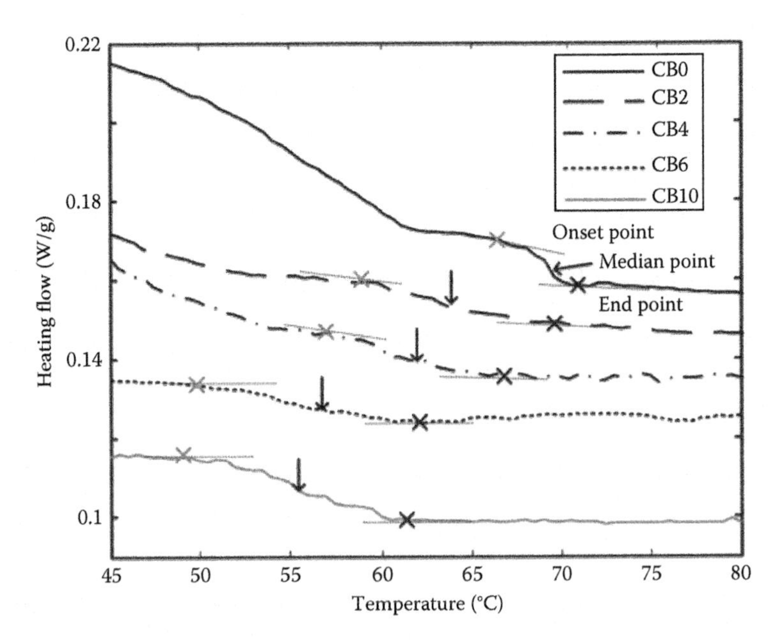

FIGURE 6.12
DSC results of SMP with various concentrations of nanocarbon powders. (Reproduced from Leng, J.S. et al., *Smart Mater. Struct.*, 18, 074003, 2009.)

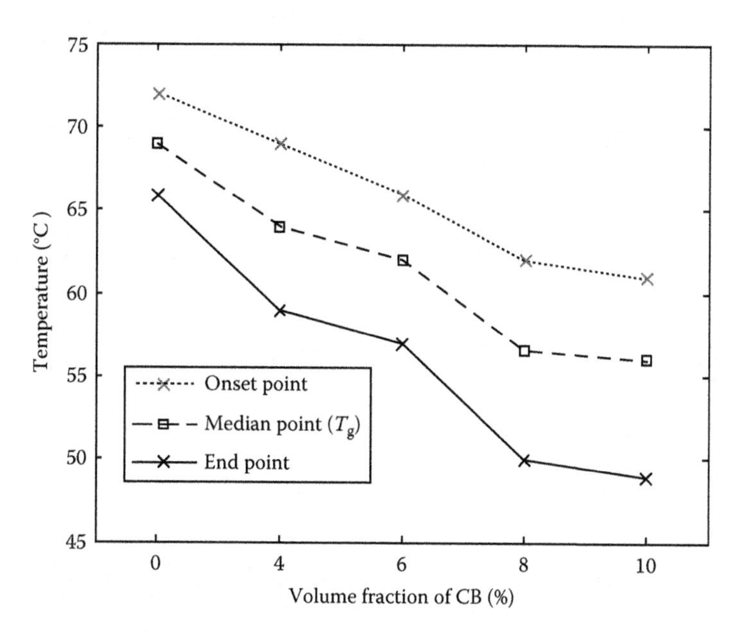

FIGURE 6.13
T_g against the content of nanocarbon powders. (Reproduced from Leng, J.S. et al., *Smart Mater. Struct.*, 18, 074003, 2009.)

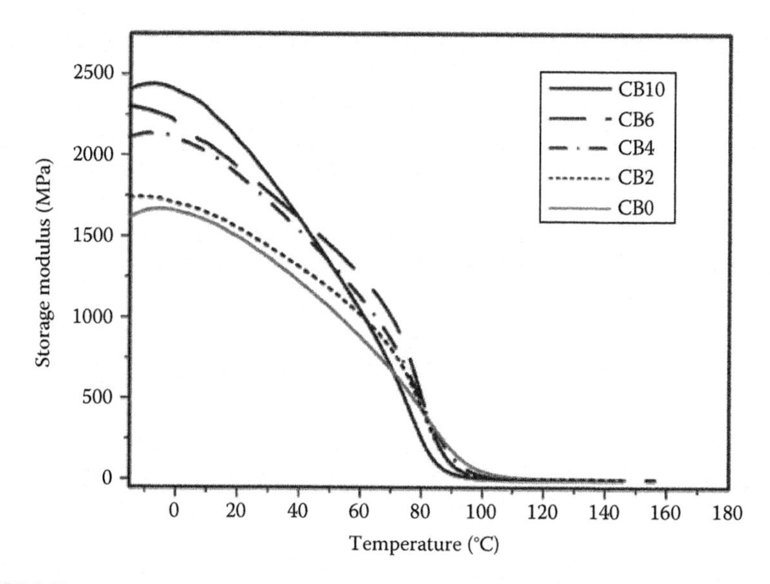

FIGURE 6.14
Storage modulus as a function of temperature. (Reproduced from Leng, J.S. et al., *Smart Mater. Struct.*, 18, 074003, 2009.)

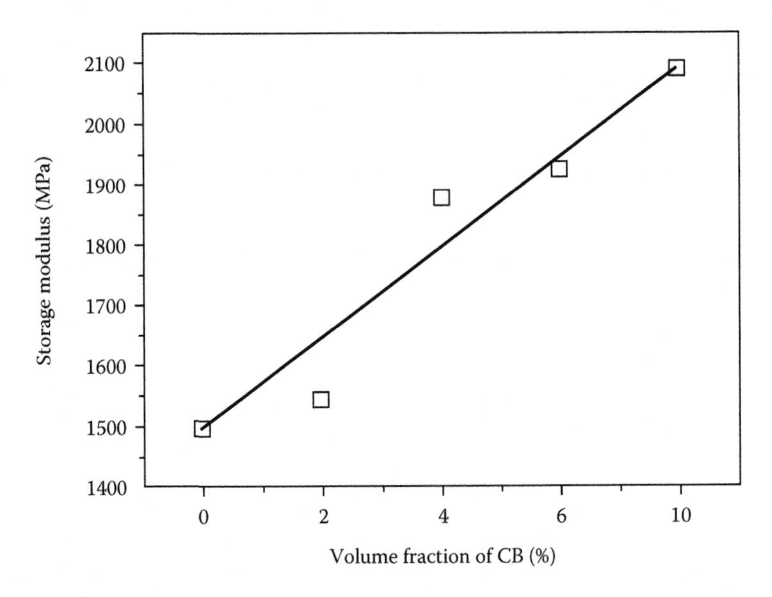

FIGURE 6.15
Storage modulus at 20°C versus the content of nanocarbon powders. (Reproduced from Leng, J.S. et al., *Smart Mater. Struct.*, 18, 074003, 2009.)

indicates the curves of measured storage modulus versus temperature of SMP composite with different nanocarbon powder concentrations, namely CB0, CB2, CB4, CB6, and CB10. For an SMP composite with a certain amount of nanocarbon powder content, the storage modulus far below T_g is about two orders of magnitude larger than that above T_g. For instant, the SMP composite CB10 exhibits an elastic modulus of 2091.27 MPa at 20°C, while it is just 6.44 MPa at 100°C. For a clearer illustration, the storage modulus at 20°C of composite is also summarized in Figure 6.15. It is found that the storage modulus at 20°C increases with the increase of the contents of nanocarbon powders. It indicates that the incorporation of nanocarbon powders reinforces the SMP.

6.4.2.3 Micro/Nanopatterns of SMP Composite

As far as the composites filled with conductive particles are concerned, the distribution, aggregation, morphology, and microporosity of particles greatly affect the conductivity. As shown in Figure 6.16a, the nanocarbon particles in CB2 distribute randomly and separate from each other in the thermoset styrene-based SMP matrix. With the increase in carbon powder content (Figure 6.16b), carbon powders in CB4 distribute uniformly in the SMP matrix, aggregating as clusters instead of absolutely separating from each other. The clusters of carbon particles get connected with each other and continuous carbon networks are formed. In Figure 6.16c, nanocarbon particles in CB6 distribute

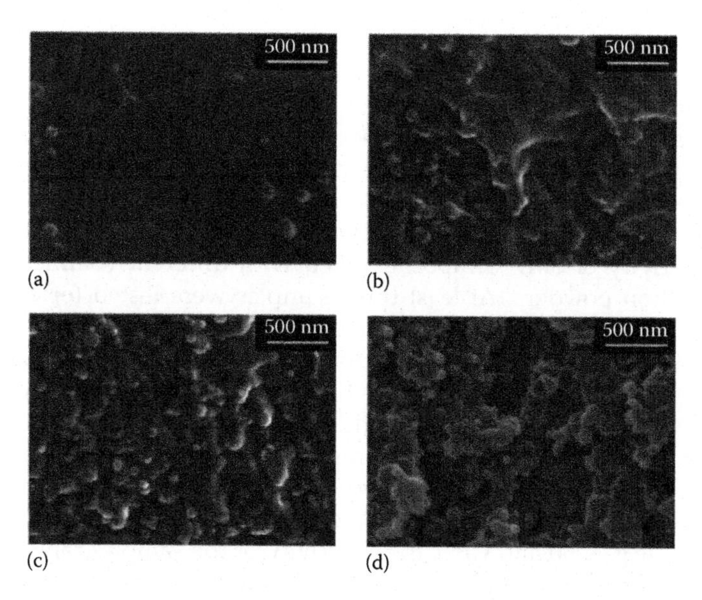

FIGURE 6.16
SEM images (SEI, 20 kV, 20000.0×) of the distributions of nanocarbon powders in the SMP matrix (a) CB2, (b) CB4, (c) CB6, (d) CB10. (Reproduced from Leng, J.S. et al., *Smart Mater. Struct.*, 18, 074003, 2009.)

more uniformly. With the continuous increase in carbon loading (up to 10%), the nanocarbon powder aggregations in CB10 distribute homogeneously and show a high density (see Figure 6.16d). These nanocarbon powders are in high structure, anisometric, as well as high microporosity shape. Based on the structure, a great amount of conductive channels are formed in CB10. Hence, the resistivity may be relatively low and stable.

6.4.2.4 Electroactive Properties for Shape-Memory Effect

In order to determine the electrical resistivity of thin films, samples were cut into the same dimensions ($0.5 \times 5 \times 20\,mm^3$) with two copper electrodes clamped tightly on both ends along the lengthwise direction. The resistance R was monitored by a digital multimeter (IDM91E). The resistivity of the SMP composite with a smaller volume fraction of nanocarbon powders is higher than the upper limit of the digital multimeter. Therefore, a digital high-resistivity determiner (RP2680) was used instead. Subsequently, given the geometrical dimensions of samples, the corresponding volume electrical resistivity ρ is calculated by

$$\rho = \frac{RS}{l} \tag{6.1}$$

where

R is the measured electrical resistance

S the cross-sectional area of sample

I the length between the two copper electrodes (refer to the top inset in Figure 6.17 for the setup of the volume-resistivity measurement)

The circle symbols in Figure 6.17 represent the experimental results of electrical resistivity of SMP composites filled with different volume fractions of nanocarbon powders. At least three samples were tested for each composition. It shows that the electrical resistivity of composite with less than 3% volume fraction of nanocarbon powders is extraordinarily high (10^{14}–10^{13} Ω cm). In contrast, a sharp transition of electrical conductivity occurs between 3% and 5%, which is called the percolation threshold range. As the fillers content is larger than 5%, the resistivity reduces to a low and stable level (10^3–10^1 Ω cm). Moreover, the error bar in Figure 6.17 reveals the reproducibility of the results for each composition. At a content of fillers below percolation (e.g., CB0 and CB2), the resistivity of the samples ranges quite a bit. In contrast, at a content of fillers above percolation (e.g., CB4, CB6, and

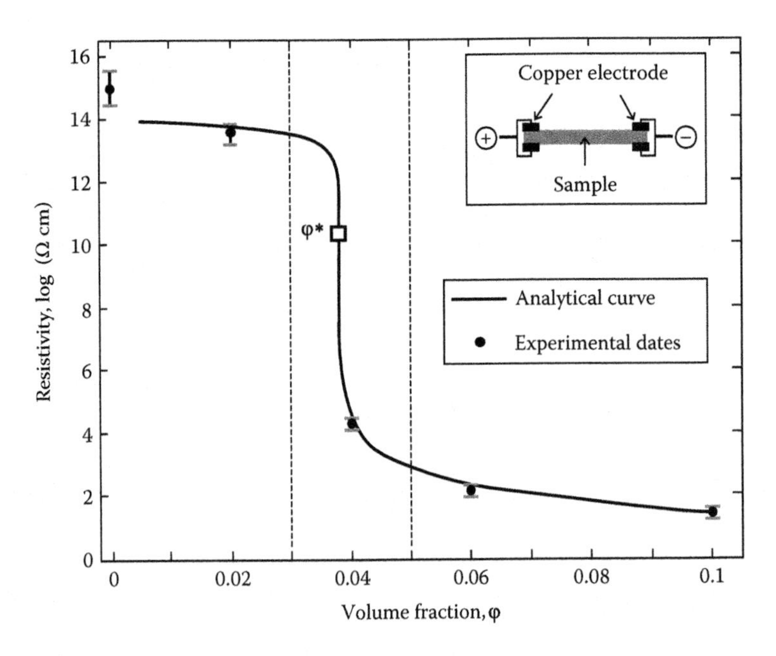

FIGURE 6.17
DC electrical resistivity of SMP composite filled with nanocarbon powders. The circle symbols denote the mean values of at least three measurements for each composition from experimental results. The error bar reveals the variability of results. The curves represent the calculated results according to the percolation theory using the exponents of Equations 6.2 and 6.3 and percolation threshold $\varphi^* \approx 3.8\%$. Inset figure illustrates how the resistance was measured. (Reproduced from Leng, J.S. et al., *Smart Mater. Struct.*, 18, 074003, 2009.)

CB10), it all hovers around the mean values. It shows that the resistivity of the samples at a content of fillers above percolation shows a higher stability than that below percolation.

According to SEM images in Figure 6.16, for the SMP composites with a low-volume fraction of nanocarbon powders (<3.8%), the carbon particles in the SMP matrix are more separated, so that almost no conductive channels can be formed in the material. Hence, the electrical resistivity is very high. As the volume fraction of conductive filler approaches the percolation range (3.8%–5%), some carbon aggregates directly contact with each other and have a higher tendency to form conductive networks in composite. Thus, the composite becomes much more electrically conductive. However, a further increase of carbon powders (>5%) reduces the gap slightly, and only a few more conductive channels are formed. Hence, the reduction in resistivity is in a very gradual manner. It is noticed that the percolation threshold (~3.8 vol%) of this new system of SMP composite is slightly lower than that of many other carbon-based conductive polymer composites, whose percolation is usually above 5% in volume fraction of conductive fillers. This is because of the high structures, high microporosity and homogenous distributions of nanocarbon powders in this SMP matrix. In particular, the BET surface area of carbon powder is about 1000 m^2/g, which is higher than that of common carbon powder (BET, 100–500 m^2/g). Therefore, the carbon powder has a relatively high surface area and high microporosity. Due to this high structure of carbon powders, the composite presents a higher tendency to form a three-dimensional network in the composite, ensuring a relatively lower percolation threshold.

In order to demonstrate the experimental results, the theory of percolation was used. The DC electrical resistivity (ρ_{dc}) of insulator–conductor binary composites is expressed as follows (for randomly distributed conductors in a nonzero conductive matrix):

$$\rho_{dc} = \rho_m(\varphi^* - \varphi)^s, \quad \varphi < \varphi^* \tag{6.2}$$

$$\rho_{dc} = \rho_f(\varphi - \varphi^*)^{-t}, \quad \varphi > \varphi^* \tag{6.3}$$

where φ is the volume fraction of the conductive filler, φ^* is the percolation threshold, and s and t are the critical exponents close to 0.73 and 2.0, respectively, which are affected by the distribution, geometry of conductive fillers, and some other factors. ρ_m and ρ_f are critical coefficients relating to the conductivity of the polymer matrix and the conductive filler. The resistivity of the matrix ($\rho_m = 10^{15}$ Ω cm), the filler ($\rho_f = 10$ Ω cm), and the percolation ($\varphi^* = 3.8\%$) have been used as fitting parameters. By data fitting with the experimental results, the fitting curve of electrical resistivity in Figure 6.17 is obtained through Equations 6.2 and 6.3. As percolation theory describes, our experimental results coincide rather well with the reservation that there

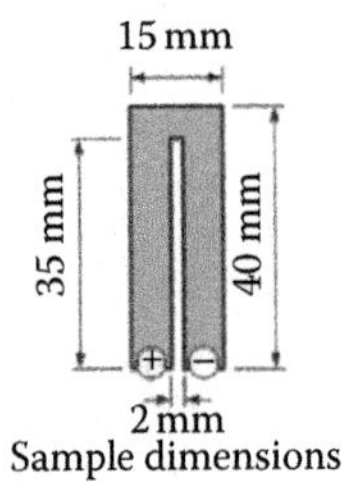

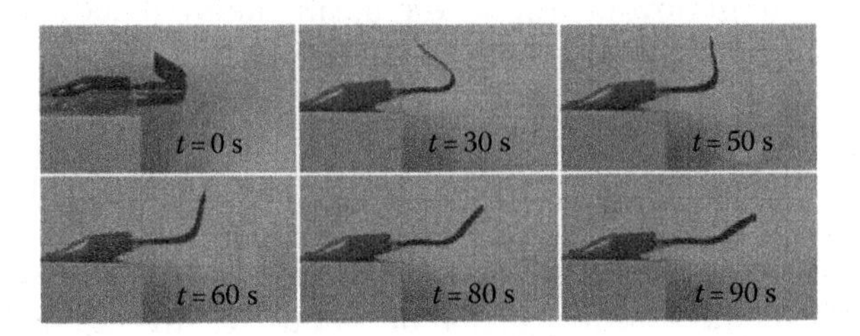

FIGURE 6.18
Sequences of the shape recovery of sample CB10 by passing an electrical current (voltage, 30 V). (Reproduced from Leng, J.S. et al., *Smart Mater. Struct.*, 18, 074003, 2009.)

are very limited numbers of different nanocarbon powders concentrations for the composite.

It is noticed that for the practical applications of SMP composite, high and stable electrical conductivity is required. Consequently, the SMP filled with nanocarbon powders at a content of 10 vol% was employed in a shape recovery test.

The electric field-triggered shape recovery of sample CB10 is shown in Figure 6.18. The SMP composite CB10 was cut into an "n" shape for demonstrating the shape-memory effect upon heating by passing an electrical current through it (Figure 6.18, left inset). The strip can be easily bent to about 135°C at 70°C (above T_g). After switching off the electrical power and cooling back to a room temperature of 22°C, the temporary curved shape (predeformed shape) was obtained (Figure 6.18, $t = 0$ s). Then, the strip had a constant voltage of 30 V applied through two conductive clampers. Figure 6.18 shows the time dependency of shape recovery. The curved strip began to deploy at about 5 s right after heating by the electrical current. The curved strip recovered spontaneously at about 30° at the end of the 90 s (Figure 6.18, $t = 90$ s). The recovered shape of the SMP strip was about 75%–80% recovered compared with the original shape. It was also obvious that the deployment velocity of the sample in the range of 40–60 s was relatively higher than those in other time ranges. Furthermore, compared with thermoplastic SMP composites, a systematic investigation of mechanical properties for this thermosetting styrene-based SMP composite is still required in a future study.

6.4.3 Shape-Memory Polymer Filled with Electromagnetic Fillers

6.4.3.1 Electromagnetic-Induced SMP Composite

By adding surface-modified super-paramagnetic Fe_3O_4 nanoparticles into oligo(e-caprolactone)dimethacrylate/butyl acrylate to fabricate shape-memory polymer composite, it has been demonstrated that the actuation of

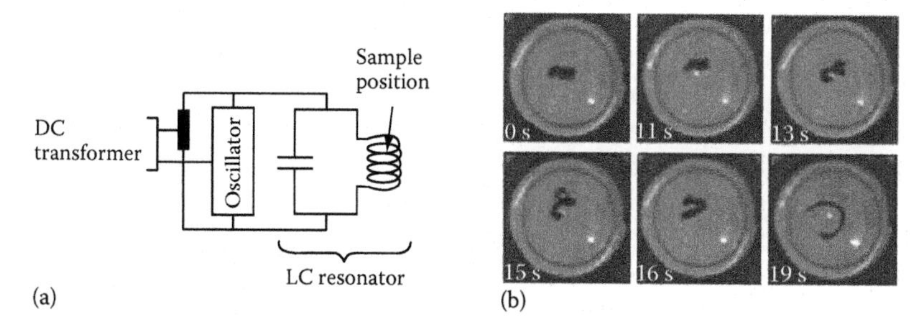

FIGURE 6.19
(a) Schematic diagram of the LC resonant circuit-based HF generator used for induction heating experiments, and sample positioning; (b) photo series demonstrating the shape-memory transition induced by the impact of an HF electromagnetic field, measured from the topside of the induction coil. The shape of the sample is changed from helical (temporary shape) to a rectangular strip (permanent shape). (Reproduced from Schmidt, A.M., *Macromol. Rapid Commun.*, 27, 1168, 2006.)

shape-memory effect by electromagnetic fields is achieved [63]. The thermosetting SMP composites contains between 2 and 12 wt% magnetite nanoparticles serving as nanoantennas for magnetic resistive heating. The specific loss of power of the particles is determined to be 30 W g^{-1} at 300 kHz and 5.0 W. During the shape transition at 43°C, no further temperature increase is observed.

In Figure 6.19a, a series of photos impressively documented the electromagnetically induced shape-memory effect of the sample. The used specimen was cut from the composite film as a rectangular strip (about $15 \times 2 \times 0.5\,mm^3$), referred to as the permanent shape. After heating the sample to 70°C, it was deformed to a helix and cooled to fix this temporary shape by the formation of oligo(e-caprolactone) crystallites. After this programming process, the sample kept the helical shape in the absence of external forces as can be seen in Figure 6.19b (0 s). The shape transition in the AC field of 300 kHz was documented with a digital camera. After 10 s, the starting conversion of the helix was observed, taking another 10 s to be completed. The final shape was close to the original rod with some remaining flexion due to the friction between the soft polymer and the glass plate. The observed time window is in good agreement with the experimental sample temperature in the induction heating experiment with the temperature control described above, and the temperature of the sample's environment (the open-aired system inside the coil is shown to be hardly affected by the process).

6.4.3.2 *Electroactive Thermoplastic SMP Composite Filled with Ni Powder Chains*

While shape-memory polymers (SMPs) can be actuated by various stimuli, an external heater is normally required for the shape recovery in thermo-responsive SMPs. In order to get rid of the external heaters, SMP composites with different types of fillers have been developed, so that they can be actuated by means of Joule heat or induction heat. While the latter approach has the advantage of wireless/remote operation, the former might be more of interest in many engineering applications in which a bulk system to generate magnetic field is not appreciated.

Electrically conductive powders, fibers, and even nanowires/tubes have been utilized as the fillers to improve the electrical conductivity of polymers [58]. Although conductive fiber and nanowire/tube can significantly enhance the stiffness and strength of polymers, their deformable strain is limited to within a few percent. Since the recoverable strain in the SMPs is normally in the order of 100%, there is a potential problem of deformation compatibility. As such, electrically conductive powders, such as carbon black and Ni powder, should be a better choice.

SEM images reveal that single chains start to be formed at 1% volume fraction of Ni [58]. With the increase of Ni content, multichains (bundles) are resulted, and eventually no clear Ni chain can be recognized (Figure 6.20, left column). In addition, after five stretching shape recovery cycles, the Ni chains still exist (Figure 6.20, right column), which indicates the possibility of using chained SMPs for cyclic actuation.

Figure 6.21 plots the electrical resistivity ρ of the random and chained samples against the volume fraction of Ni powders. Although limited by the measuring range of the multimeter, as it is expected, ρ of the chained sample in the transverse direction (i.e., perpendicular to the chain direction) is the highest, while ρ of random samples is lower. ρ of the chained samples in the chain direction is always the lowest. For example, at 10% volume fraction of Ni powder, ρ of the random sample is $2.36 \times 10^4\ \Omega$ cm; while in the chained sample, it is $2.93 \times 10^6\ \Omega$ cm in the transverse direction, and only $12.18\,\Omega$ cm in the chain direction. However, at a high Ni content, ρ of all types of samples is close. This is due to the fact that the Ni chains become unrecognizable at a high Ni content, as revealed in Figure 6.20. At 10% volume fraction of Ni, the chained sample (with a dimension of $16 \times 0.6 \times 5\,\text{mm}^3$) can be heated from room temperature (20°C) to 55°C by applying a voltage of 6 V (refer to infrared image in Figure 6.21, bottom-left inset), which is enough to trigger the shape recovery. However, for the same setup and configuration, it is only about 26°C in the random sample. No shape recovery can be actuated, as it is far below the actuation temperature. The recovery sequence of a chained sample upon Joule heating is shown in Figure 6.22. According to the DSC results in Figure 6.23, with the increase of Ni content, the glass transition shifts a little bit toward the low temperature range, which indicates that there is a slight chemical interaction between Ni powders and SMP.

From the measured storage modulus versus temperature curves of random and chained samples (Figure 6.24), the storage modulus is plotted (at 0°C) against the volume fraction of Ni for both random samples and

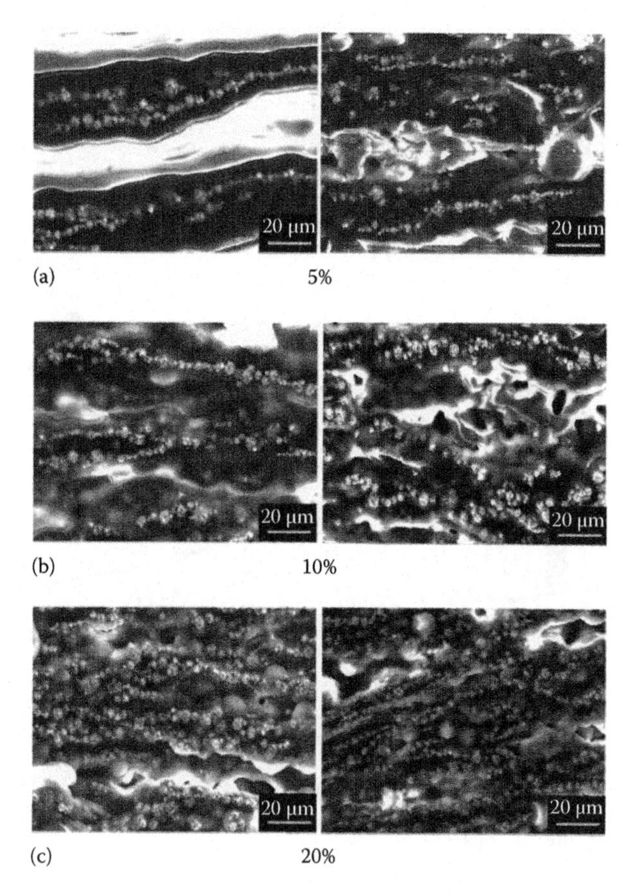

FIGURE 6.20
Typical SEM images before (left column) and after (right column) five stretching (at 50% strain) shape recovery cycles. (Reproduced from Leng, J.S. et al., *Appl. Phys. Lett.*, 92, 014104, 2008.)

chained samples (chain direction only) in Figure 6.25. At around 10% volume fraction of Ni content, the composite is already significantly strengthened. For the same amount of Ni content, the storage modulus of chained samples in the chain direction is always much higher than that of the random sample. This indicates that the reinforcement is more effective by forming Ni powder chains.

T_g is one of the major characteristics of SMPs. Traditionally, T_g can be defined in various ways. Here, it is the differential storage modulus curve with regard to temperature, and then the peak temperature is defined as T_g. Figure 6.26 plots T_g against the volume fraction of Ni powders for both random samples and chained samples (chain direction only). As can be seen, with the increase of Ni content, T_g decreases in a more or less same linear fashion in both types of samples. At 20% Ni content (this is extremely

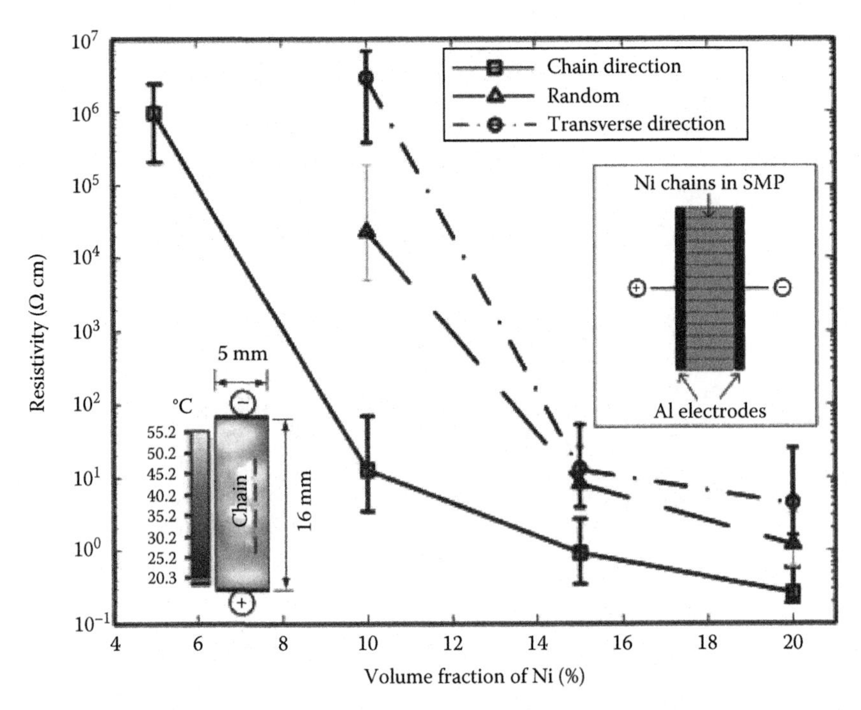

FIGURE 6.21
Electrical resistivity versus volume fraction of Ni powder. Right inset: illustration of setup for the resistivity measurement along the chain direction. Bottom-left inset: infrared image of temperature distribution in chained sample (10% Ni, 6 V). (Reproduced from Leng, J.S. et al., *Appl. Phys. Lett.*, 92, 014104, 2008.)

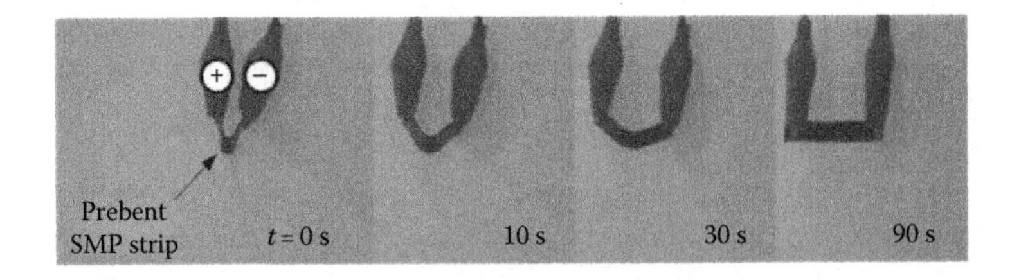

FIGURE 6.22
Shape recovery sequence in a chained sample ($30\times7\times1\,mm^3$, 10% Ni, 20 V). (Reproduced from Leng, J.S. et al., *Appl. Phys. Lett.*, 92, 014104, 2008.)

high), T_g drops by about 15°C to 35°C, which confirms the slight chemical interaction between SMP and Ni powders.

In recent years, it has been reported that the shape-memory effect in SMPs is not only a macroscopic phenomenon, but also occurs even on the 10s of nm scale in bulk SMPs [64]. As such, SMPs have great potential for micron

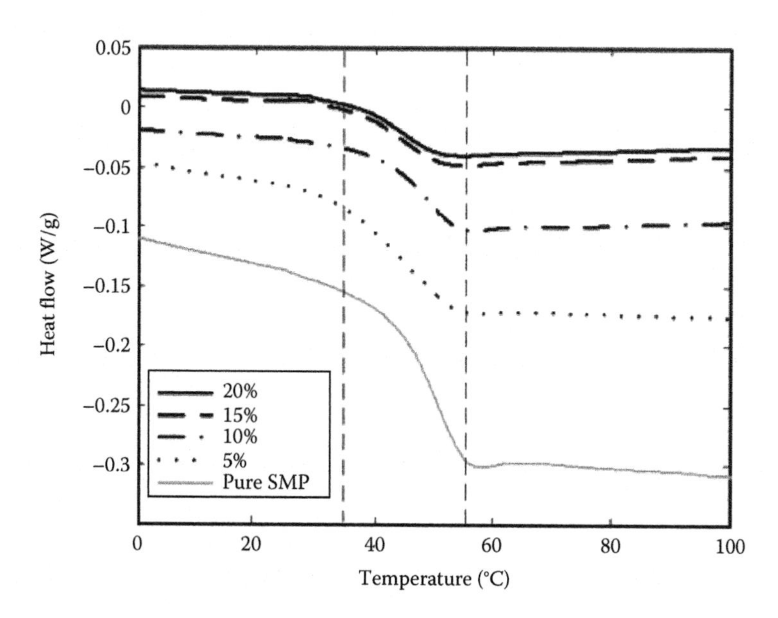

FIGURE 6.23
DSC results of random samples. (Reproduced from Leng, J.S. et al., *Appl. Phys. Lett.*, 92, 014104, 2008.)

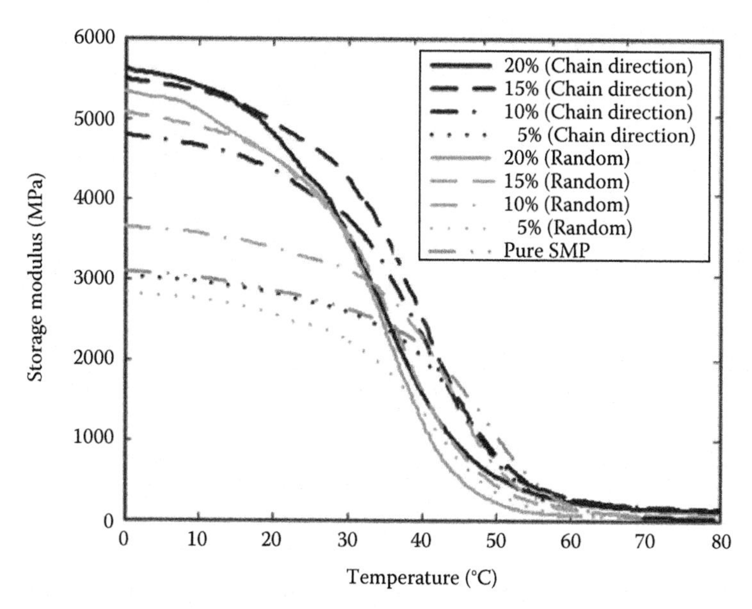

FIGURE 6.24
Typical storage modulus versus temperature curves obtained from DMA test. (Reproduced from Leng, J.S. et al., *Appl. Phys. Lett.*, 92, 014104, 2008.)

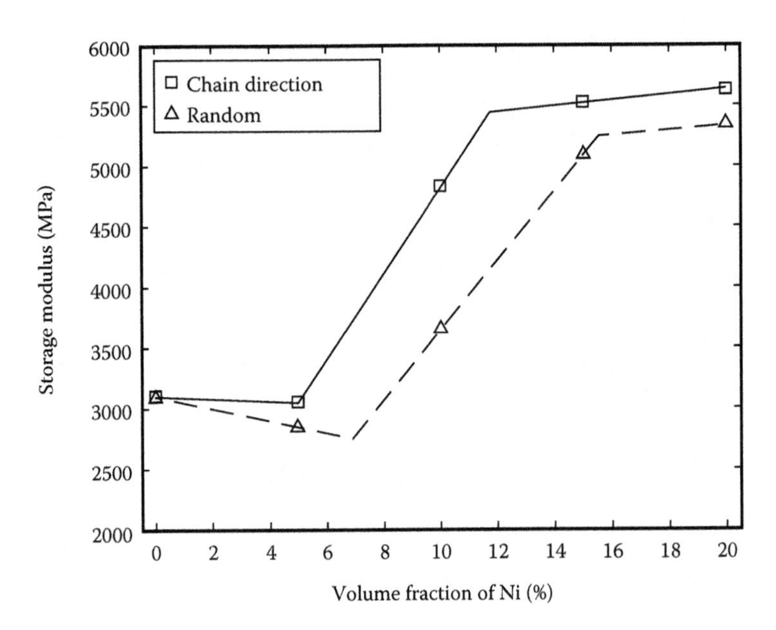

FIGURE 6.25
Storage modulus versus volume fraction of Ni at 0°C. (Reproduced from Leng, J.S. et al., *Appl. Phys. Lett.*, 92, 014104, 2008.)

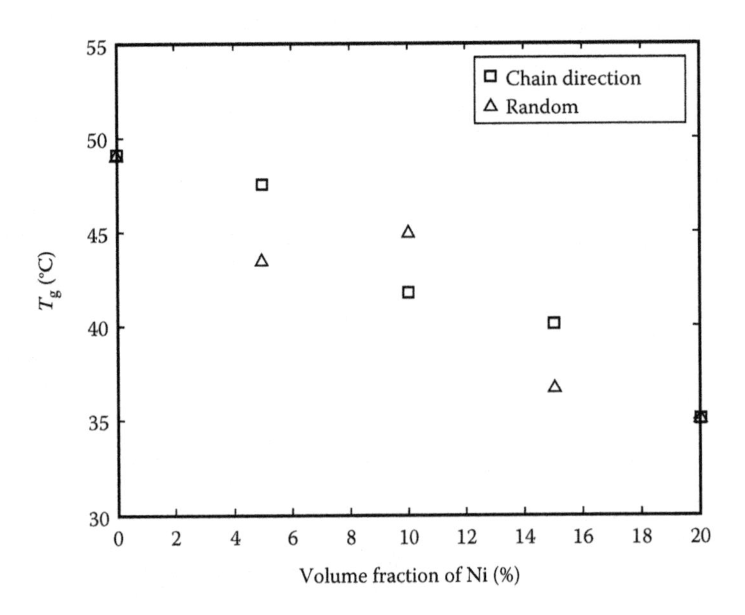

FIGURE 6.26
T_g of chained and random samples. (Reproduced from Leng, J.S. et al., *Appl. Phys. Lett.*, 92, 014104, 2008.)

and even submicron scale actuation for MEMS and NEMS applications, micro/nano patterning [65,66] and biomedical device [55,67], etc.

Micron sized thermo/moisture-responsive PU SMP/Ni powder chains were fabricated and their shape recovery was investigated as an indirect way to demonstrate the SME in micron/submicron sized SMPs. Vertical protrusive SMP/Ni chains were fabricated using the following procedure. First, a certain volume percentage of Ni powders was mixed with the SMP/DMF solution and stirred to create a uniform dispersion. The highly viscous mixture was subsequently poured into a petri dish (30 mm diameter, 8 mm height) until a height of about 4 mm (Figure 6.27). Two magnets ($20 \times 20 \times 40$ mm^3) were placed in the positions illustrated in Figure 6.27 with a gap of 10 mm between them. Note that the petri dish was placed 1 mm above the bottom magnet. The magnetic density at the bottom of petri dish was measured as 0.4 ± 0.05 T. The whole setup (including the petri dish) was placed in an air-tight box and kept in an oven for 48 h at a constant temperature of 80°C for solidification. After the volatilization of DMF, samples with vertical protrusive SMP/Ni chains were obtained. The resulting samples contained an array of vertical chains along the direction of the magnetic field atop the SMP substrate Figure 6.28.

Figure 6.29 presents the morphology of the array of flattened chains atop the 2 vol% Ni sample and the morphology after subsequent heating for shape recovery. The zoomed-in view of a small chain reveals that in the flattened shape, the initially vertical chain was bent around 90°. However, after heating, the bending angle was largely recovered, which reveals the shape recovery phenomenon in the chain and indirectly demonstrates

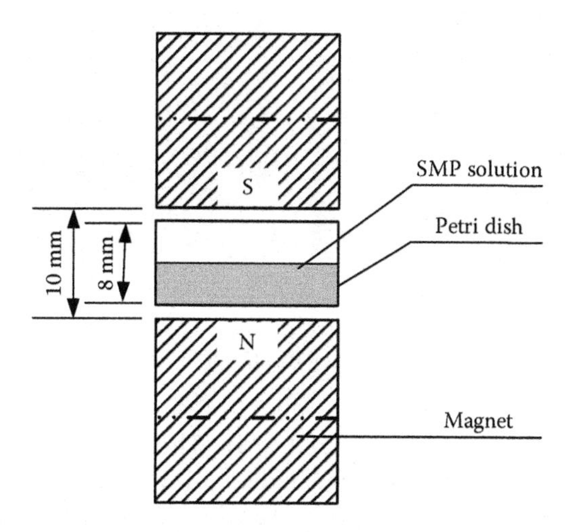

FIGURE 6.27
Illustration of setup for sample preparation. The strength of magnetic field was measured as 0.4 ± 0.05 T within the area of petri dish. Dimensions of magnet: $20 \times 20 \times 40$ mm^3. Gap between magnet and petri dish: about 1 mm.

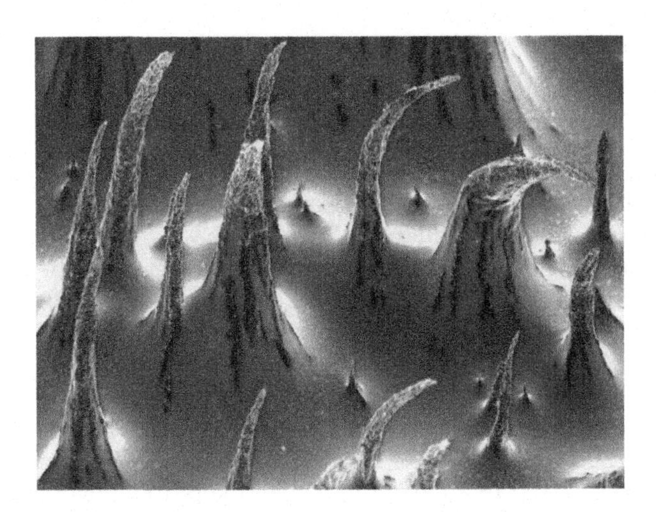

FIGURE 6.28
Samples contained an array of vertical chains atop the SMP substrate.

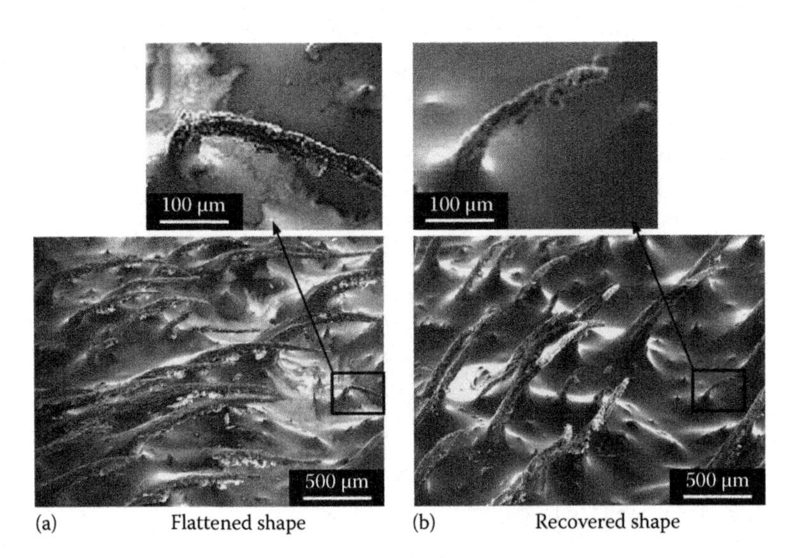

FIGURE 6.29
Typical shape-memory effect in protrusive chains (2 vol % Ni).

the SME in micron/submicron sized SMP. The SME in an array of such chains (like a micro brush) may provide a simple mechanism and convenient approach to achieve reversible surface morphology for a significant change in many surface related properties such as friction and wetting ability.

6.4.3.3 *Electroactive Thermoplastic SMP/CB Composite Filled with Ni Powder Chains*

Furthermore, an approach to significantly reduce the electrical resistivity in a polyurethane shape-memory polymer (SMP) filled with randomly distributed carbon black (CB) is presented [59]. With an additional small amount of randomly distributed Ni microparticles (0.5 vol%) in the SMP/CB composite, its electrical resistivity is only reduced slightly. However, if these Ni particles are aligned into chains (by applying a low magnetic field on the SMP/CB/Ni solution before curing), the drop in electrical resistivity is significant. This approach, although demonstrated in a SMP, is applicable to other conductive polymers.

Three types of thin films were fabricated, namely SMP/CB/Ni (chained), SMP/CB/Ni (random), and SMP/CB (i.e., without Ni). Within each type of thin film, the volume fraction of CB varies from 4%–10%. But in all thin films with Ni particles, Ni is always 0.5 vol%. At a low Ni content, 0.5 vol%, many short Ni single chains, i.e., the Ni particles aligned one after another in one line, are formed instantly upon applying a weak magnetic field. After solidification, these chains are fixed (Figure 6.30). Given that the length of chains ranges from 100 to 200 μm and the size of Ni particle is about 3–7 μm, the aspect ratio of Ni chains is about 14–70, which is comparable to short microfibers.

In order to determine the resistivity of these composites, samples ($1 \times 5 \times 20\,mm^3$) were cut out of the thin films and connected to aluminum electrodes (refer to inset in Figure 6.31). The resistance was measured by a digital multimeter (IDM91E). Given the geometrical dimensions of the samples, the corresponding electrical resistivity was derived. Four readings from four different pieces of samples of the same type of film were obtained. One month later, the samples were remeasured in order to check the stability.

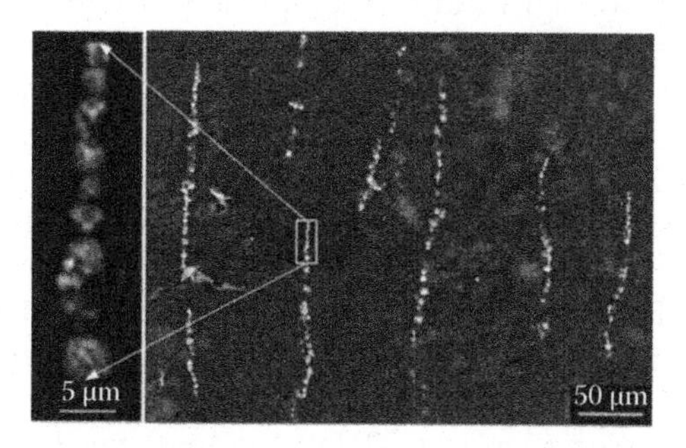

FIGURE 6.30
SEM images of conductive SMP with 10 vol% of CB and 0.5 vol% of chained Ni. Inset: zoom-in view of one Ni chain. (Reproduced from Leng, J.S. et al., *Appl. Phys. Lett.* 92, 204101, 2008.)

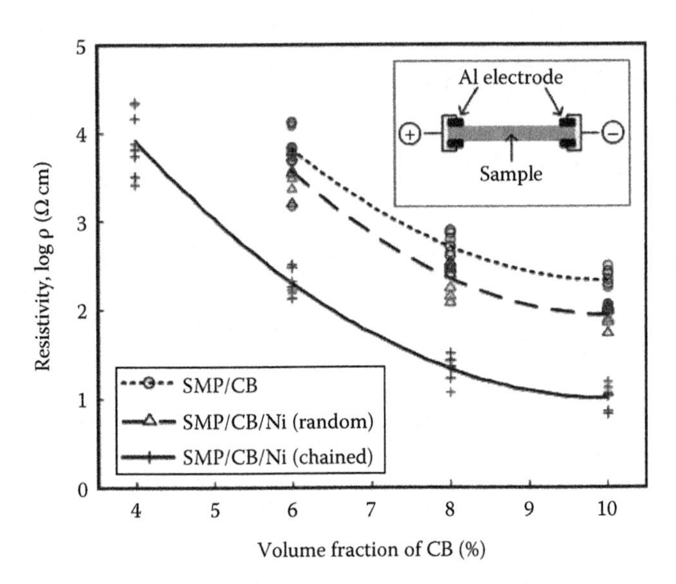

FIGURE 6.31
Resistivity versus volume fraction of CB with/without 0.5 vol% of Ni. Inset figure illustrates how the resistance was measured. (Reproduced from Leng, J.S. et al., *Appl. Phys. Lett.*, 92, 204101, 2008.)

Figure 6.31 reveals the relationship of CB content versus electrical resistivity of SMP/CB/Ni (chained), SMP/CB/Ni (random), and SMP/CB. It is clear that the additional 0.5 vol% of Ni, if distributed randomly, only slightly reduces the resistivity of the composites. However, the same amount of Ni particles, if well aligned to form chains, can significantly reduce the electrical resistivity by more than 10 times. Obviously, the remarkable reduction in the electrical resistivity is the result of the conductive chains, which serve as conductive channels to bridge those small isolated CB aggregations. This bridging effect is more significant in composites loaded with a low amount of CB in which the CB aggregates are relatively small in size and more isolated. It should be pointed out that for the same sample the resistance measured one month later is about the same as before. Hence, the resistivity of these samples is stable.

In order to demonstrate the shape recovery by Joule heating, three samples [namely SMP/CB/Ni (chained), SMP/CB/Ni (random) and SMP/CB], all with 10 vol% of CB, 1 mm thick and in a shape as illustrated in Figure 6.32 (left-top inset) were bent by about 150° at 80°C, and then cooled to room temperature (22°C) (refer to Figure 6.32, left-middle inset). Subsequently, a 30 V power was applied (Figure 6.32, left-top inset for the setup). An infrared video camera (AGEMA, Thermo-vision 900) was used to monitor the temperature distribution and shape recovery simultaneously. Figure 6.32 (right) presents four snap shots of each sample. It is clearly revealed that Sample (c) [SMP/CB/Ni (chained)] reaches the highest temperature (about

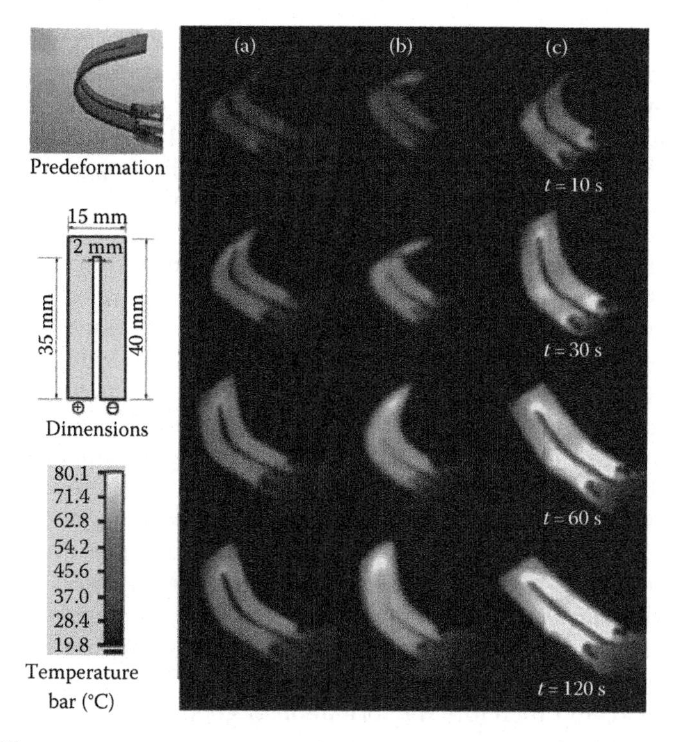

FIGURE 6.32
Sequence of shape recovery and temperature distribution. Top-left inset: Prebent shape; middle-left inset: Dimensions of sample. Bottom-left inset: Temperature bar (in °C). Sample (a) 10 vol% of CB only; Sample (b) 10 vol% of CB, 0.5 vol% of randomly distributed Ni; and Sample (c) 10 vol% of CB, 0.5 vol% of chained Ni. The tests were repeated for more than five times on each sample. (Reproduced from Leng, J.S. et al., *Appl. Phys. Lett.*, 92, 204101, 2008.)

80°C everywhere, which is much higher than the T_g of the SMP, so that almost full recovery is observed in 120 s), while the temperature of Sample (a) [SMP/CB] is the lowest (about 45°C only, slightly higher than the T_g of the SMP). Hence, the shape recovery is small. Sample (b) [SMP/CB/Ni (random)] reaches around 65°C and the shape recovery is not completed after 120 s. In term of power consumption, it is about 1.2 W for Sample (c). In addition, the evolution of resistivity after up to 20 shape recovery cycles (20% pre-strained) is investigated.

6.4.3.4 Electroactive Thermoset SMP Composite Filled with Ni Powder Chains

The former discussions are based on thermoplastic polyurethane SMPs with embedded micron-sized Ni particle chains. It demonstrates that the alignment geometry of Ni chains in SMP presents higher electrical conductivity than SMP filled with the same content of random-distributed Ni powders. However, the required amount of Ni particles for good conductivity is still high (15%–20% in volume fraction) and the conductivity is unstable. Moreover, as a thermoplastic polymer, this SMP composite shows poor mechanical properties and bad moisture resistance. Instead of using micro-sized Ni particles in thermoplastic SMP previously, Jinsong Leng's group at Harbin Institute of Technology fabricated a thermoset styrene-based SMP with embedded nano-sized Ni powder chains.

Results demonstrate that the resistivity of the chained sample in the chain direction is lower than that of random samples (see Figure 6.33). For example, at 6% volume fraction of Ni powders, ρ of chained sample is about 20 times lower than that of random samples. However, at a high Ni content, ρ of the two types of samples is relatively close. This is due to the fact that the Ni chains become unrecognizable at a high Ni content. It is to be noted that the improvement of the electrical conductivity along the chain direction in the thermoset SMP is not so apparent compared with those in our previous study with thermoplastic polyurethane SMP. From Figure 6.33, the raw data of resistivity in chained samples and random samples are sometimes overlapped, and therefore, in this case, the improvement of conductivity in the chain direction cannot even be demonstrated. This may be because of the types, diameters, microporosity and distributions of Ni particles, and the types of SMP. The chain geometry is also found to be very sensitive to the magnetic field, and therefore the formation of good chain geometry is very difficult sometimes.

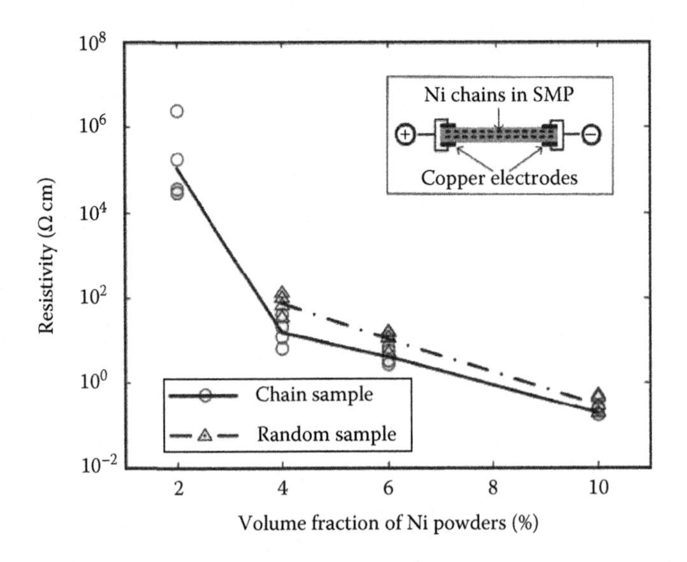

FIGURE 6.33
Electrical resistivity versus volume fraction of Ni powder. Four raw experimental points are given for each composition. Top-right inset: illustration of setup for the resistivity measurement along the chain direction.

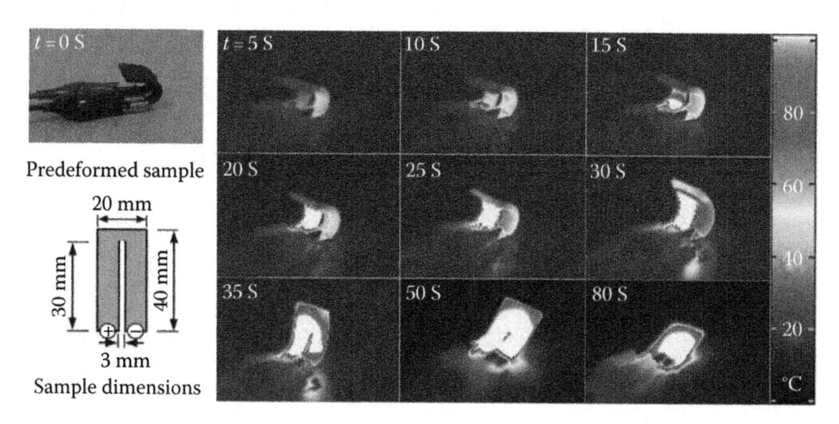

FIGURE 6.34
Shape recovery sequences in a chained sample (10% Ni, 12 V, 3.6 W).

In order to demonstrate the shape recovery by Joule heating, a chained sample with 10 vol% of Ni powders, 1 mm thick, and in an "n" shape as illustrated in Figure 6.34 (bottom-left inset) was bent by about 180°C at 90°C. After switching off the electrical power and cooling back to a room temperature of 15°C, the temporary curved shape (predeformed shape) was obtained (refer to Figure 6.34, top-left inset). Subsequently, the strip was applied a voltage of 12 V through two conductive clampers. An infrared video camera (InfraTec, VarioCam) was used to monitor the temperature distribution and shape recovery sequence of the sample upon Joule heating simultaneously. Figure 6.34 (right) presents nine snap shots of the shape recovery of the sample. The curved strip began to deploy about 5 s right after heating by the electrical current. The curved strip recovered spontaneously to about 30° in 80 s (Figure 6.34, $t = 80$ s). The recovered shape of the SMP strip was about 80%–85% recovered compared with the original shape. The stable current in the shape recovery process is about 0.3 A at a power of 3.6 W.

6.4.3.5 SMP Composite Actuated by External Magnetic Force

Jinsong Leng's group at Harbin Institute of Technology investigated the activated approach of SMP composite filled with magnetic filler is achieved by external magnetic force. The magnetic power is blended into SMP matrix to prepare multifunctional magnetic SMP composites. Two kinds of magnetic powders (Ni powder and Fe powder) were uniformly dispersed into the matrix material to investigate the magnetic field strength of the composite material. In the magnetic field, the stretching experiments of the magnetic shape-memory composite were also conducted.

During the stretching experiment in a magnetic field, Ni powder (diameter: 15 μm) and Fe powder (diameter: 80 μm) were incorporated into the SMP to evaluate the magnetic performances. The experimental result is shown in Figure 6.35. The weight contents of the magnetic powder are 22.5%, 27.5%,

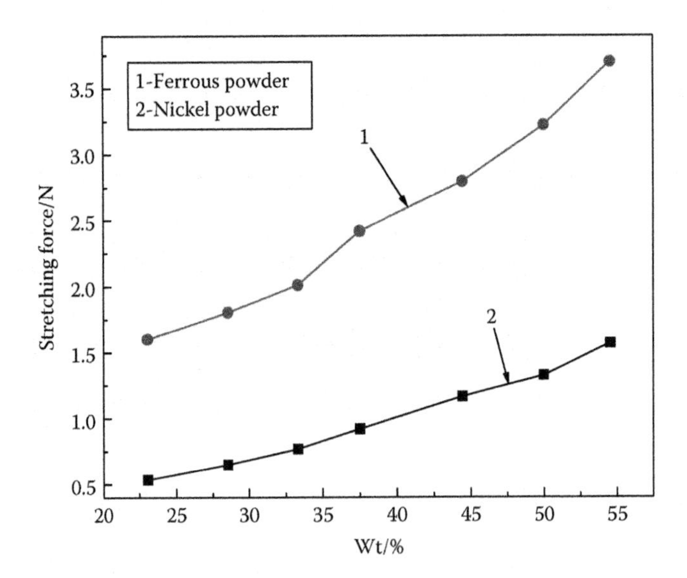

FIGURE 6.35
Relationship of the magnetic power and stretching force.

33.5%, 37.5%, 45%, 50%, and 55%, respectively. Apparently, the residual deformation of the magnetic SMP composites increases with increment in magnetic powder content initially.

While the magnetic shape-memory composite filled with Fe powder and Ni powder can only be unidirectionally attracted under the magnetic field, composite based on Nd-Fe-B powder can be enabled to give a bidirectional response, including both attraction and rebellion. Figures 6.36 and 6.37 show the deformation process of magnetic shape-memory composite (doped with 20 wt% Nd-Fe-B powder) under attractive and repelling magnetic force, respectively. In Figure 6.36, the magnetic shape-memory composite is prebent to 40° ($t = 0$ s). Upon applying the magnetic field ($t = 1$ s), the free end of the strip began

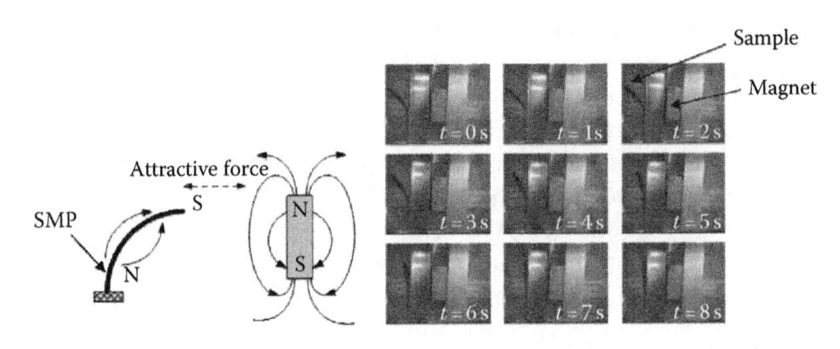

FIGURE 6.36
Deformation process of magnetic shape-memory composite under magnetic attractive force.

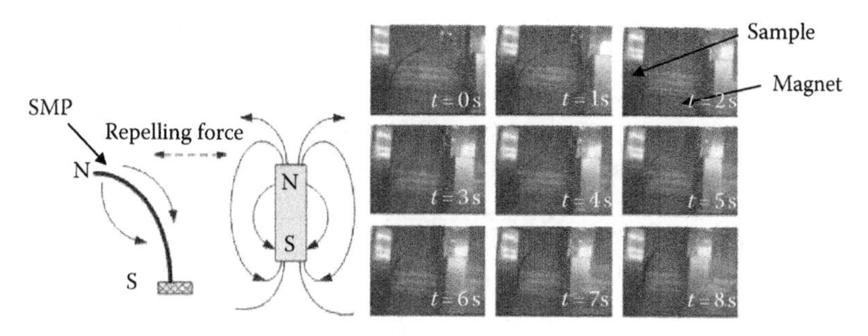

FIGURE 6.37
Deformation process of magnetic shape-memory composite under repelling force.

to move toward the magnet block. In this process, the bent strip with a predeformed angle of about 40° deformed to a straight configuration within 8 s.

As shown in Figure 6.37, the magnetic shape-memory composite is bent to 30°, namely predeformed angle. In the magnetic field, the predeformed strip moved away from the magnet block upon applying the external repelling force, and the residual angel was about 10° within 8 s. Note that the strip is not fully recovered, as the repelling force is relatively small compared to the attractive force.

This magnetic shape-memory polymer shows a potential to achieve the two-way shape-memory effect just like that of shape-memory alloy. The magnetic SMP composite is proposed to be used for designing novel actuator in the future.

6.4.4 Shape-Memory Polymer Filled with Hybrid Fibers

The continuously conductive network of hybrid fibers that contains particulate and fibrous fillers has been demonstrated in the commercial styrene-based SMP. In the conductive network, a fibrous filler is considered as a long pathway to electrons, and the particulate filler works as a cross-linked point. Synergistic effect could be found to significantly improve the electrical conductive property of SMP composites. In this section, hybrid filler of CB and SCF was selected to achieve SMP nanocomposites driven by electrical resistive heating. Figure 6.38 shows the distinct relationship between the electrical resistivity of composites with respect to conductive filler content. The data of SMP filled with micro-carbon powder and hybrid filler of micro carbon powder and SCF were referenced to [68] for comparison purposes. Experimental results reveal that the composites containing nanoparticles have a better conductivity than that blended with microdimension conductive filler. That is may be attributed to the nanosized effect. As the amount of filler content increased, the electrical resistivity of both composites and nanocomposites is decreased, and a synergistic effect occurring in the CB/SCF system with lowest resistivity was expected. As presented in the curves,

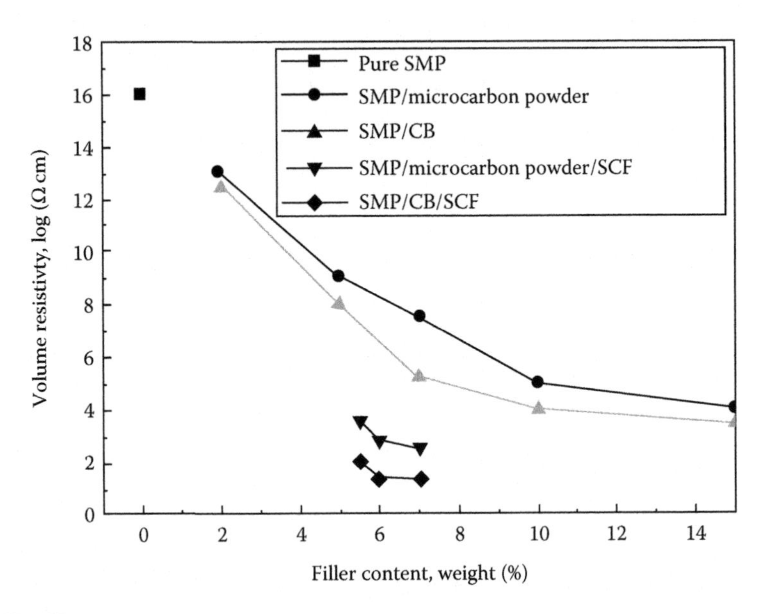

FIGURE 6.38
Resistivity of SMP matrix filled with MCP, CB, MCP/SCF, and CB/SCF systems versus filler content. (Reproduced from Leng, J.S. et al., *Appl. Phys. Lett.*, 91, 144105, 2007.)

the electrical resistivity of the composite filled with CB is 10^4 times higher than that of the composite with CB/SCF at the same content of 7wt%. This could be attributed to the fact that the inherent fibrillar form of SCF has a higher tendency to support long pathway to electrical current in the composites, ensuring better electrical response than that of particulate fillers.

The characteristic volume resistivity curves for the SMP/CB/SCF composite (shown in Figure 6.38) indicate that the SMP filled with 5wt% CB and 0.5wt% SCF, whose resistivity is 128.32 Ω cm, has turned into a semiconductor from an insulator; while the resistivity of composite containing 5wt% CB and 2wt% SCF is 2.32 Ω cm and can be defined as conductor.

The electrically induced shape-memory effect is exemplarily demonstrated for SMP composite with a dimension of $112 \times 23.2 \times 4\,mm^3$ which is filled with 5wt% CB and 2wt% SCF as shown in Figure 6.39, where a change in shape from temporary flexural shape to permanent plane stripe shape occurring within 50s is shown, when a constant voltage of 24V is applied. However, the shape recovery behaviors such as recovery time, recovery ratio, and so on, are strongly dependent on the magnitude of the applied voltage and the electrical resistivity of SMPC.

As shown in Figure 6.40, the short fibers disperse randomly in the SMP matrix. There are many interconnections between fibers. These interconnections form the conductive networks which can be used to explain the excellent electrical conductivity of composites filled with SCF. However, the dispersion of SCF is normally inhomogeneous within composites. As

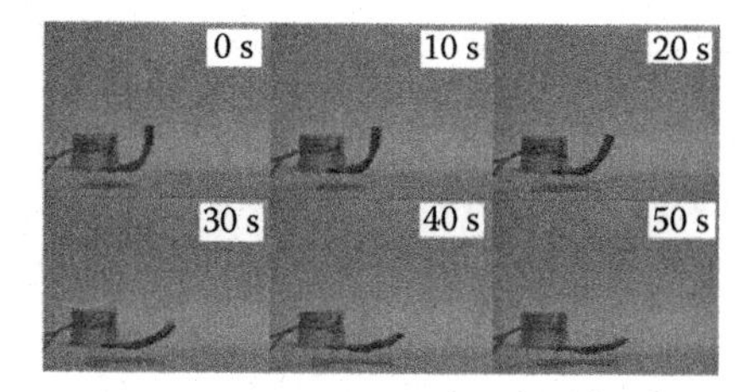

FIGURE 6.39
Electro-activated SMP filled with CB/SCF induced by applying 25 V. (Reproduced from Leng, J.S. et al., *J. Appl. Phys.*, 104, 104917, 2008.)

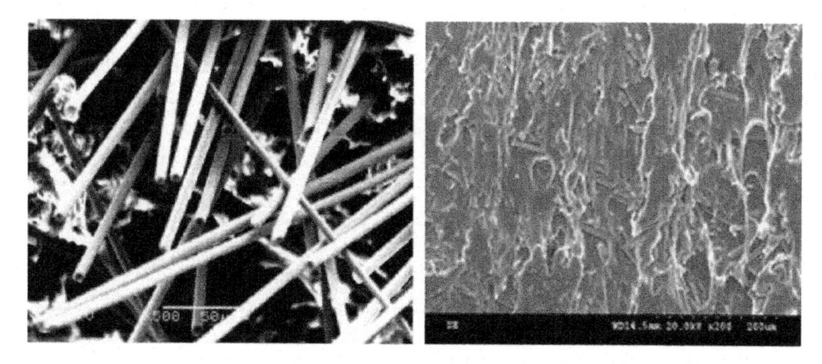

FIGURE 6.40
Morphology of short carbon fibers (the left one) and sectional observation (the right one). (Reproduced from Leng, J.S. et al. *Appl. Phys. Lett.*, 91, 144105, 2007.)

a result, the electrical conductivity of composites filled with only SCF may not be good.

Two microstructure images in Figure 6.40 show the formation of cosupporting conductive networks. Such conductive networks could improve the electrical properties of SMP composites. The dispersion of particles and short fibers in the SMP composites was evaluated through investigating the cross-section surface of samples.

From Figure 6.40, it is found that there is a distinct difference between fibrous and particulate filler in influencing electrical conductivity of the composites. This may be contributed to the inherent tendencies of these two fillers. SCFs may be considered as a rigid long aggregate of carbon, leading to easy formation of continuous conductive networks, and the amount of the networks generally determine the effectiveness of conductivity. The increase of filler content would increase the number of filler particles in the composite interacting with incident conductive networks, although normally the aggregates are separate from each other. Thus, the conductivity of composites including particulate filler is poor, unless, very high filler content is used. However, the particulate filler plays an important role in the SMPCs.

There are two major reasons are being used to account for it. First, its aggregate works as the node of conductive network and makes orientations of short fiber improved; second, there are many particles and their aggregates adsorbed on the surface of SCF, enlarge the surface of the conductive fillers and improve the electrical properties of the polymer which act as one of the worst conductors.

In the past decade, numerous researches have been conducted on the mechanism, shape recovery, and electrical properties of SMP filled with conductive filler and the corresponding applications. The maturity of electro-activated SMP has been attributed to the works done at the early stage. This part regards of the-art-of-status of SMP composites filled with electrical conductive filler and summarizes the growth of electro-activated SMP composites investigation by our group and other academic groups. Electricity as a stimulus enables resistive actuation of SMP composites filled with conductive fillers. In this way, external heating, which is unfavorable for many applications and is used to stimulate conventional SMP matrix, can be avoided. The electrical triggering approach of SMP composites would enlarges their development and technological potential.

6.5 Solution-Driven Shape-Memory Effect

SMP material in response to solution (namely, solvent or mixture), the mechanism behind this phenomenon is solution molecule firstly diffuses into polymer network, and has a plasticizing effect on it, makes flexibility of macromolecule chains increase. These effects lead to the reduction of the transition temperature of materials till shape recovery occurs. This effective actuation approach is firstly carried out in commercial polyurethane SMPs. On the basis of this outcome, actuation of styrene-based SMP driven by its interactive solvent has been demonstrated and achieved. Both these two findings clearly figure out the relationship of the plasticizing effect and the recovery behavior. Therefore, shape recovery of SMP materials driven by a physical swelling effect is predicted and exemplified.

6.5.1 Water-Driven Shape-Memory Effect

Shape recovery actuation of shape-memory polyurethanes are incorporated of two parts, namely, hard segment (with a relative higher transition temperature) and soft segment (with a relative lower transition temperature), has been carried out by inductively lowering T_{trans} [69–72]. When immersed in water, water molecules have hydrogen interaction with the soft segment of the polymer chain. This interaction leads to the significant lowering of the transition temperature of the soft segment. As we know, the switch

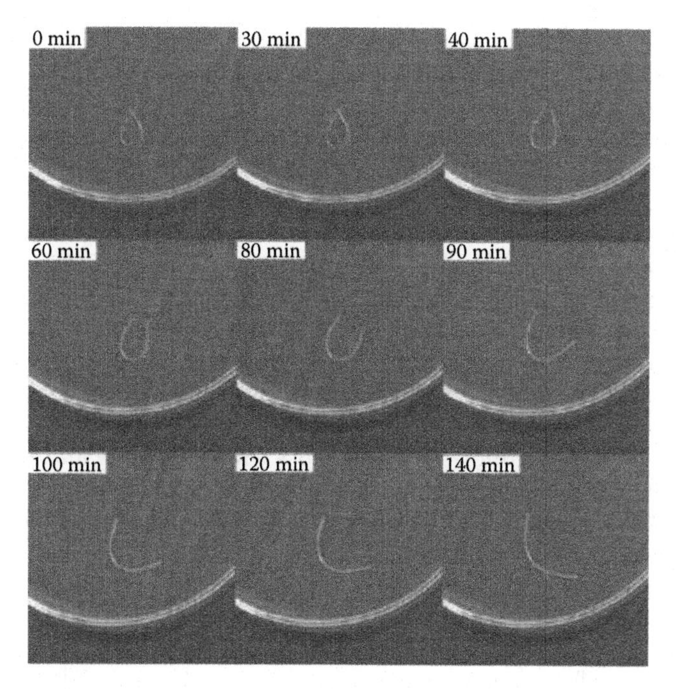

FIGURE 6.41
Water-driven actuation of a shape-memory effect. A shape-memory polyurethane with a circular temporary shape is immersed into water. After 30 min, the recovery of the linear permanent shape begins. (From Huang, W.M. et al., *Appl. Phys. Lett.*, 86, 114105, 2005.)

temperature of the shape recovery for SMP materials is dominated by the transition temperature of the soft segment. Moreover, as the switch temperature of SMP is reduced to or below ambient temperature, shape recovery behavior occurs (Figure 6.41). With immersion time increase, more and more water molecules are involved into interaction with the soft segment. In time-dependent immersion tests, the relationship between water uptake and T_g was studied and analyzed. This results revealed that the absorption process of water molecules obeyed with the diffusion rule, and the hydrogen interaction between polymer and solvent is reversible. And these absorbed water can be completely dried out. Alternatively, another strategy for water-actuated SMPs has been carried out in polyetherurethane polysilesquisiloxane block copolymers [73]. Here, low molecular weight poly(ethylene glycol) (PEG) has been used as the polyether segment. Upon immersion in water, the PEG segment is dissolved, leading to the disappearance of T_m, resulting in recovery of the permanent shape.

6.5.1.1 Glass Transition Temperature after Immersion

In this part, a systemic evaluation between thermal properties and shape recovery behavior is figured out. DSC tests were performed for these samples on a DSC 2920 (TA Instruments) to study the dependence of T_g on immersion time in water. The samples for testing are approximate to 10 mg in weight and the constant heating/cooling rate is 20°C/min. T_g was defined as the median point in the range of glass transition temperature during the heating process.

Figure 6.42 plots the DSC results of samples with various immersion times. This experimental result reveals that T_g decreases remarkably with the increase of immersion time. For further investigation, T_g (together with the onset and end of T_g) was summarized against the immersion time in Figure 6.43. As shown, with an increment in immersion time, T_g decreases gradually (dropped by about 35°C after 240 h immersion). In the immersion process, the decrease is rapid at the beginning. Then it becomes more moderate by prolonging the immersion time. By increasing immersion time to more than 168 h, the samples are close to a saturated state and the dependence of T_g on immersion time becomes minor. Moreover, it also reveals that the temperature range of glass transition is widened from about 10°C to 40°C with the increase of immersion time.

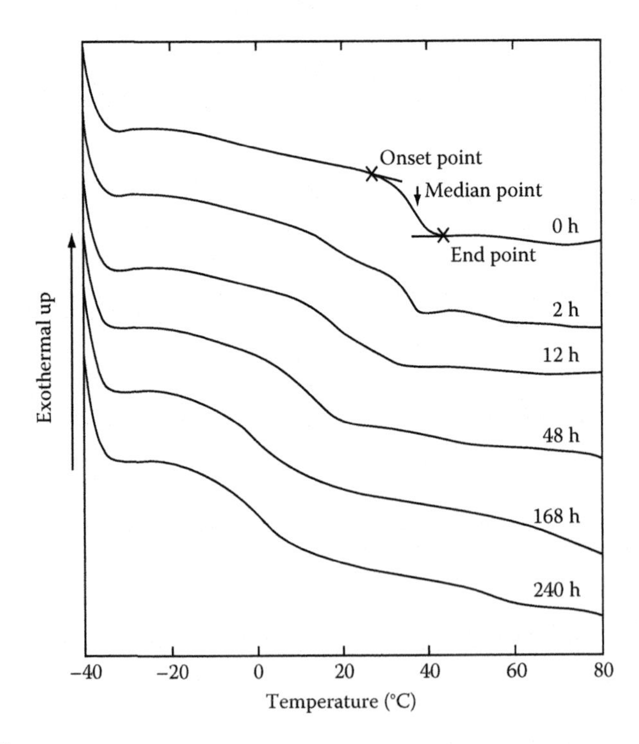

FIGURE 6.42
DSC results after immersion in water for different hours. (From Leng, J. et al., *Appl. Phys. Lett.*, 91, 144105, 2007. With permission.)

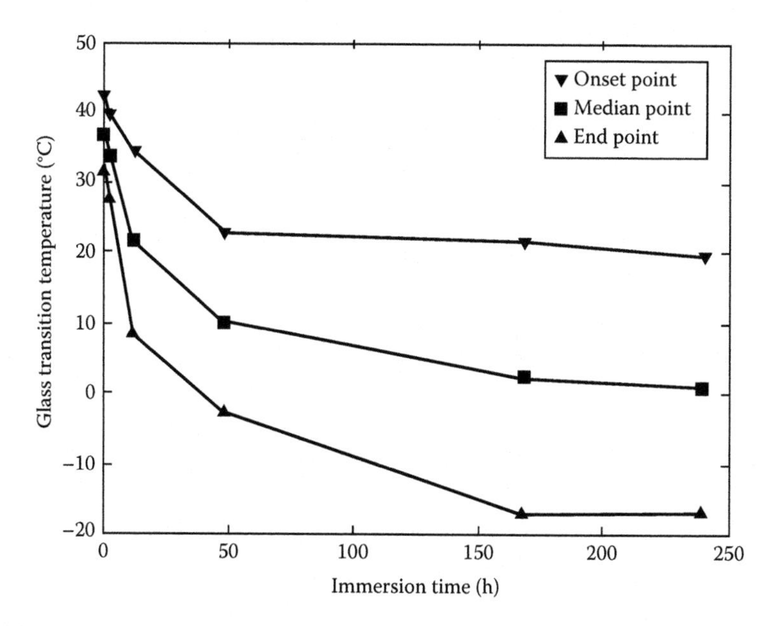

FIGURE 6.43
Changes of T_g with the immersion time. (From Leng, J.S. et al., *J. Appl. Phys.*, 104, 104917, 2008. With permission.)

6.5.1.2 Correlation among Main Factors

The dependence of T_g on the ratio of moisture to SMP in weight percentage for all SMP samples in immersing and heating processes was shown in Figure 6.44. Similar slanted L-shaped curves for T_g against water content were obtained for samples upon different immersion times and then heating to different temperatures. There are two distinct roles in playing on transition temperature, they are free water. The change in T_g of polymer can be obviously divided into two stages. At a lower heating temperature, the transition temperature remains almost as a constant, despite the water content is continuously reduced. This part of water is named as free water, which has no effect on the transition temperature reduction. However beyond a critical temperature, T_g starts to increase linearly with further decrease in water content. This part of water named as bound water, which directly influence the transition temperature reduction. Note that there is a turning point in the L-shaped curve during heating process in Figure 6.44.

The heating temperature of 120°C presented in Figure 6.44 has a physical significance owing to the total absorbed water in the polyurethane SMP which can be divided into two parts: the free water and the bound water. The bound water can be rid of from the polymer only at a relatively higher heating temperature. On this background, the critical heating temperature

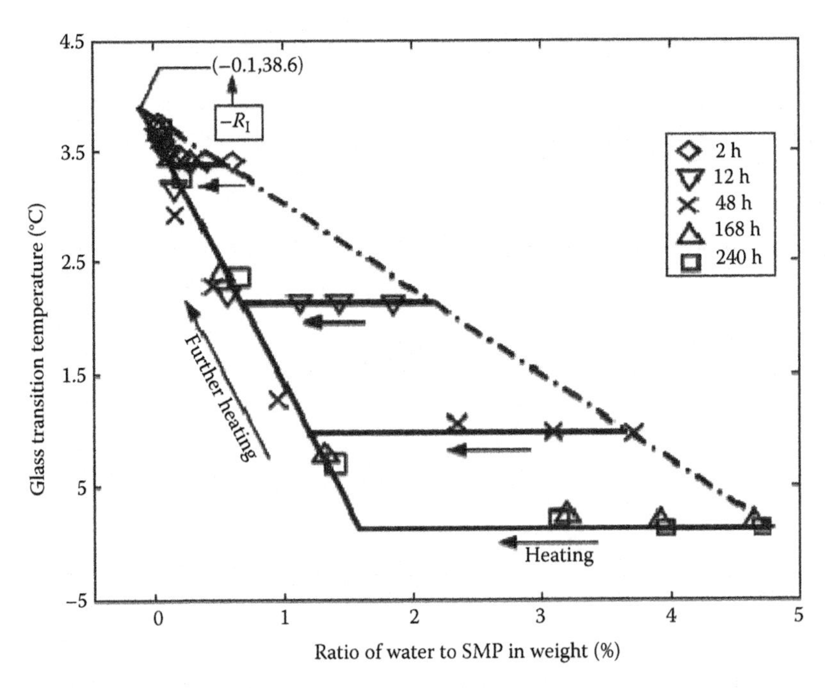

FIGURE 6.44
T_g versus ratio of water to SMP in weight percentage. (From Leng, J. et al., *Appl. Phys. Lett.*, 91, 144105, 2007. With permission.)

is around 120°C and the bound water is removed from the polymer in an approximate linear fashion by increasing the heating temperature. Regarding the free water, all horizontal segments in Figure 6.44 indicate that free water has negligible effect on T_g [74].

As shown in Figure 6.44, the lines for the immersion history and the heating processes have an intersection in the minus zone of the moisture content. This point corresponds to the real dry state of polyurethane SMP, namely, with 240°C being taken as reference, there is still some moisture trapped in the material. The real ratio of moisture to SMP in weight, R, can be obtained as

$$R = (R_t + R_e) / (1 - R_t) \tag{6.5}$$

where

R_t presents the ratio of moisture to SMP composite in weight at the intersection point (positive value, refer to Figure 6.44)

R_e is defined as the measured ratio of moisture to SMP in weight (based on the weight of sample at 240°C)

Now the amounts of free and bound water during immersion process can be further identified by Equation 6.1. Using the two segments in those slanted L-shape curves, the ratios of the free, bound, and total absorbed water in polymer can be obtained as functions of immersion time as plotted in Figure 6.45. It suggests that the absorbed moisture increases dramatically at the beginning of immersion. And experimental results present that more amount of free water than bound water is involved at any instant.

Through FT-IR measurement, it is found that the characteristic groups of N–H and C=O are involved into chemical interaction with water molecules. This hydrogen bonding is directly revealed as the stretching characteristic peak shift to a lower wavenumber. It is implied that a plasticizing effect occurs and transition temperature is subsequently decreased. The infrared bands of the hydrogen bonded N–H and C=O stretchings as a function of immersion time in water are plotted in Figure 6.46. The shifts of infrared bands in hydrogen-bonded N–H and C=O groups is obvious in the first 48 h of immersion and then flatten out. The water amount in the polyurethane SMP also increases with the immersion time increase in a similar manner (Figure 6.45). This phenomenon reveals that hydrogen-bonding interaction between water and polyurethane SMP occurs, which can be illustrated by a model as in Figure 6.47. Some absorbed water molecules in the polyurethane SMP upon immersion provide bridges between the hydrogen

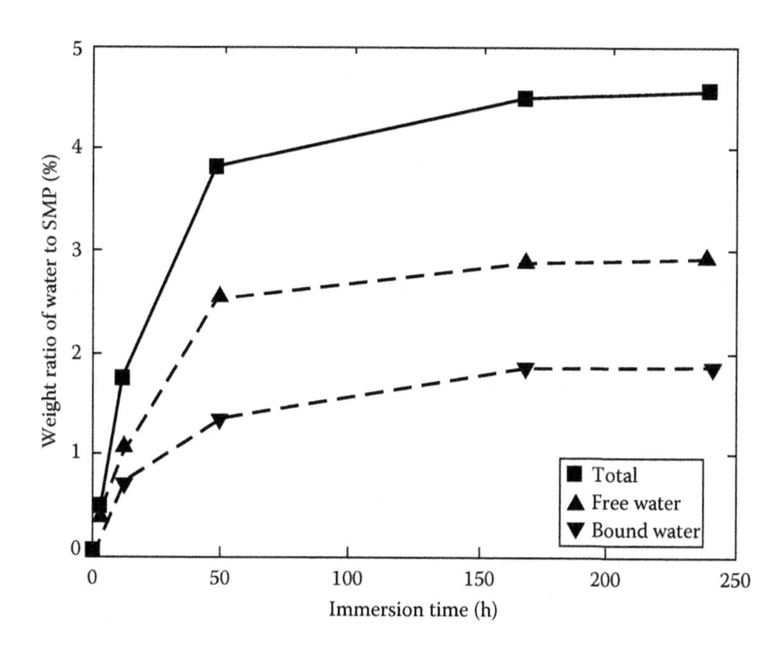

FIGURE 6.45
Ratio of water to SMP in weight versus immersion time. (From Huang, W.M. et al., *Appl. Phys. Lett.*, 86, 114105, 2005. With permission.)

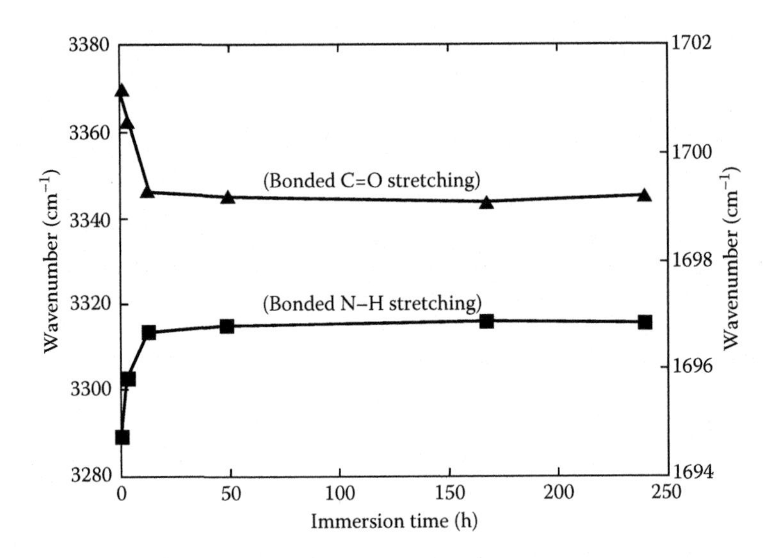

FIGURE 6.46
Infrared band of bonded C=O and N–H stretching versus immersion time in water. (From Yang, B. et al., *Polymer*, 47, 1348, 2006. With permission.)

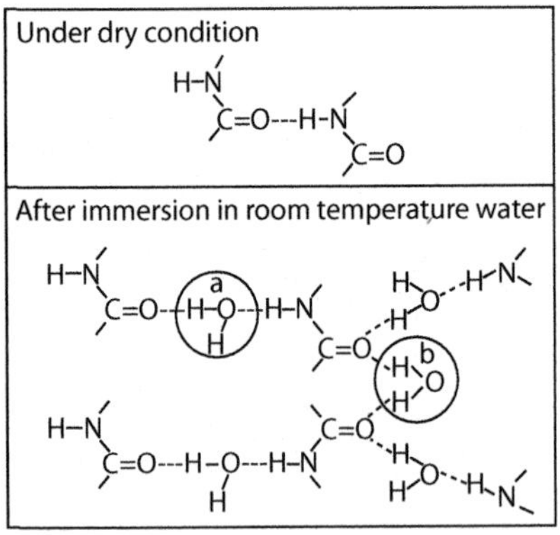

Under dry condition

After immersion in room temperature water

FIGURE 6.47
Effects of water on the hydrogen bonding in the polyurethane SMP. (From Yang, B., Influence of moisture in polyurethane shape-memory polymers and their electrical conductive composites, A thesis of Nanyang Technological University, Singapore, 2005. With permission.)

bonded N–H and C=O groups (site "a" in Figure 6.47). In this model, there is only one possible interaction between water and bonded N=O (site "a"), which directly relates to the hydrogen bonding in SMP. Thus, the change of bonded N–H infrared band induced by water can be interpreted by the hydrogen-bonding effect between water molecules and polymer groups. The loosely bound water directly weakens the hydrogen bonding, which can be evidenced by the shift of infrared band of the hydrogen-bonded

N–H to a higher frequency. Together with the function of water working as plasticizer, T_g is reduced. At the mean time, some absorbed water molecules will form double hydrogen bonds with two already hydrogen-bonded C=O groups (site "b" in Figure 6.47). Due to the hydrogen bonding in site "a," the infrared band of bonded C=O stretching is shifted to a higher frequency while the hydrogen bond in site "b" brings it down to a lower frequency. These two hydrogen bonds may work and counteract at the same time [76–78].

In the heating progress, water molecules absorbed into polymer will be evaporated and dried out. Through weight change versus temperature, it is helpful to test the remaining water content in polymer. From the above-mentioned DSC and thermal stability results, free water and bound water play a distinct different role in transition temperature reduction. Thus, different roles of water on polymer will be qualitatively separated and analyzed. By monitoring the heating process, the free water is first removed, and the weight loss of polymer is significant. With further heating, the weight of polymer is not changed, and it is implied that the bound water is dried out totally. As shown in Figures 6.50 and 6.51, the influenced characteristic groups of N–H and C=O gradually return to their originate state, and the effect of absorbed water is removed completely, leading to the plasticizing effect being eliminated. The polymer regains its physical properties and chemical structures such as transition temperature and hydrogen-bonding structure. Finally, the reversible shape-memory behavior of polyurethane SMP driven by water is exemplified.

A novel approach of polyurethane SMP triggered by water or moisture explores and demonstrates that recovery actuation of thermal responsive shape-memory material can be carried out by its interaction solvent. It opens a generation for SMP materials, breaks in traditional recognitions on them. Based on above-mentioned results and outcomes, the development and application will be extremely extended such as in chemical engineering, materials science and engineering, surgery medicine, and bioengineering.

6.5.2 Solution-Driven Shape-Memory Effect by Chemical Interaction

It is well known that phase transition often accompanies great changes in physical properties of polymeric materials, like a large decrease of modulus, on which solution-responsive SMP is based. The mechanism is that the plastic effect between the polymeric macromolecule and the micromolecule of absorbed solution. There are three major steps carried out to achieve solution-driven SMP. Firstly, the plastic effect enlarges the flexibility of polymeric chains and monomers [59,60]. Secondly, the interaction results in a volume change of the polymer that reduces the elastic modulus of the polymer explained by the continuum theories of rubber elasticity, the Mooney–Rivlin Equation and the volume change refinement theory. Finally, when the temperature of the solution (or the ambient) is lower than

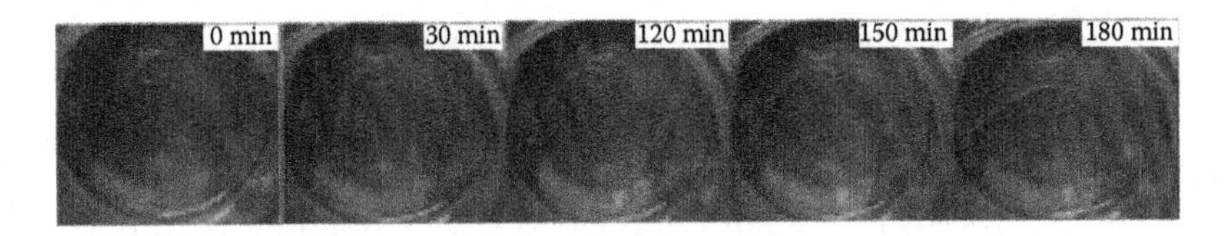

FIGURE 6.48
Shape recovery of a 2.88 mm diameter SMP wire in DMF. The wire was bent into "n"-like shape. (From Yang, B. et al., *Polymer*, 47, 1348, 2006. With permission.)

the glass transition temperature (T_g) of the polymer, the solution firstly diffuses and intenerates the polymer till the T_g is lowered down to the temperature of the solution, followed the plastic effect. Thus, the actuation of the SMPs can be triggered by their solutions. The styrene-based thermosetting SMP can recover from the predeformed "n" shape sequentially after being immersed in *N,N*-dimethylformamide (DMF) has been demonstrated in Figure 6.48.

In order to quantify the content of solution absorption in the samples, thermogravimetric analysis (TGA) was conducted using a TGA 2950 (TA Instruments). The styrene-based SMP sheets were first immersed in DMF at room temperature for different periods. The sheets were then cut into pieces, and heated and heated at the temperature of 80°C to remove the remaining solution molecules on polymer's surface. In subsequence, the immersed samples were tested from 25°C to 450°C at a constant rate of 20°C/min. Figure 6.49 reveals the TGA results of the samples after immersion in the solution. All the samples start to decompose at about 350°C, as evidenced by the rapid decrease in weight. Before decomposition, the samples that have been immersed in DMF lose dramatic amount of weight between 125°C and 250°C. This should be the result of the evaporation of the absorbed solution molecules in the polymer. As expected, weight loss becomes significant with the increase in immersion time. For convenience, the weight fraction at 250°C was chosen as the reference for the comparison of the results in the following study. The weight loss of the polymer with various amounts of solution absorbed at 250°C are 9.47 wt%, 10.51 wt%, 11.68 wt% and 14.32 wt% for 10 min, 30 min, 60 min, and 120 min immersion time, respectively. That is to say, with immersion time increasing, the amount of molecular weight of the absorbed solution increases, and the degree of interaction between the polymeric molecules and the solution molecules is augmented. Thus, the physical properties of the polymer are influenced more obviously.

In the following part, thermomechanical properties of SMPs immersed into DMF solvent is studied and analyzed. Dynamic mechanical analysis (DMA) test was carried out at a constant frequency of 1 Hz on a DMA Q800 (TA Instruments). After different immersion minutes in DMF at room temperature, the rectangular samples were heated at a rate of 5°C/min.

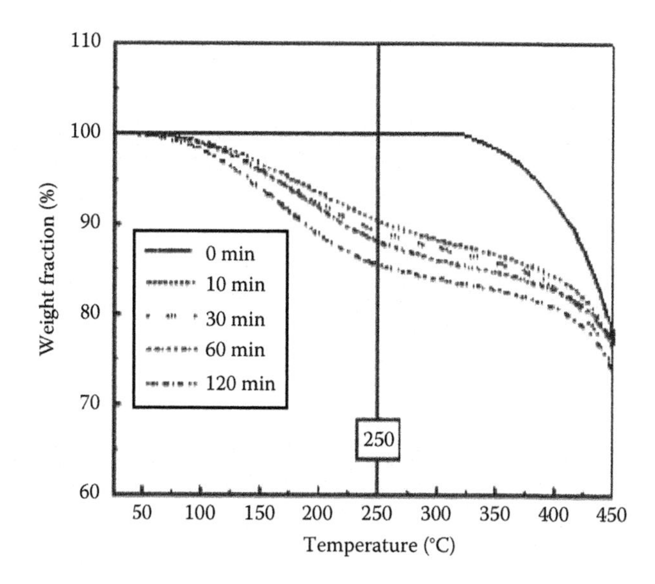

FIGURE 6.49
TGA results of pure SMP after different immersion times in DMF. (From Yang, B., Influence of moisture in polyurethane shape-memory polymers and their electrical conductive composites, A thesis of Nanyang Technological University, Singapore, 2005. With permission.)

Typical results of the storage modulus, the loss modulus, and the tangent delta as a function of temperature are plotted in Figure 6.50. The storage modulus is the modulus of the elastic portion of the material; it reveals the transition velocity from the glass state to the rubber state. Figure 6.50a shows that, upon gradually absorbing solution, the storage modulus decreases from 1475 MPa (0 min) to 604 MPa (after 120 min immersion) at the room temperature; and the transition temperature determined by the storage modulus curves reveals that it decreases gradually with the immersion time increasing. At last, the transition temperature determined from the storage modulus curves decreases with increasing immersion time.

Tangent delta is defined as the ratio of the loss modulus over the storage modulus, and the peaks of the tangent delta curve is an alternative glass transition temperature (T_g). T_g defined in such a way is determined as 84.29°C, 74.04°C, 65.30°C, 60.72°C, and 55.05°C for immersion time 0, 10, 30, 60, and 120 min, respectively. As shown in Figure 6.50b, upon gradually absorbing the solution, the maximum peaks of tangent delta curves shifted to a relative low temperature. It reveals that the tangent delta curves reach their maximum at lower temperatures with increasing immersion time.

The DMA results show that the thermomechanical properties of SMP are reduced by its interactive solution for chemical interaction. That's because the bonded hydrogen between the polymer and the solution results in improving the flexibility of the polymeric molecule, which plays a negative role on

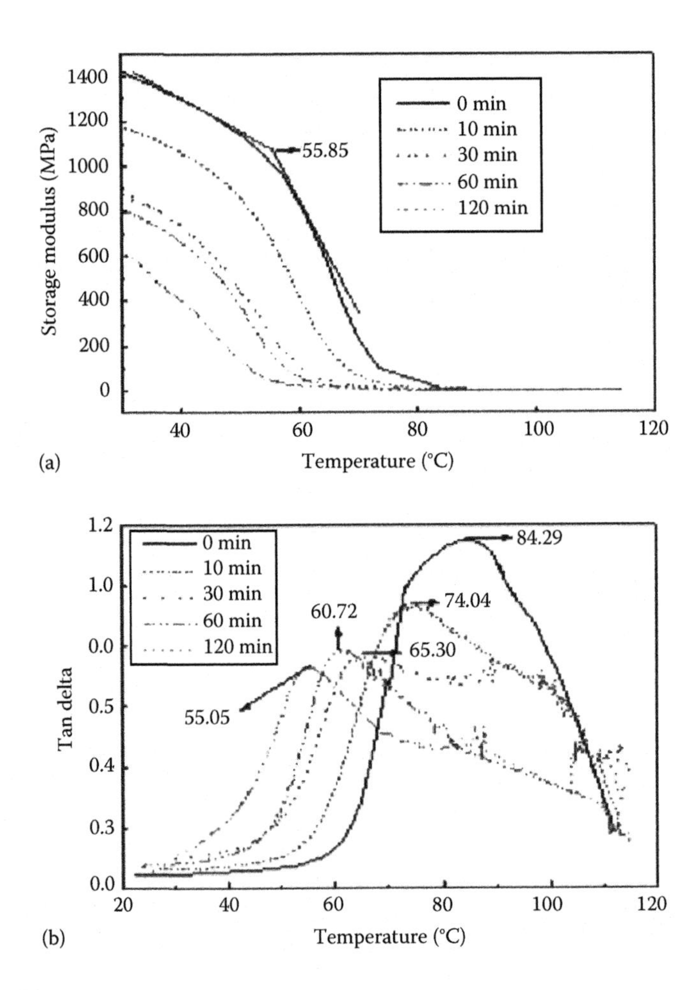

FIGURE 6.50
Storage modulus, loss modulus, and tangent delta curves of samples with different immersion times at the oscillation frequency of 1 Hz. (From Yang, B., Influence of moisture in polyurethane shape-memory polymers and their electrical conductive composites, A thesis of Nanyang Technological University, Singapore, 2005. With permission.)

the polymeric mechanical properties. All the above-mentioned can be cited from the physics of polymer. From the DMA results, it is found the chemical interaction between SMP and solvent makes transition temperature of polymer is lowered. Then, Differential scanning calorimeter (DSC) test was carried out to further support above mentioned phenomenon.

SMP samples, with 10 ~ 15 mg in weight, were performed on DSC test (DSC 2920, TA Instruments), at a heating rate of 20°C/min.

Figure 6.51 shows the DSC results of the samples with various immersion history. The T_g of each sample was determined as the peak point in the glass transition region as marked out on the scanning curves. The T_gs defined in

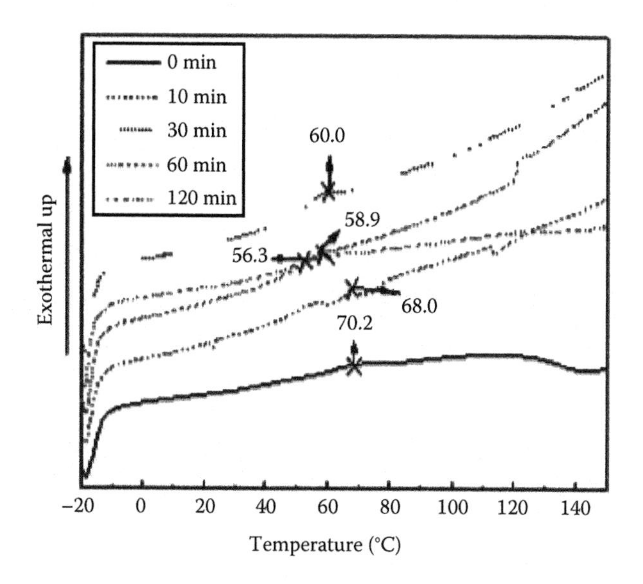

FIGURE 6.51
DSC results after immersion in toluene for different times. (From Yang, B. et al., *Smart Mater. Struct.*, 13, 191, 2004. With permission.)

such a manner are 70.2°C, 68.0°C, 60.0°C, 58.9°C, and 56.3°C. As can be seen, with increment in immersion time the T_g decreases significantly. Thus the DSC result is further testified the thermal, and thermomechanical properties of SMPs are extremely reduced by the DMF solvent.

Now the amounts of swelling the volume ratio during immersion can be further identified according to the Figure 6.52. It reveals that the micromolecular solution absorption increases dramatically in the first 60 min of immersion. With the swelling ratio increasing, the T_g drops by about 56.3°C after 120 min immersion.

To find out the mechanism behind the interaction between polymer and solvent, FTIR spectroscopy was used to study the interaction between the SMP and the DMF solvent, and separate from other factors. The full FTIR spectra of the SMPs (at the temperature of 20°C) after different immersion minutes in DMF are presented in Figure 6.53. The peak of bonded C=O that is relevant to this study has been marked out and identified. The infrared band of free C=O stretching at 1724 cm^{-1} shifts to that of the bonded one at 1740 cm^{-1}. As such, the hydrogen bonding in the SMP is apparent.

Figure 6.54 shows that after immersion the infrared band of the hydrogen-bonded C=O stretching shifts slightly to a lower frequency. Furthermore, with the increase of immersion time the infrared band intensity of the bonded C=O stretching becomes more striking compared with that of free C=O stretching, which indicates that a longer immersion time triggers more C=O groups in hydrogen bonding.

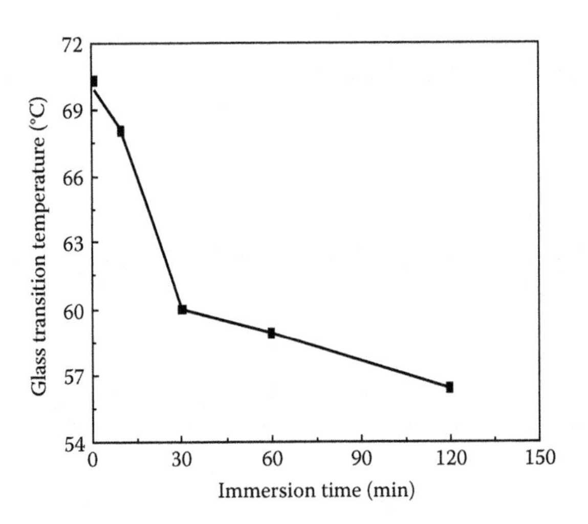

FIGURE 6.52
Glass transition temperatures versus immersion time. (From Yang, B. et al., *Polymer*, 47, 1348, 2006. With permission.)

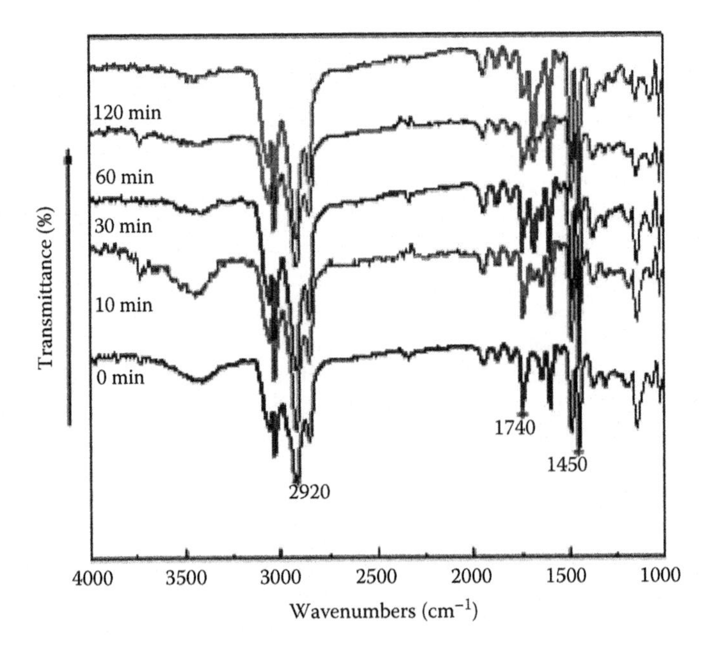

FIGURE 6.53
FTIR spectra of styrene-based SMP after different immersion minutes. (From Yang, B. et al., *Polymer*, 47, 1348, 2006. With permission.)

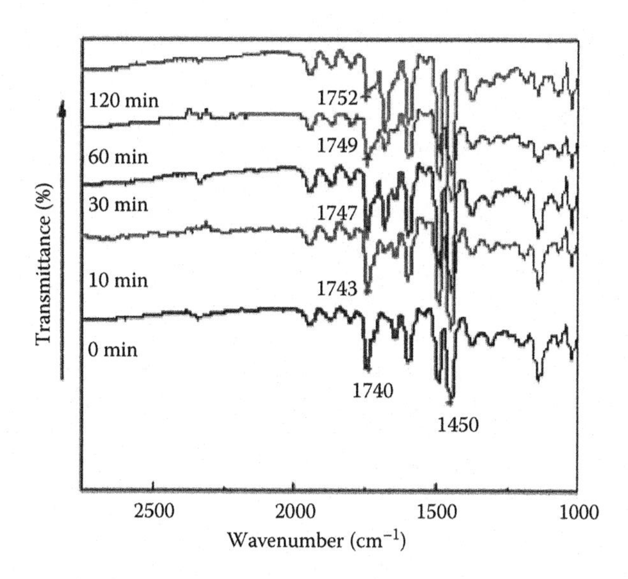

FIGURE 6.54
FTIR spectra of styrene-based SMP after different immersion hours for C=O stretching region. (From Yang, B. et al., *Polymer*, 47, 1348, 2006. With permission.)

Figure 6.54 presents the FTIR spectra in the C=O stretching regions in samples after different immersion hours, respectively. After immersion the infrared band of the hydrogen-bonded C=O stretching shifts slightly to a higher frequency. Moreover, with the immersion time increasing, the intensity of the bonded C=O stretching becomes more unapparent, which indicates that a longer immersion time triggers more C=O groups involved in hydrogen bonding.

Chemical interaction-triggered SMP provides an alternative approach for activation, and so heating approach becomes unnecessary. The actuation can be triggered upon immersing the material into the solution. That is to say, instead of heating the material above its T_g, shape recovery can be achieved by means of reducing the T_g of the material itself upon immersing the SMP into a solution that is the same as water-driven PU SMP in principle. This outcome enlarges the water-driven polyurethane SMP result, to explore a new feature for SMP materials triggered by another approach, which must extend the applications in medicine and bio-engineering.

6.5.3 Solution-Driven Shape-Memory Effect by Physical Swelling Effect

Polymeric materials have a distinct feature that can be swollen by its interactive solvent or solution. In this section, work has been carried out to make SMP to be actuated by physical swelling effect. Based on the free-volume theory, rubber elasticity theory, and Mooney–Rivlin Equation, it is theoretically predicted that the feasibility of SMP is activated by swelling effect. The mechanism is that the solvent acts as plasticizer to reduce the glass transition temperature (T_g) and melting temperature (T_m) of the polymers due to the easier movement after volume swelling effect. In addition to this physical action, the intermolecular interactions among the chains are weakened, because interactions are hindered at the points where the plasticizer is located.

It is found that this solvent had an intensive swelling effect on the polymer. Then, the motion ability (usually be defined as flexibility) of the macromolecule chains increased after being swollen. Subsequently, the increased flexibility led to the decrease in transition temperature, and resulted in the shape recovery occurring at a low temperature. All the above-mentioned phenomena obey the continuum theories of rubber elasticity, the Mooney–Rivlin Equation, and the volume change refinement theory [74–77].

6.5.3.1 Theoretical Basis and Analysis

A suitable approximation for analyzing the glass transition behavior of polymer materials concerns the free volume [50,78].

The free-volume hole hypothesis supposed by Cohen and Turnbull states that the volume of the polymer includes occupied volume, which is occupied by the macromolecule itself, and free-volume hole, which is distributed in all the polymers without occupied by anything. The configuration of the macromolecular chain can change through turning and moving due to the existence of free-volume holes. When the polymer is stored at the temperature lower than the phase transition temperature, the polymer chains are frozen, the micro-Brownian motion of macromolecules is limited, and the free-volume hole size is too small to free the gas or vapor, and so low gas/vapor permeability is obtained at low temperature. In the temperature above the phase transition temperature, the micro-Brownian motion enhances and the size of free-volume hole increases. Therefore, the vapor/liquid permeability of the polymers increases dramatically [77].

The influenced factors such as chemical structure, molecular weight, crosslinking, and plasticizer type in the glass transition of polymers can be related to the changes that they provoke on the free-volume fraction, which reaches a critical value at the glass transition temperature. The factors affecting the glass transition can be classified into two types: inherent molecular structure and external factors.

The T_g of polymers decreases when a liquid such as a solvent or solution is mixed with them. Solvent molecules are added to polymers to make them softer and more flexible at ambient temperature, are known as plasticizers. These substances are generally low molecular weight organic compounds, weakly polar, with lower T_g and higher boiling points to prevent evaporation losses. These compounds behave to weaken the intermolecular interactions by means of diffusion action. Plasticizers reduce the transition temperature of polymers, making them softer and more flexible, by distributing their molecules through the polymer and separating the chains. In addition to this physical action, the intermolecular interactions among the chains are weakened, because interactions are hindered at the points where the plasticizer is located. Therefore, the swelling effect also is originated from plastic effect between polymer and its interactive solvent or solution.

One way to deform the polymeric network is to put it in contact with an interactive solvent. In this case, the molecules of the solvent are absorbed in the network, giving rise to a phenomenon known as volume swelling. Swelling of a network by the action of a solvent constitutes a three-dimensional deformation, and the network absorbs the solvent until the swelling equilibrium is reached. The interaction between the polymer and the solvent molecules obeys the diffusion theorem. There are two diffused types: when the temperature of the solution is higher than the T_g, the diffusion of the solution molecules obeys Fickian first law; when the temperature is lower than the T_g, the diffusion obeys the Fickian second law. In the condition of the former, the solution first intenerates the polymer to make T_g of the polymer lowered down to the ambient temperature or more lower than it, followed with swelling effect. All mentioned phenomena can be explained by polymer solution theory and rubber elasticity theory. And the swelling behavior is expressed as rubber elastic equation and its amendment.

Based on the theory of rubber elasticity, the force per unit area is expressed as:

$$\sigma = nRT\frac{\overline{r_i^2}}{\overline{r_0^2}}\left(\alpha - \frac{1}{\alpha^2}\right) \tag{6.6}$$

In this equation, L_0 is increased to L, and $\alpha = L/L_0$ is the elongation ratio, and RT is the gas constant multiplying with absolute temperature. The quantity $\overline{r_i^2}$ represents the isotropic, unstrained, end-to-end distance of the chain. The two quantities $\overline{r_i^2}$ and $\overline{r_0^2}$ represent the same chain in the network and the uncross-linked states, respectively. And under many circumstances, the quantity $\overline{r_i^2}/\overline{r_0^2}$ approximately equals unity. The quantity n represents that the network is made up of n chains per unit volume.

Several other relationships may be derived immediately, since Young's modulus is a function of state:

$$E = L\left(\frac{\partial \sigma}{\partial L}\right)_{T,V}. \tag{6.7}$$

Therefore,

$$E = nRT\frac{\overline{r_i^2}}{\overline{r_0^2}}\left(\alpha + \frac{2}{\alpha^2}\right) \cong 3n\frac{\overline{r_i^2}}{\overline{r_0^2}}RT. \tag{6.8}$$

If a polymer network is swollen with a "solvent" (it does not dissolve), the detailed effect has two parts:

1. Effect on the $\overline{r_i^2}/\overline{r_0^2}$. The quantity $\overline{r_i^2}$ increases with the volume V to the two-thirds power, while $\overline{r_0^2}$ remains constant. The work done on the polymer network is

$$\left(\frac{\overline{r_i^2}}{\overline{r_0^2}}\right) = \left(\frac{V}{V_0}\right)^{2/3}\left(\frac{\overline{r_i^2}}{\overline{r_0^2}}\right)^* = \frac{1}{v_2^{2/3}}\left(\frac{\overline{r_i^2}}{\overline{r_0^2}}\right)^* \tag{6.9}$$

where
"*" presents the swollen state
v_2 is the volume ratio of the polymer in the no swollen and swollen stages

Of course, v_2 is less than unity, as it is commonly experienced.

2. Effect on the number of network chain segments concentration, n. The quantity n gives the expression:

$$\left(\frac{V}{V_0}\right)n = v_2 n = n_s \tag{6.10}$$

where
n is the chain segment concentration in the swollen state
By substituting Equations 6.9 and 6.10 into Equation 6.8 we obtain

$$E = 3n \cdot v_2^{1/3}\frac{\overline{r_i^2}}{\overline{r_0^2}}RT. \tag{6.11}$$

The elasticity modulus, defined as the force per unit original cross-section, is decreased by $v_2^{1/3}$. Since the number of chains occupying a given volume has decreased. Thus, the modulus of the swollen polymer must be lower than that of the no swollen polymer due to that the volume ratio value is always bigger than 1. So the solution-responsive SMP is feasible based on rubber elasticity theory. Furthermore, the recovery behavior of SMP driven by swelling effect must obey with the above mentioned rubber elastic equation.

6.5.3.2 Experimental Tests and Demonstration

Successful swelling effect-induced styrene-based SMP (i.e., filled with thermo-responsive dye, the color altering from purple to pink occurs at the transition temperature of 45°C) was conducted in Toluene solvent. Shape recovery activation driven by swelling effect had been shown in Figure 6.55, where accompanying shape recovery, volume change occurred. In all cases, the temporary shape was programmed by conventional methods for thermo-responsive SMPs. The permanent straight SMP recovered from the predeformed "n" shape in sequence after immersion in solvent at a temperature similar to the human body's temperature of 35°C. With immersion time increase, shape recovered step by step along with volume swelling ratio was also enhanced. After 60 min, SMP almost fully regained its original shape. Finally, through the comparison of length change in non-absorption and 60 min immersion samples, it is found that swelling effect occurred in the process of shape recovery.

In addition to shape recovery, the volume swelling of SMP accompanies simultaneously. A different strategy for solution-triggered actuation has been realized in styrene-based SMP. However, the swelling effect is applicable for almost all polymeric materials. Swelling effect actuation must be suitable for all the SMPs and their relational materials.

6.5.3.2.1 Dynamic Mechanical Analysis

From the theoretical analysis, it is revealed that the swelling effect makes elastic modulus of polymeric materials reduced. Therefore, in the following part, thermomechanical properties of SMP was studied and exemplified. The dynamic mechanical properties of the SMP with various immersion times

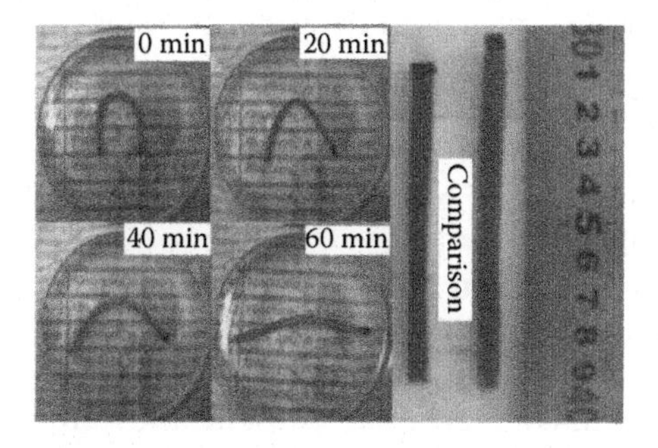

FIGURE 6.55
Series photos of shape recovery of rectangle SMP sample (bent into "n"-like shape) in toluene; comparison of length change between unswollen and swollen samples. (From Yang, B. et al., *Scripta Mater.*, 53, 105, 2005. With permission.)

have been obtained using dynamic mechanical analysis (DMA NETZSCH 204F1) instrument in a stretch mode over a range of 0°C to 100°C. The rectangular specimens of each material were prepared with dimension of $16.5\times13.0\times3.5\,mm^3$. The storage modulus and tangent delta curves have been recorded as a function of temperature. The storage modulus is the modulus of the elastic portion of the materials that reveal the transition velocity from glassy state to rubber state. Tangent delta is defined as the ratio of the loss modulus over the storage modulus, and the peaks of the curves indicate the transition temperature.

DMA tests were carried out at a heating rate of 5.0°C/min and a constant frequency of 1 Hz. The effect of the immersion history of SMPs with immersion time of 0, 20, 40, 60, 80, and 100 min is captured in Figure 6.56. The storage modulus in the glassy phase decreases by a factor of the immersion time increase from 1204 to 434 MPa. In addition, there is a shift to lower T_g from 55.77°C (0 minute immersion time) to 35.82°C after 100 min immersion, and also a decrease in the tangent delta peak height, indicating an increase in the chain movability. Figure 6.56a reveals that the modulus of the samples decreases dramatically with increasing immersion time. And the tangent delta curves shown in Figure 6.56b display the interrelation between T_g and the immersion time.

On the basis of the experimental results, it is demonstrable that the SME can be induced by swelling effect, indeed through reducing the elastic modulus (owing to swelling effect arising from the interaction between the polymer and the solvent molecules), which has a direct effect on the transition temperature.

6.5.3.2.2 FTIR Studies

At this point, it has been demonstrated that in the progress T_g depression accompanies physical swelling effect from the macroscopic angle. However, the main challenge in the development of the swelling effect actuation at the molecular level behind this phenomenon remains unknown.

The FTIR spectroscopy was used to study the interaction of Toluene molecule with the styrene-based polymer chains and identify other possible factors. The samples used for FTIR test were thin polyurethane SMP sheets with a thickness of 1.0 mm. The FTIR spectra were collected by averaging 70 scans at a resolution of 4^{-1} cm in a reflection mode from a FTIR spectrometer (AVATAR 360, Nicolet).

The full FTIR spectra of the SMPs (at the temperature of 20°C) after different immersion minutes in Toluene solvent are presented in Figure 6.57. Some peaks of this spectrum that are relevant to this study have been marked out and identified. Figure 6.57 presents the FTIR spectra of the O–H (3430 cm^{-1}) and C=O (1740 cm^{-1}) stretching regions in the samples after different immersion times, respectively. Figure 6.57a shows that the infrared band intensity of the bonded O–H stretching have not been changed compared with that of the sample without immersion. Even if the immersion time is

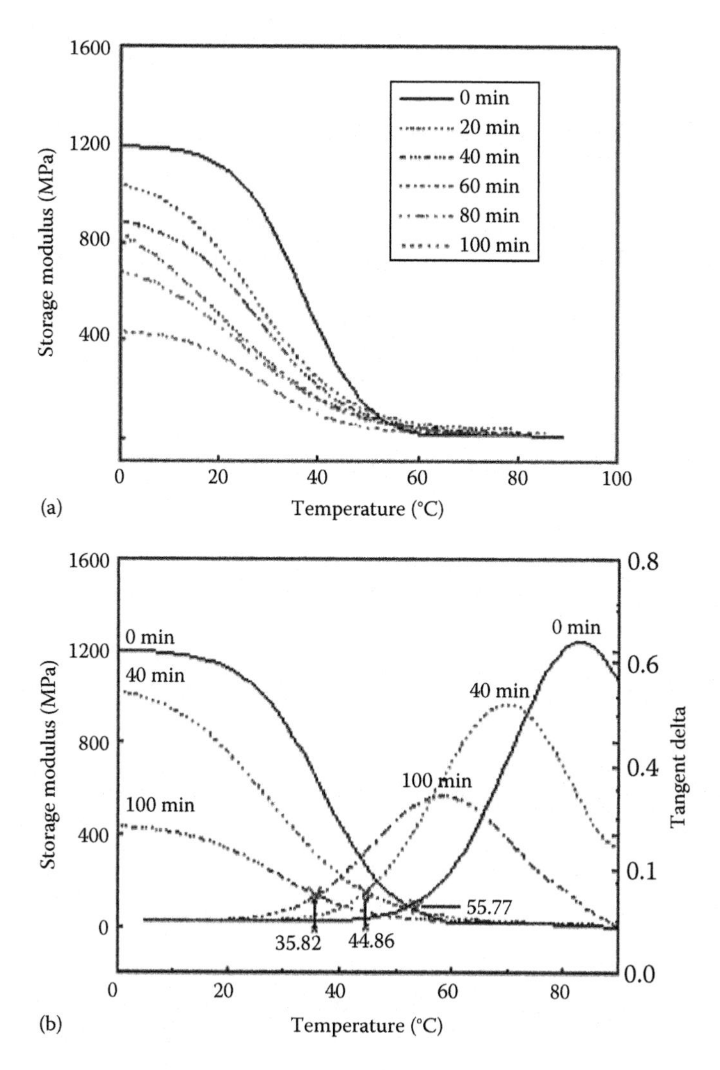

FIGURE 6.56
Storage modulus (a) and tangent delta curves (b) of samples with different immersion times at the oscillation frequency of 1Hz. (From Yang, B. et al., *Scripta Mater.*, 53, 105, 2005. With permission.)

increased, there are no O–H groups involved in the interaction with solvent molecule, leading to the characteristic peak of O–H groups not shifted any after more than 100min immersion. Figure 6.57b shows the infrared band of the C=O stretching region of the samples with different immersion times. Furthermore, with the increase of immersion time the location of infrared band of the bonded C=O stretching shows no shift compared with that of the sample 0min immersed, which indicates that there are no C=O groups involved in the chemical interaction with solvent molecule.

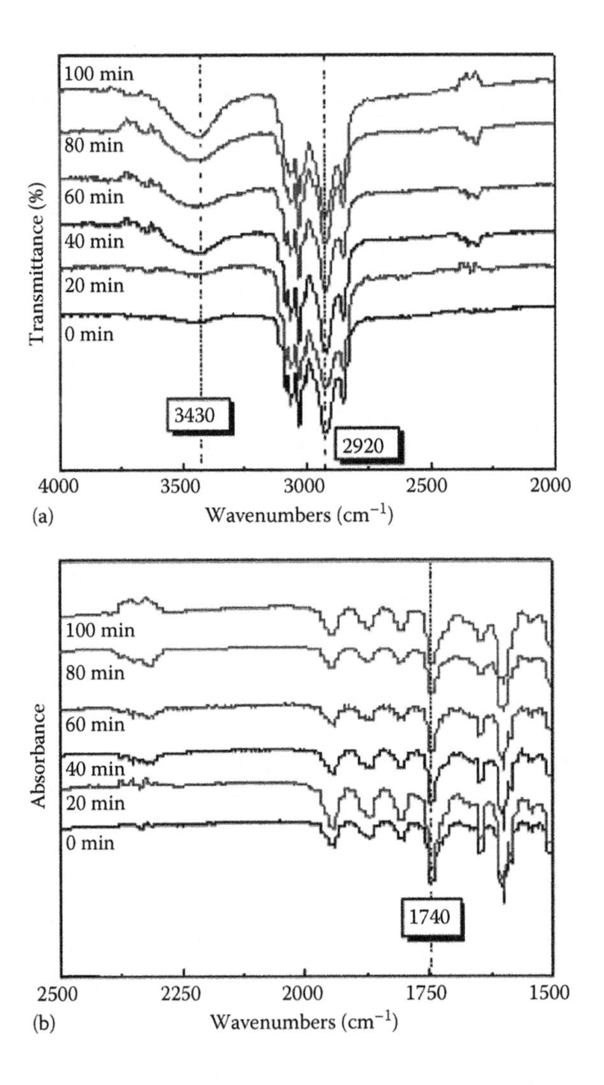

FIGURE 6.57
FTIR spectra of styrene-based SMP after different immersion minutes. (a) O–H stretching region, (b) C=O stretching region. (From Yang, B., Influence of moisture in polyurethane shape-memory polymers and their electrical conductive composites, A thesis of Nanyang Technological University, Singapore, 2005. With permission.)

6.5.3.2.3 Correlations among Volume Swelling Ratio, Elastic Modulus, and T_g

In order to quantify the content of the solvent absorption in the samples, the time-dependent volume study was conducted to observe the relationship between the volume swelling ratio and the immersion time. Figure 6.58a shows the results of the samples after immersion in Toluene for upto 100 min. The solvent molecules were absorbed into the samples, as evidenced by the

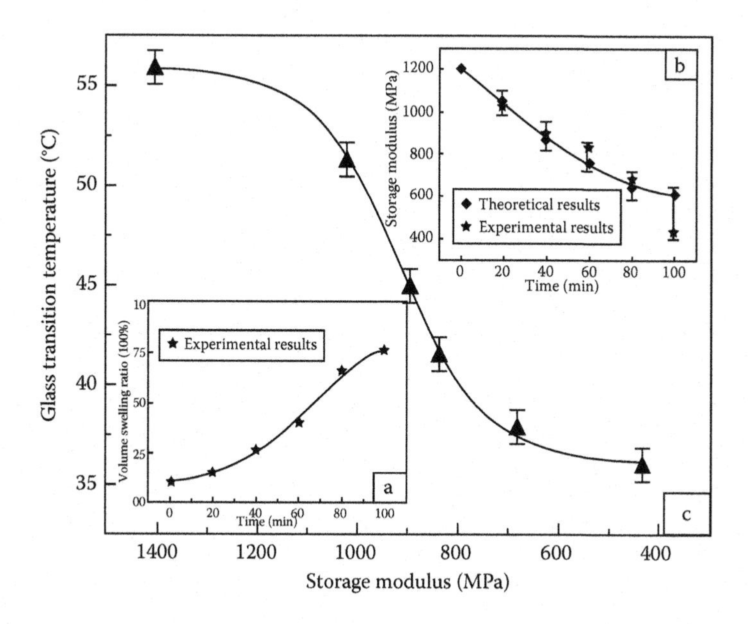

FIGURE 6.58
Correlations among immersion time, volume swelling ratio, elastic modulus, and glass transition temperature. (Modified from Yang, B. et al., *Polymer*, 47, 1348, 2006.)

increase in the volume. As expected, the volume swelling ratio became significant with the increase in immersion time.

In the view of the above considerations, a measure of the storage modulus at the room temperature (~20°C) is chosen between the rubber elasticity theory profile and DMA test profile. A comparison between the theory and test data is presented in Figure 6.58b. From the storage modulus curves, it is shown that the theoretical result is quite in agreement with the experimental result. However, it can also be seen in Figure 6.55b that the analytical result gives a slightly stiffer behavior than that of the test. Overall, the theoretical depiction of the swelling effect that is being used to predict the feasibility of the actuation for the SMP is of great value, especially for the initiation of the test.

Utilizing the data obtained from the previous DMA test, namely the storage modulus curves and the Tangent delta curves, it is can be used to determine the relationship between T_g and elastic modulus. The evolution of the T_g indicating the relationship between the transition temperature and the elastic modulus is shown in Figure 6.58c. It reveals that the depression of T_g is found increasingly dependent on the progressive storage modulus. Consequently, the T_g depression should be attributed to the plasticizing effect of the solvent on T_g. This result provides the evidence to support our previous statement that the solvent molecules absorption affects the mobility of the polymer chains and increases the chain flexibility.

Now we can qualitatively identify the role played by swelling effect in influencing the transition temperature. Swelling effect occurs due to the interaction between the macromolecules and the solvent molecules, leading to the increase in the free volume of polymeric chains (namely the flexibility of polymer chains increasing), and resulting in T_g decreasing. All the above-mentioned investigation can be used to account for the shape recovery induced by swelling effect.

Polyurethane SMP's response to water or, to the extension, SMP's response to solution (namely, solvent or mixture) is due to the plasticizing effect of the solution molecule on the polymeric materials which then increases the flexibility of macromolecule chains. These two effects reduce the transition temperature of materials until shape recovery occurs. These studies are based on polymer physics and chemical interaction explaining the mechanism. It is summarized that this approach makes tremendous process in the actuation of SMP. Many extension results and achievements are based on it.

6.6 Summary and Outlook

As a smart material, both the SMPs and their composites have shape-memory effect. In addition to thermal actuation, SMP composites filled with functional fillers can also be actuated by other external stimuli. Till now, there has been a maturity in SMP composites for which the shape recovery can be carried out by electrical-resistive heating and the magnetic field. These novel actuation approaches play a critical role in development and applications. The SMP materials are being, or will be, served in many important fields like aerospace, automobile, textiles, and medical science. Actuations of their corresponding composites have been carried out under controllable and programmable conditions. Finally, these SMPs composites are filled with functional fillers for multifunction. They have the capability of shape-memory effect and respond in various ways to stimuli, besides Joule heating.

Furthermore, thermo-responsive SMPs are always defined to be triggered by direct or indirect heating. It is declared that traditional thermo-response SMPs can also be driven by corresponding solvent. SMP in response to water or solution (namely, solvent or mixture) is due to the plasticizing effect of the solution molecule on polymer materials. That is because the solvent increases the flexibility of macromolecule chains and reduces the transition temperature. These studies are based on polymer physics and chemical interactions explaining the mechanism. It is summarized that this approach makes tremendous process in the actuation of SMP. Many extensive results and achievements are based on it.

References

1. Vaia, R. 2005. Nanocomposites—Remote-controlled actuators. *Nature Materials* 4:429–430.
2. Razzaq, M. Y., Anhalt, M., Frormann, L., and Weidenfeller, B. 2007. Mechanical spectroscopy of magnetite filled polyurethane shape memory polymers. *Materials Science and Engineering A—Structural Materials Properties Microstructure and Processing* 471:57–62.
3. Maitland, D. J., Metzger, M. F., Schumann, D., Lee, A., and Wilson, T. S. 2002. Photothermal properties of shape memory polymer micro-actuators for treating stroke. *Lasers in Surgery and Medicine* 30:1–11.
4. Finkelmann, H., Nishikawa, E., Pereira, G. G., and Werner, M. 2001. A new opto-mechanical effect in solids. *Physics Review Letters* 87:15501.
5. Camacho-Lopez, M., Finkelmann, H., Muhoray, P. P., and Shelley, M. 2004. Fast liquid-crystal elastomer swims into the dark. *Nature Materials* 3:307–310.
6. Lendlein, A., Jiang, H., Junger, O., and Langer, R. 2005. Light-induced shape-memory polymers. *Nature (London)* 434:879–882.
7. Akabori, S. and Tsichiya, S. 1990. The preparation of crown ethers containing dicinnamoyl groups and their complexing abilities. *Bulletin of the Chemical Society of Japan* 63:1623–1628.
8. Tanka, H. and Hunda, K. 1977. Photoreversible reactions of polymers containing cinnamylideneacetate derivatives and the model compounds. *Journal of Polymer Science Part A: Polymer Chemistry* 15:2685–2689.
9. Choi, N. Y. and Lendlein, A. 2007. Degradable shape-memory polymer networks from oligo[(L-lactide)-ran-glycolide]dimethacrylates. *Soft Matter* 3:901–909.
10. El Feninat, F., Laroche, G., Fiset, M., and Mantovani, D. 2002. Shape memory materials for biomedical applications. *Advanced Engineering Materials* 4:91–104.
11. Sankaran, V., Walsh, J. T., and Maitland, D. J. 2002. Comparative study of polarized light propagation in biologic tissues. *Journal of Biomedical Optics* 7:300–306.
12. Small, W., Wilson, T. S., Benett, W. J., Loge, J. M., and Maitland, D. J. 2005. Laser-activated shape memory polymer intravascular thrombectomy device. *Optical Express* 13:8204–8213.
13. Lendlein, A. and Kelch, S. 2005. Degradable, multifunctional polymeric biomaterials with shape-memory. In *Functionally Graded Materials VIII*, Van der Biest, O. et al. (eds.), Technology Transition Publications, Leuven, Belgium, vol. 219, pp. 492–493.
14. Bellin, I., Kelch, S., Langer, R., and Lendlein, A. 2006. Polymeric triple-shape materials. *Proceedings of the National Academy of Sciences of the United States of America* 103:18043–18047.
15. Lendlein, A. and Kelch, S. 2002. Shape-memory polymers. *Angewandte Chemie—International Edition* 41:2034–2057.
16. Kim, B. K., Lee, S. Y., and Xu, M. 1996. Polyurethanes having shape memory effect. *Polymer* 37:5781–5793.
17. Li, F. K., Hou, J. N., Zhu, W., Zhang, X., Xu, M., Luo, X. L., Ma, D. Z., and Kim, B. K. 1996. Crystallinity and morphology of segmented polyurethanes with different soft-segment length. *Journal of Applied Polymer Science* 62:631–638.

18. Ma, Z. L., Zhao, W. G., Liu, Y. F., and Shi, J. R. 1997. Intumescent polyurethane coatings with reduced flammability based on spirocyclic phosphate-containing polyols. *Journal of Applied Polymer Science* 66:471–475.
19. Jeong, H. M., Lee, S. Y., and Kim, B. K. 2000. Shape memory polyurethane containing amorphous reversible phase. *Journal of Materials Science* 35:1579–1583.
20. Jeong, H. M., Kim, B. K., and Choi, Y. J. 2000. Synthesis and properties of thermotropic liquid crystalline polyurethane elastomers. *Polymer* 41:1849–1855.
21. Sun, X. and Isayev, A. L. 2007. Ultrasound devulcanization: Comparison of synthetic isoprene and natural rubbers. *Journal of Materials Science* 42:7520–7529.
22. Cho, J. W. and Jung, H. 1997. Electrically conducting high-strength aramid composite fibers prepared by vapor-phase polymerization of pyrrole. *Journal of Materials Science* 32:5371–5376.
23. Booth, C. J., Kindinger, M., McKenzie, H. R., Handcock, J., Bray, A. V., and Beall, G. W. 2006. Copolyterephthalates containing tetramethylcyclobutane with impact and ballistic properties greater than bisphenol A polycarbonate. *Polymer* 47:6398–6405.
24. Perez-Foullerat, D., Meier, U. W., Hild, S., and Rieger, B. 2004. High-molecular-weight polyketones from higher alpha-olefins: A general method. *Macromolecular Chemistry and Physics* 205:2292–2302.
25. Charlesby, A. 1960. *Atomic Radiation and Polymers*, Pergamon Press, New York, p. 198.
26. Kleinhans, G., Starkl, W., and Nuffer, K. 1984. Structure and mechanical properties of shape memory polyurethane. *Kunststoffe* 74:445–449.
27. Kleinhans, G. and Heidenhain, F. 1986. Actively moving polymers. *Kunststoffe* 76:1069–1073.
28. Zhu, G., Liang, G., Xu, Q., and Yu, Q. 2003. Shape-memory effects of radiation crosslinked poly(epsilon-caprolactone). *Journal of Applied Polymer Science* 90:1589–1595.
29. Zhu, G., Lu, G. W., Zhang, J., Lan, J. H., Li, Y. F., Wang, X. Q., Xu, D., and Xia, H. R. 2006. Spectra analysis of [MnHg(SCN)(4) (H2O)2]center dot 2C(4)H(9)NO (MMTWD) crystals. *Spectroscopy and Spectral Analysis* 26:267–270.
30. Luo, X. L., Zhang, X. Y., Wang, M. T., Ma, D. H., Xu, M., and Li, F. K. 1997. Thermally stimulated shape-memory behavior of ethylene oxide ethylene terephthalate segmented copolymer. *Journal of Applied Polymer Science* 64:2433–2440.
31. Liu, C. D., Chun, S. B., Mather, P. T., Zheng, L., Haley, E. H., and Coughlin, E. B. 2002. Chemically cross-linked polycyclooctene: Synthesis, characterization, and shape memory behavior. *Macromolecules* 35:9868–9874.
32. Li, F. K. and Larock, R. C. 2002. New soybean oil-styrene-divinylbenzene thermosetting copolymers. V. Shape memory effect. *Journal of Applied Polymer Science* 84:1533–1543.
33. Kagami, Y., Gong, J. P., and Osada, Y. 1996. Shape memory behaviors of crosslinked copolymers containing stearyl acrylate. *Macromolecular Rapid Communication* 17:539–543.
34. Goethals, E. J., Reyntjens, W., and Lievens, S. 1998. Poly(vinyl ethers) as building blocks for new materials. *Macromolecular Symposia* 132:57–64.
35. Reyntjens, W. G., Du-Prez, F. E., and Goethals, E. J. 1999. Polymer networks containing crystallizable poly(octadecyl vinyl ether) segments for shape-memory materials. *Macromolecular Rapid Communication* 20:251–255.

36. Lee, S. H., Kim, J. W., and Kim, B. K. 2004. Shape memory polyurethanes having crosslinks in soft and hard segments. *Smart Materials and Structures* 13:1345–1350.
37. Rousseau, I. A. and Mather, P. T. 2003. Shape memory effect exhibited by smectic-c liquid crystalline elastomers. *Journal of the America Chemical Society* 125:15300–15301.
38. Behl, M. and Lendlein, A. 2007. Actively moving polymers. *Soft Matter* 3:58–67.
39. Jiang, H. Y., Kelch, S., and Lendlein, A. 2006. Polymers move in response to light. *Advanced Materials* 18:1471–1475.
40. Yu, Y. L. and Ikeda, T. 2005. Crystallization and fluorescence properties of Nd3+-doped transparent oxyfluoride glass ceramics. *Macromolecular Chemistry Physics* 206:1705–1708.
41. Zhang, D., Liu, Y., and Leng, J. S. 2008. Infrared laser-activated shape memory polymer. 15th SPIE International Conference on Smart Structures/NDE, San Diego, CA, March 9–13, 2008. SPIE 6932.
42. Zhang, D., Liu, Y., and Leng, J. S. 2007. Shape memory polymer networks from styrene copolymer. SPIE International Conference on Smart Materials and Nanotechnology, Harbin, China, July 1–4, 2007. SPIE 6423.
43. Liu, C. D., Wu, J., and Mather, P. T. 2003. Nonisothermal crystallization kinetics of polycyclooctene: Characterization using thermal and optical methods. *Abstract of Papers of the American Chemical Society* 226:U521.
44. Biercuk, M. J., Monsma, D. J., Marcus, C. M., Becker, J. S., and Gordon, R. G. 2003. Low-temperature atomic-layer-deposition lift-off method for microelectronic and nanoelectronic applications. *Applied Physics Letters* 83:2405–2407.
45. Li, F. K. and Larock, R. C. 2000. Thermosetting polymers from cationic copolymerization of tung oil: Synthesis and characterization. *Journal of Applied Polymer Science* 78:1044–1056.
46. Liang, C. and Rogers, C. A. 1997. One-dimensional thermomechanical constitutive relations for shape memory materials. *Journal of Intelligent Materials Systems and Structures* 8:285–302.
47. Gall, K., Mikulas, M., Munshi, N. A., Beavers, F., and Tupper, M. 2000. Carbon fiber reinforced shape memory polymer composites. *Journal of Intelligent Materials Systems and Structures* 11:877–886.
48. Ash, B. J., Schadler, L. S., and Siegel, R. W. 2001. Thermal and mechanical properties of PMMA/alumina nanocomposites: Effect of strong and weak interfaces. *Abstract of Papers of the American Chemical Society* 222:U286.
49. Leng, J. S., Lan, X., Liu, Y. J. and Du, S. Y. 2009. Electroactive shape-memory polymer composite filled with nanocarbon powders. *Smart Materials and Structures* 18(074003):1–7.
50. Lv, H. B., Leng, J. S., Liu, Y. J., and Du, S. Y. 2008. Shape-memory polymer in response to solution. *Advanced Engineering Materials* 10:592–595.
51. Langer, R. and Tirrell, D. A. 2004. Designing materials for biology and medicine. *Nature (London)* 428:487–492.
52. Lv, H. B., Liu, Y. J., Zhang, D., Leng, J. S., and Du, S. 2008. Solution-responsive shape-memory polymer driven by forming hydrogen bonding. *Advanced Materials Research* 47–50:258–261.
53. Lendlein, A. and Langer, R. S. 2004. Self-expanding device for the gastrointestinal or urogenital area. WO 2004073690 A1.

54. Marco, D., 2006. Biodegradable self-inflating intragastric implants for curbing appetite. WO 2006092789 A2.
55. Huang, W. M., Lee, C. W., and Teo, H. P. 2006. Thermomechanical behavior of a polyurethane shape memory polymer foam. *Journal of Intelligent Materials Systems and Structures* 17:753–760.
56. Hampikian, J. M., Heaton, B. C, Tong, F. C., Zhang, Z. Q., and Wong, C. P. 2006. Mechanical and radiographic properties of a shape memory polymer composite for intracranial aneurysm coils. *Materials Science and Engineering C—Biomimetic and Supramolecular* 26:1373–1379.
57. Fare, S., Valtulina, V., Petrini, P., Alessandrini, E., Pietrocola, G., Tanzi, M. C., Speziale, P., and Visai, L. 2005. In vitro interaction of human fibroblasts and platelets with a shape-memory polyurethane. *Journal of Biomedical Materials Research Part A* 73A:1–11.
58. Leng, J. S., Lan, X., Huang, W. M., Liu, Y. J., Liu, N., Phee, S. Y., Du, S. Y. et al., 2008. Electrical conductivity of shape memory polymer embedded with micro Ni chains. *Applied Physics Letters* 92:014104.
59. Leng, J. S., Huang, W. M., Lan, X., Liu, Y. J., Liu, N., Phee, S. Y., and Du, S. Y. 2008. Significantly reducing electrical resistivity by forming conductive Ni chains in a polyurethane shape-memory polymer/carbonblack composite. *Applied Physics Letters* 92:204101.
60. Sahoo, N. G., Jung, Y. C., Goo, N. S., and Cho, J. W. 2005. Conducting shape memory polyurethane-polypyrrole composites for an electroactive actuator. *Macromolecular Materials and Engineering* 290:1049–1055.
61. Koerner, H., Price, G., Pearce, N. A., Alexander, M., and Vaia, R. 2004. Remotely actuated polymer nanocomposites—Stress-recovery of carbon-nanotube-filled thermoplastic elastomers. *Nature Materials* 3:115–120.
62. Leng, J.S., Lv, H., Liu, Y., and Du, S. 2008. Investigation of shape-memory polymer nanocomposite. *Journal of Applied Physics* 104:104917.
63. Schmidt, A. M. 2006. Electromagnetic activation of shape memory polymer networks containing magnetic nanoparticles. *Macromolecular Rapid Communication* 27:1168–1172.
64. Huang, W. M., Liu, N., Lan, X., Lin, J. Q., Pan, J. H., and Leng, J. S. 2009. Formation of protrusive micro/nano patterns atop shape memory polymers, *Materials Science Forum* 14:243–248.
65. Liu, N., Xie, Q., Huang, W. M., Phee, S. J., and Guo, N. Q. 2008. Formation of micro protrusion arrays atop shape memory polymer. *Journal of Micromechanics and Microengineering* 18:027001.
66. Liu, N., Huang, W. M., Phee, S. J., Fan, H., and Chew, K. L. 2008. The formation of micro-protrusions atop a thermo-responsive shape memory polymer. *Smart Materials and Structures* 7(5):057101
67. Sokolowski, W., Metcalfe, A., Hayashi, S., Yahia, L., and Raymond, J. 2007. Medical applications of shape memory polymers. *Biomedical Materials* 2:S23–S27.
68. Leng, J., Lv, H., Liu, Y., and Du, S. 2007. Electroactivate shape-memory polymer filled with nanocarbon particles and short carbon fibers. *Applied Physics Letters* 91:144105.
69. Yang, B., Huang, W. M., Li, C., Lee, C. M., and Li, L. 2004. On the effects of moisture in a polyurethane shape memory polymer. *Smart Materials and Structures* 13:191–195.

70. Huang, W. M., Yang, B., An, L., Li, C., and Chan, Y. S. 2005. Water-driven programmable polyurethane shape memory polymer: Demonstration and mechanism. *Applied Physics Letters* 86:114105.
71. Yang, B., Huang, W. M., Li, C., and Li, L. 2006. Effects of moisture on the thermomechanical properties of a polyurethane shape memory polymer. *Polymer* 47:1348–1356.
72. Yang, B., Huang, W. M., Li, C., Li, L., and Chor, J. H. 2005. Qualitative separation of the effects of carbon nanopowder and moisture on the glass transition temperature of polyurethane shape memory polymer. *Scripta Materialia* 53:105–107.
73. Jung, Y. C., So, H. H., and Cho, J. W. 2006. Water-responsive shape memory polyurethane block copolymer modified with polyhedral oligomeric silsesquioxane. *Journal of Macromolecular Science Part B—Physics* 45:441–451.
74. Yang, B. 2005. Influence of moisture in polyurethane shape memory polymers and their electrical conductive composites. A thesis of Nanyang Technological University, Singapore.
75. Morton, M. 1987. *Solution Theory of Polymer*, Wiley-VCH, Weinheim, Germany, vol. 1, p. 7.
76. Mark, J. E. and Erman, B. 1998. *Rubberlike Elasticity—A Molecular Primer*, vol. 1, Wiley-Interscience, New York, p. 5.
77. Hudgin, D. E. 2000. *Rubber Elasticity*, vol. 3, Marcel Dekker, Inc. New York, p. 1.
78. Leng, J. S., Lv, H., Liu, Y., and Du, S. 2008. Comment on "water-driven programmable polyurethane shape memory polymer: Demonstration and mechanism." [*Applied Physics Letters* 86, 114105, (2005)]. *Applied Physics Letters* 92:206105.

7

Shape-Memory Polymer Composites

Jinsong Leng, Xin Lan, and Shanyi Du

Centre for Composite Materials and Structures, Harbin Institute of Technology, Harbin, P.R. China

CONTENTS

7.1 Introduction

The pure SMPs usually exhibit lower modulus and recovery forces than the shape-memory alloys or shape-memory ceramics, and therefore the unreinforced SMPs are not suitable for many applications that require special performances (e.g., high strength and high recovery force) [1,2]. In order to improve their mechanical properties, the incorporation of various reinforcing fillers within SMPs has been investigated. SMPs reinforced by particles or fibers could achieve high strength, high Young's modulus, and diversify the application of SMPs.

Depending on the type of filler, the SMP composites can be classified as particle or fiber reinforced. The SMP composites filled with particles (e.g., carbon powders, carbon nanotubes, carbon nanofibers, and SiC nanoparticles) or short fibers (e.g., carbon, glass, and Kevlar fiber) develop some particular functions, such as high stiffness and high electrical conductivity. Hence, the type of SMP composites is studied as a functional material. The SMP composites filled with particles may meet various requirements in practical applications. While the SMP composites have been discussed as functional materials in Chapter 6, this section will focus on their mechanical and thermomechanical properties.

In general, SMP composites reinforced with particles or short fibers cannot be used as structural materials, because the improvement in their mechanical properties is quite limited and their mechanical properties are relatively poor. In contrast, continuous-fiber-reinforced SMPs exhibit a significant improvement in mechanical properties including strengths and stiffnesses. Liang et al. [3] and Gall et al. [4] have investigated the carbon, glass, or Kevlar fiber-reinforced SMP composites. As a result of the incorporation of the fiber reinforcement, an improvement in stiffness but a decrease in recoverable strain was obtained. In addition, Ohki et al. [5,6] investigated the mechanical and shape recovery properties of thermoplastic SMP composites reinforced by strand glass fiber. They found that, with the incorporation of 50 wt% of glass fiber, the failure stress increased by 140% while the recovery rate decreased by 62%. As both a functional and structural material, these fiber-reinforced SMPs exhibit good potential for many applications [6]. When used as actuator materials, they require no moving parts. Substantial interest has therefore been generated for the use of fiber-reinforced SMPs in the deployable structures including hinges, reflectors, beams, and solar arrays, which will be discussed in Chapter 8 [7].

7.2 Constitutive Relationship of Shape-Memory Polymers

Engineering applications demand effective and simple theoretical models that present the mechanism of shape-memory effect for SMP. It is difficult to construct the mechanical constitutive behavior model of SMP theoretically because the mechanical behaviors of polymer depend extensively on both time and temperature. Generally, there are two kinds of approaches to develop the constitutive model of SMP: the first is based on the standard linear viscoelastic theorem whereas the second is based on the micromechanical theorem of phase transformation.

In 1997, Tobushi et al. [8] proposed a linear constitutive equation of SMP by taking account of a slip mechanism due to internal friction and the thermal expansion in the standard linear viscoelastic model. In this model, the stress–strain–temperature relationship of SMP is expressed as follows:

$$\begin{cases} \dot{\varepsilon} = \dfrac{\dot{\sigma}}{E} + \dfrac{\sigma}{\mu} - \dfrac{\varepsilon - \varepsilon_s}{\lambda} + \alpha \dot{T} \\ \varepsilon_s = C(\varepsilon_c - \varepsilon_l) \end{cases} \tag{7.1}$$

where

σ, ε, and T denote stress, strain, and temperature, respectively
The dot denotes time derivative
ε_s denotes irreversible strain

The temperature-dependent parameters of E, μ, and λ present elastic modulus, viscosity, and retardation time, respectively. The temperature-dependent C and ε_l are related to the process of shape fixity. The dependence of the temperature-dependent parameters on temperature is expressed by the following material parameter function:

$$x = x_g \exp\left[a\left(\frac{T_g}{T} - 1\right)\right] \quad (T_l \le T \le T_h) \tag{7.2}$$

where

x_g is the value of x at $T = T_g$
T_l and T_h are the temperatures at the starting and finishing points of the glass transition from glassy state to rubbery state in SMP

The shape-memory thermomechanical cycle of polyurethane-based SMP can be predicted by the constitutive equation, Equation 7.1, coupled with the material parameter function, Equation 7.2. The temperature-dependent material parameters at various temperatures of T_g, T_h, and T_l are given in Table 7.1. The stress–strain curves, stress–temperature curves, and strain–temperature

TABLE 7.1

Material Parameters of SMP for Figures 7.1 through 7.3

	E/MPa	μ/GPa s	λ/s	C	ε_l/%
$T_l = T_g - 15$ K	907	116	2840	0.716	0.0184
T_g	146	14	521	0.112	0.3
$T_h = T_g + 15$ K	27.6	2.03	111	0.0206	3.83
	a_E	A_μ	A_λ	a_c	a_ε
	38.1	44.2	35.4	38.7	58.3

Source: From Tobushi, H. et al., *J. Intell. Mater. Syst. Struct.*, 8, 711, 1997.

curves are plotted in Figures 7.1 through 7.3, respectively. In these figures, symbols ①–④ denote the loading path. In process ①, the maximum ε_m was applied at high temperature $T_h = T_g + 20\,\text{K}$. In process ②, maintaining ε_m constant, the specimen was cooled to $T_l = T_g - 20\,\text{K}$. In process ③, the specimen was unloaded at T_l. In process ④, it was heated from T_l to T_h under free-load conditions.

In 2001, a nonlinear constitutive relation by adding two nonlinear parameters in the linear constitutive equation had been developed by Tobushi et al. [9], Equation 7.1. In this model, the stress–strain–temperature relationship is expressed as follows:

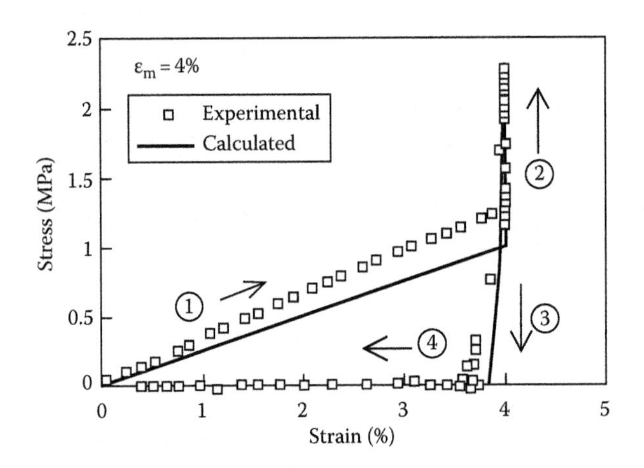

FIGURE 7.1
Relationship between stress and strain of SMP by Equations 7.1 and 7.2. (From Tobushi, H. et al., *J. Intell. Mater. Syst. Struct.*, 8, 711, 1997. With permission.)

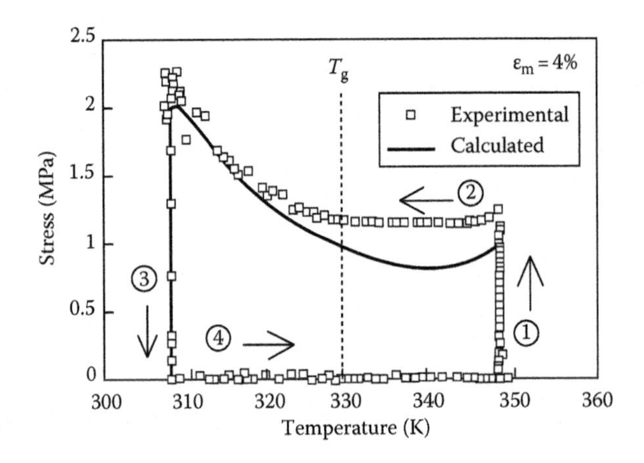

FIGURE 7.2
Relationship between stress and temperature of SMP by Equations 7.1 and 7.2. (From Tobushi, H. et al., *J. Intell. Mater. Syst. Struct.*, 8, 711, 1997. With permission.)

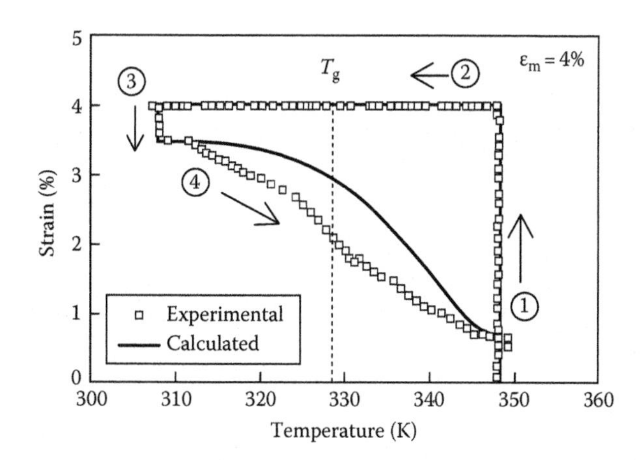

FIGURE 7.3
Relationship between strain and temperature of SMP by Equations 7.1 and 7.2. (From Tobushi, H. et al., *J. Intell. Mater. Syst. Struct.*, 8, 711, 1997. With permission.)

$$\begin{cases} \dot{\varepsilon} = \dfrac{\dot{\sigma}}{E} + m\left(\dfrac{\sigma - \sigma_y}{k}\right)^{m-1} \dfrac{\dot{\sigma}}{E} + \dfrac{\sigma}{\mu} + \dfrac{1}{b}\left(\dfrac{\sigma}{\sigma_c} - 1\right)^{n} - \dfrac{\varepsilon - \varepsilon_s}{\lambda} + \alpha \dot{T} \\ \varepsilon_s = S(\varepsilon_c + \varepsilon_p) \end{cases} \tag{7.3}$$

where

σ_y and σ_c denote the elastic and viscous proportional limits of SMP

The thermomechanical cycle of the polyurethane-based SMP was simulated by using the nonlinear constitutive equation, Equation 7.3, coupled with the material parameter function, Equation 7.2. The stress–strain curves, stress–temperature curves, and strain–temperature curves were plotted in Figures 7.4 through 7.6, respectively.

In 2006, a three-dimensional constitutive model with internal state variable based on the experimental results and the molecular mechanism of shape-memory effect in SMP was developed by Liu et al. [10]. Based on this model there are two kinds of extreme phases—frozen and active phases—in an SMP at an arbitrary temperature. The fractions of frozen and active phases are defined as follows:

$$\phi_f = \frac{V_{frz}}{V}, \quad \phi_a = \frac{V_{act}}{V}, \quad \phi_f + \phi_a = 1 \tag{7.4}$$

where

ϕ_f and ϕ_a denote the fractions of frozen and active phases in SMP, respectively

V, V_{frz}, and V_{act} stand for the total volume, the volumes of the frozen and active phases, respectively

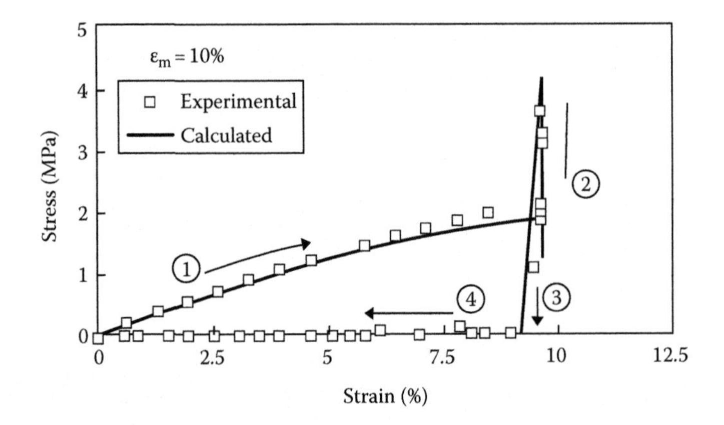

FIGURE 7.4
Relationship between stress and strain of SMP by Equations 7.3 and 7.2. (From Tobushi, H. et al., *Mech. Mater.*, 33, 545, 2001. With permission.)

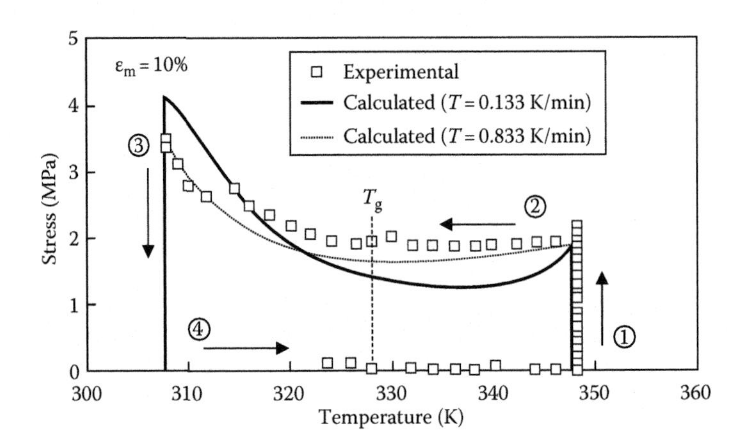

FIGURE 7.5
Relationship between stress and temperature of SMP by Equations 7.3 and 7.2. (From Tobushi, H. et al., *Mech. Mater.*, 33, 545, 2001. With permission.)

The strain of SMP is composed of three parts: the stored strain, the mechanical elastic strain, and the thermal expansion strain, expressed as

$$\varepsilon = \varepsilon_s + \varepsilon_m + \varepsilon_T \tag{7.5}$$

where

ε denotes second-order total strain tensor

ε_s, ε_m, and ε_T present second-order stored strain, mechanical elastic strain, and thermal expansion strain, respectively

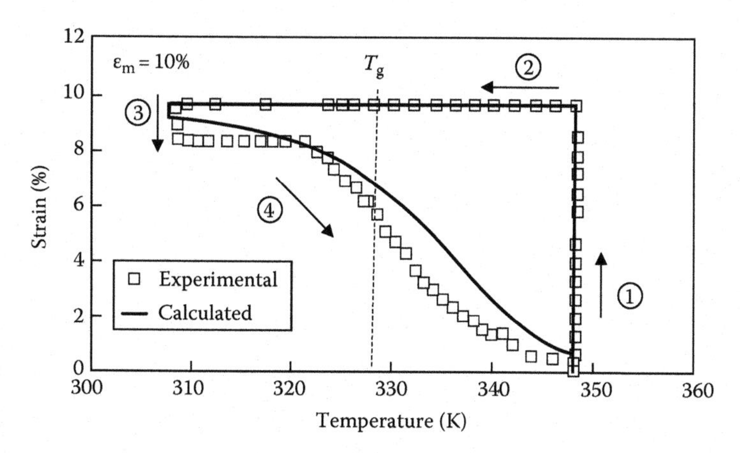

FIGURE 7.6
Relationship between strain and temperature of SMP by Equations 7.3 and 7.2. (From Tobushi, H. et al., *Mech. Mater.*, 33, 545, 2001. With permission.)

The stored strain of SMP is expressed as

$$\varepsilon_s = \int_0^{\phi_f} \varepsilon_f^e(x)\mathrm{d}\phi \tag{7.6}$$

where ε_f^e denotes the entropic frozen strain. The mechanical elastic strain of SMP is expressed as

$$\varepsilon_m = [\phi_f S_i + (1-\phi_f)S_e] : \sigma \tag{7.7}$$

where
- σ denotes second-order stress tensor
- S_i is the elastic compliance fourth-order tensor corresponding to the internal energetic deformation
- S_e is the elastic compliance fourth-order tensor corresponding to the entropic deformation

The thermal expansion strain is expressed as

$$\varepsilon_T = \left\{ \int_{T_0}^{T} \left[\phi_f \alpha_f(\theta) + (1-\phi_f)\alpha_a(\theta) \right] \mathrm{d}\theta \right\} I \tag{7.8}$$

where
- α_f and α_a are the thermal expansion coefficients of the frozen phase and the active phase
- I is the second-order identity tensor

In subsequence, a nonlinear constitutive model for SMP was constructed by Chen and Lagoudas [11,12] to describe the thermomechanical properties under large deformation. The model is originated from the idea that is developed on the basis of Liu et al. [10], that the coexisting active and frozen phases of SMP and the transition between them provide the underlying mechanisms for strain storage and recovery during a shape-memory cycle.

In 2009, Zhou et al. [13–15] developed a macromechanical constitutive equation to describe the shape-memory cycle of SMP from a practical viewpoint that is based on the viscoelastic theorem for solid materials. According to this model, the strain rate of SMP is composed of three parts: the elastic strain rate, the viscous strain rate, and the thermal expansion strain rate. The stress–strain–temperature relationship is expressed as follows:

$$\dot{\varepsilon} = \left[\frac{\dot{\sigma}}{E} + m\theta\left(\frac{\langle\sigma - \sigma_p\rangle}{E}\right)^{m-1}\frac{\dot{\sigma}}{E}\right] + \left[\frac{\sigma}{\mu} + \varphi\left(\frac{\langle\sigma - \sigma_p\rangle}{\sigma_p}\right)^{n}\frac{\sigma}{\mu} - \frac{\varepsilon}{\lambda}\right] + \alpha\dot{T} \tag{7.9}$$

where the first item with square brackets expresses the elastic strain rate, the second item with square brackets denotes the viscous strain rate, and the last item is the thermal expansion rate. The temperature-dependent material parameters of E, μ, λ, and σ_p are elastic modulus, viscous modulus, retardation time, and proportional limit, respectively. The temperature-independent material parameters of m, n, θ, and φ are elastic power-number, viscous power-number, elastic nonlinear-coefficient, and viscous nonlinear-coefficient, respectively. The singular function $\langle\sigma - \sigma_p\rangle$ is defined as follows:

$$\langle\sigma - \sigma_p\rangle = \begin{cases} \sigma - \sigma_p & (\sigma > \sigma_p) \\ 0 & (\sigma \le \sigma_p) \end{cases} \tag{7.10}$$

The relationships between the temperature-dependent parameters and temperature are expressed by the following material parameter function:

$$\Phi(T) = \begin{cases} \Phi_s & (T < T_s) \\ \dfrac{(T - T_s)(T - T_g)}{(T_f - T_s)(T_f - T_g)}\Phi_f + \dfrac{(T - T_s)(T - T_f)}{(T_g - T_s)(T_g - T_f)}\Phi_g + \dfrac{(T - T_f)(T - T_g)}{(T_s - T_f)(T_s - T_g)}\Phi_s & (T_s \le T \le T_f) \\ \Phi_f & (T > T_f) \end{cases} \tag{7.11}$$

where T_s and T_f denote the glass transition starting and finishing temperatures, respectively. Φ denotes any one of the temperature-dependent material parameters, such as E, μ, λ, and σ_p. Φ_f, Φ_s, and Φ_g stand for the values of Φ at the temperatures of $T = T_f$, $T = T_s$, and $T = T_g$, respectively

The shape-memory thermomechanical cycle of the styrene-based SMP, which includes the high-temperature deformation by loading, shape fixity through cooling with displacement restraint, and shape recovery upon heating, is numerically simulated by the constitutive equation, Equation 7.9 coupled with Equation 7.12. During numerical simulations, the temperature-independent material parameters are $n = 2$, $\varphi = 0.0001$, $m = 2$, and $\theta = 2$. The temperature-dependent material parameters at the glass transition starting temperature, glass transition temperature, and glass transition finishing temperature are listed in Table 7.2.

As shown in Figure 7.7, the curve presents the relationship between stress and strain of the styrene-based SMP. During the process of high-temperature deformation through loading at the constant temperature of 40°C, the stress–strain relationship is nonlinear due to the material properties of nonlinear elasticity and viscosity. During the process of shape fixity through cooling with strain restraint, the material is cooled down from 40°C to 0°C at a rate of 1.0°C/s. The material stress increases from 11.26

TABLE 7.2

Material Parameters of the Styrene-Based SMP at Different Temperatures

Temperature	E/MPa	σ_p/MPa	λ/s	μ/GPa s
$\leq T_s$ (30°C)	940	6.74	2400	215
$= T_g$ (40°C)	520	4.50	390	35
$\geq T_f$ (50°C)	200	1.35	75	7

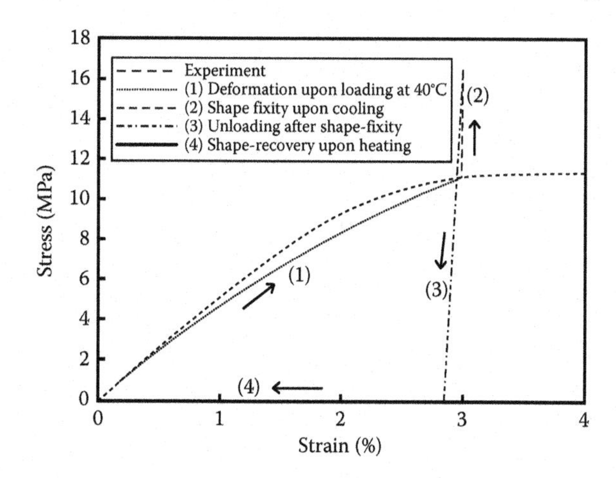

FIGURE 7.7

Stress–strain curves of SMP by Equations 7.10 and 7.11. (From Liu, Y. et al., *Int. J. Plast.*, 22, 279, 2006. With permission.)

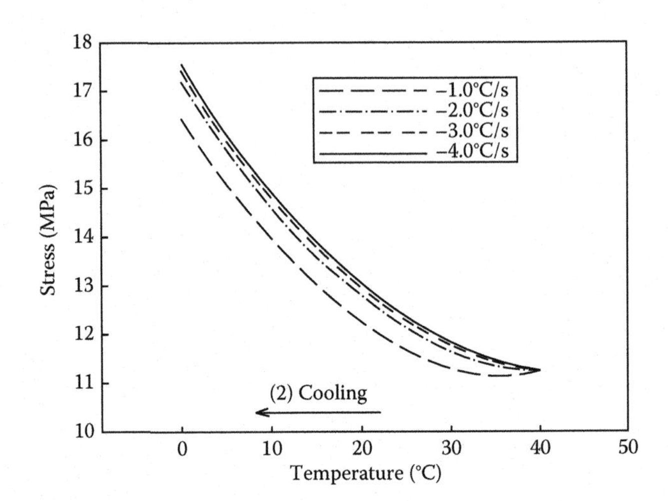

FIGURE 7.8
Curves of stress versus temperature of SMP by Equations 7.10 and 7.11. (From Liu, Y. et al., *Int. J. Plast.*, 22, 279, 2006. With permission.)

to 16.48 MPa, and the material strain is kept at a constant of 0.03. During unloading after shape fixity, the material strain decreases from 0.03 to 0.0285, and so the shape fixity ratio is 95%. During the process of shape recovery upon heating, the material is heated from 0°C to 60°C at a rate of 0.06°C/s. The material strain decreases from 0.0285 to 0.00075, and so the shape recovery ratio is 97%.

Figure 7.8 shows the curves of stress versus temperature at various cooling rates during the shape fixity process upon cooling with strain restraint. The initial values of stress and constant strain are 11.26 MPa and 0.03, respectively. The cooling rates are −1.0°C/s, −2.0°C/s, −3.0°C/s, and −4.0°C/s, respectively. The curves reveal that the stress nonlinearly increases with the decrease in temperature during the shape fixity process. This is mainly because the temperature-dependent material parameters, E, λ, and μ, nonlinearly increase with a decrease in temperature. These curves also reveal that the stress at the same temperature becomes smaller when the absolute value of cooling rate increases, which indicates that the shape-fixity behavior of SMP has a strong relationship with the cooling rate.

Figure 7.9 shows curves of strain versus temperature at different heating rates during the shape recovery process upon heating. The material is heated from 0°C to 60°C. The initial value of strain is 0.0285, and the heating rates are 0.06°C/s, 0.08°C/s, 0.10°C/s, and 0.12°C/s, respectively. Results show that material strain nonlinearly decreases with an increase in temperature during heating. This is because the retardation time, λ, nonlinearly decreases with an increase in temperature. Results also reveal that the shape recovery ratio increases with a decrease in heating rate, which explains why the shape recovery behavior of SMP depends on the heating rate.

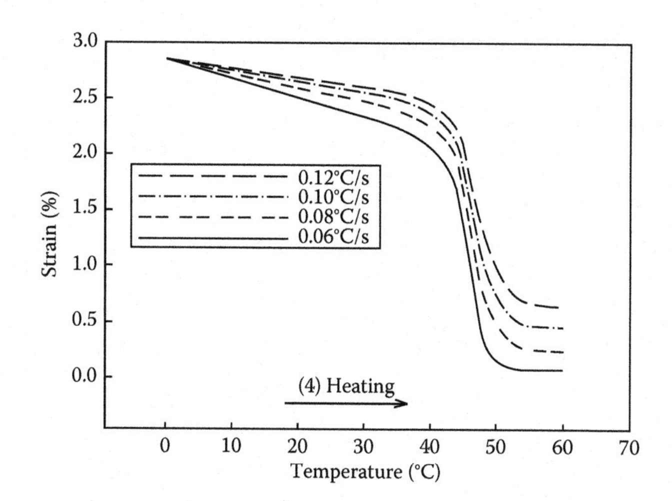

FIGURE 7.9
Curves of strain versus temperature of SMP by Equations 7.10 and 7.11. (From Liu, Y. et al., *Int. J. Plast.*, 22, 279, 2006. With permission.)

7.3 Shape-Memory Polymer Composites

7.3.1 Shape-Memory Polymer Filled with Carbon Nanotubes

SMPs are characterized by their remarkable recoverability and shape-memory effect; however, their mechanical properties such as strength and elastic modulus are low. CNTs, on the other hand, present remarkable mechanical and conductive properties, including high elastic modulus, strength, and thermal and electrical conductivity [16–20]. CNT-filled polymer composites are expected to improve the mechanical properties of the matrix polymer. Many advances in the improvement for the mechanical, thermal, and electric properties of CNT-reinforced polymers have been reported, such as CNT/polystyrene [21–23], CNT/PVA [24], CNT/PVDF [25], CNT/PP [26–32], CNT/nylon [33], and CNT/epoxy [34–36]. Recently, as a typical SMP nanocomposite [37], CNT/SMP nanocomposite were proposed by Ni et al. [38] through incorporating the CNTs of the vapor growth carbon fibers (VGCFs). A fine and homogeneous dispersion of the VGCF in the SMP matrix has been obtained. The specimens with different VGCF weight fractions, namely pure SMP bulk, 1.7 wt%, 3.3 wt%, and 5.0 wt%, were prepared.

7.3.1.1 Static Tensile Property

As shown in Figure 7.10, a typical stress–strain curve for three types of SMP nanocomposites that were incorporated with different weight fraction VGCFs

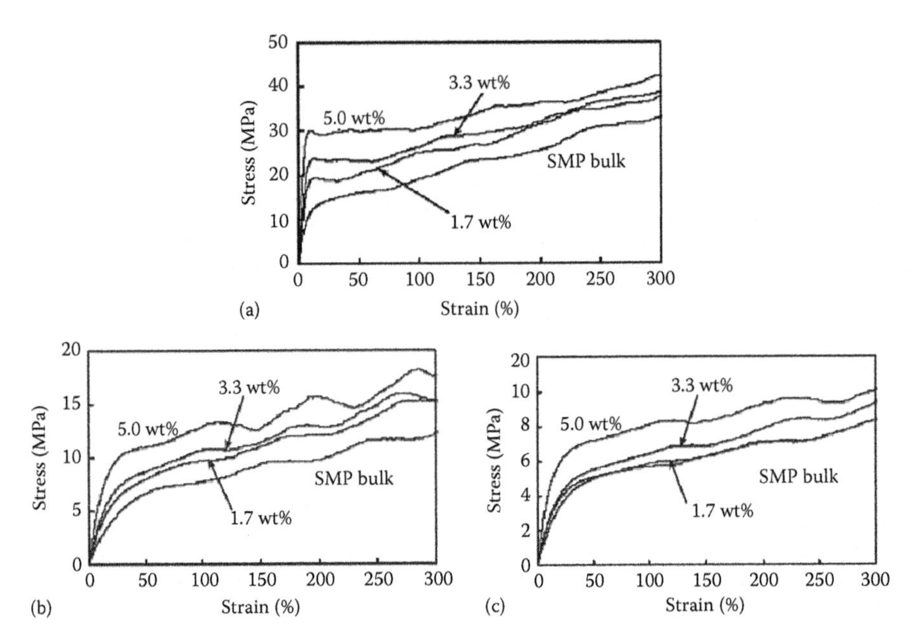

FIGURE 7.10
Stress–strain relationship in static tensile tests for four materials—SMP bulk, 1.7 wt%, 3.3 wt%, and 5.0 wt% at testing temperatures: (a) 25°C, (b) 45°C, and (c) 65°C, respectively. (Reprinted from Ni, Q.Q. et al., *Compos. Struct.*, 81, 176, 2007. With permission.)

were determined at three testing temperatures 25°C, 45°C, and 65°C [38]. It is apparent that upon increasing the weight fraction of VGCF CNTs, the mechanical properties of the SMP were improved. The SMP nanocomposite was not broken within the strain range of 300%. This result shows that these composites have excellent ductility similar to that of the pure SMP bulk. At the temperature of 25°C, which was lower than the T_g, the yield phenomenon was clearly presented in all specimens. The yield region became long when the VGCF's weight fraction increased. Beyond the yield point, the inclination of the stress–strain curve increased with increasing strain. That implied that a strain-hardening phenomenon occurred. On the other hand, when the testing temperature was 45°C and 65°C (higher than the T_g of SMP), the yield phenomenon also appeared slowly even if at a lower stress level.

Figure 7.11 indicates the dependence of Young's modulus on the VGCFs weight fraction at various testing temperatures, while Figure 7.12 reveals the relationship between the VGCF weight fraction and the yield stress. For the 5.0 wt% VGCF nanocomposite, compared with pure SMP bulk, the Young's modulus were improved with the increment of 125%, 216%, and 186% at 25°C, 45°C, and 65°C, while the yield stress had the increment of 87%, 132%, and 138%, respectively [38].

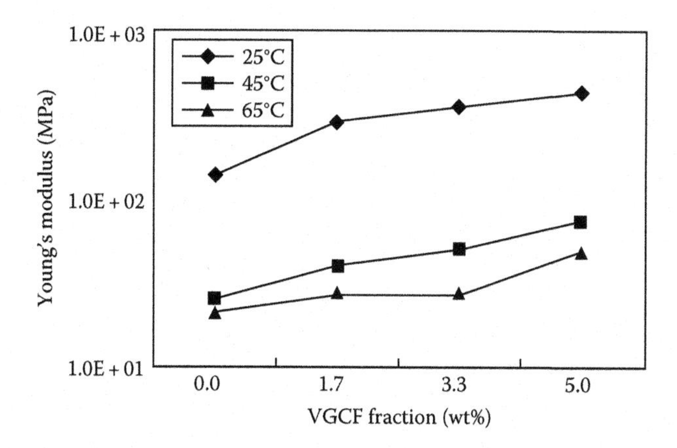

FIGURE 7.11
Young's modulus as a function of VGCFs weight fraction for four materials—SMP bulk, 1.7 wt%, 3.3 wt%, and 5.0 wt% at testing temperatures: 25°C, 45°C, and 65°C. (Reprinted from Ni, Q.Q. et al., *Compos. Struct.*, 81, 176, 2007. With permission.)

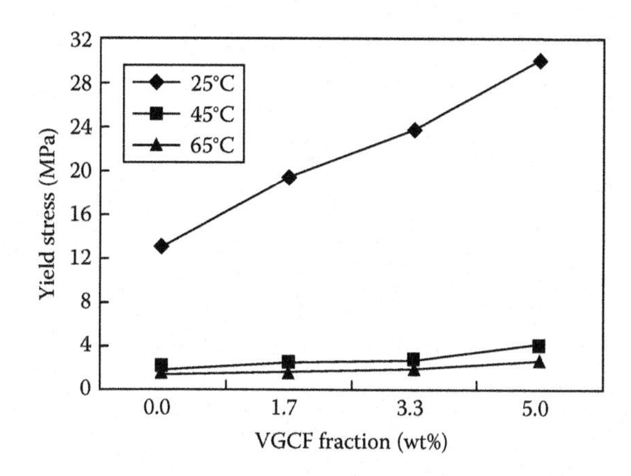

FIGURE 7.12
Yield stress as a function of VGCF's weight fraction for four materials—SMP bulk, 1.7 wt%, 3.3 wt%, and 5.0 wt% at testing temperatures: 25°C, 45°C, and 65°C. (Reprinted from Ni, Q.Q. et al., *Compos. Struct.*, 81, 176, 2007. With permission.)

7.3.1.2 Shape Recovery Force Measurement

The shape recovery performance was also evaluated. Figure 7.13 exhibits the stress–strain relationship obtained in the shape recovery test for the CNT/SMP nanocomposite and the pure SMP. It is found that the recovered strain decreased with increasing the VGCF weight fraction. The recovered strain for 3.3 wt% and 5.0 wt% VGCF was normally identical. The nanocomposite performs better shape-fixation property than SMP at low temperature. To measure the recovery stress for an application in sensors or actuators, such as temperature sensors

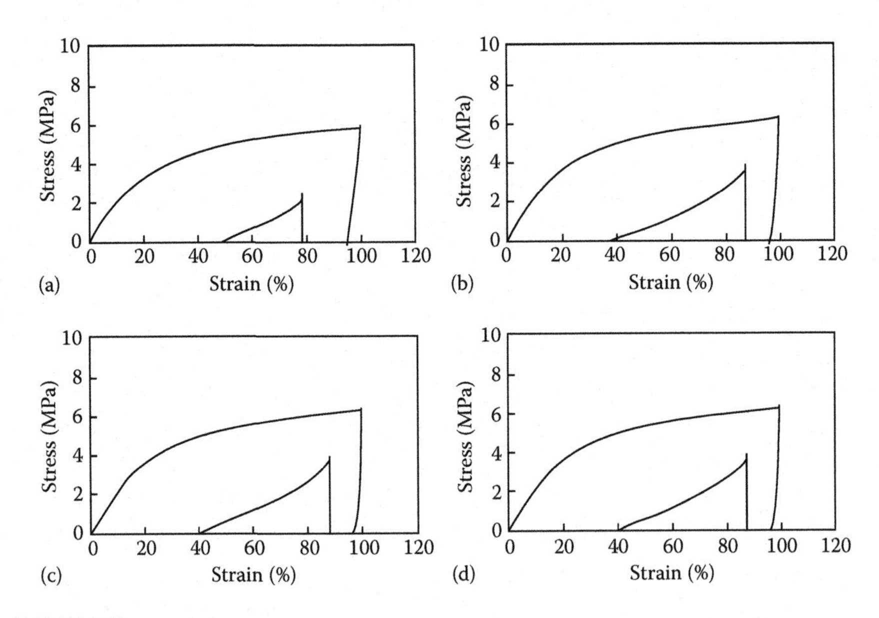

FIGURE 7.13
Relationship between stress and strain in the recovery stress cycle test for four materials (a) SMP bulk, (b) 1.7 wt%, (c) 3.3 wt%, and (d) 5.0 wt%, respectively. (Reprinted from Ni, Q.Q. et al., *Compos. Struct.*, 81, 176, 2007. With permission.)

and strain sensors, the strain can be restrained by fixing the specimen dimensions, and then the recovery stress can be measured [37,38].

7.3.2 Shape-Memory Polymer Filled with Carbon Nanofibers

Carbon nanofibers (CNFs) have attracted much attention due to their exceptional properties in the recent years. With extremely outstanding mechanical properties and high aspect ratio, CNFs are one of the best candidates for improving mechanical strength and thermomechanical properties for polymer materials. Therefore, to utilize the characteristic properties of a single CNF in macroscopic scale, macroscopic assemblies of CNFs have been explored. These attempts are inspired by the successful manufacture of CNF nanocomposites, which may significantly improve the performance of polymer, ceramic, and metal matrix. Furthermore, CNF as a carbon material possesses excellent electrical conductive properties that can be incorporated into SMPs for multifunction. This method is used to carry out shape recovery actuation by an electrical current and infrared light, respectively.

There are three types of CNTs with respect to fabrication processing: as-grown CNFs without any postproduction thermal processing, pyrolytically stripped CNFs with removing polyarocarbons from surface, and heat-treated CNFs with removing iron catalyst at temperature of 3000°C. The dimensions of CNFs vary from 5 to 300 microns in length and are between 5 to 150 nm in diameter. Young's modulus of nanofibers sometimes can approach 1.2 TPa, and some samples have been reported to perform tensile strengths up to 64 GPa. Thus, there is an expectation that composites created on the bases of nanofibers may exhibit better properties than the conventional carbon-fiber-reinforced materials. Other benefits provided by CNFs include improved heat distortion temperatures and increased electromagnetic shielding [39].

Manpreet [39] investigated overall performance of the CNF-filled SMP composite. During the dynamic mechanical analysis (DMA) testing, the storage modulus of pure SMP is 2200 MPa at the temperature of 31.8°C. With the temperature increased, the pure SMP loses its stiffness quickly as the temperature increased during the transition from glassy state to rubbery state between 37°C and 50°C. The SMP composite filled with relatively low filler content of 5% CNFs almost shows the identical storage modulus as that of pure SMP. In samples with 10% nanofiber, the storage modulus is obviously increased. As it is well known, the length of CNFs sometimes ranges from 5 to 100 microns, and the size of polymer chains are is of the identical range with CNFs. Hence, the mechanical properties of polymer will be significantly influenced by the CNF filler. As the filler material increased to 15%, there was a remarkable increase in the modulus of the material. These imply that at higher loading percentages, there is a better interaction between the fiber and the matrix. More fibers are sticking with the matrix and the applied load can be shared with the fiber and the matrix. As shown in the Figure 7.14, at higher loadings, transition from glassy state to rubbery state occurs at higher temperatures [39].

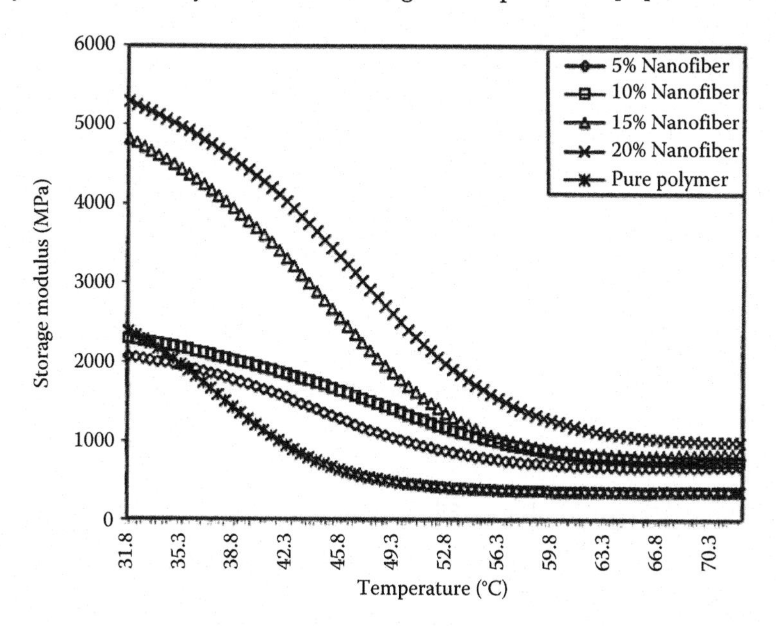

FIGURE 7.14
Effect of nanofiber-reinforcement on storage modulus. (Reprinted from Ni, Q.Q. et al., *Compos. Struct.*, 81, 176, 2007. With permission.)

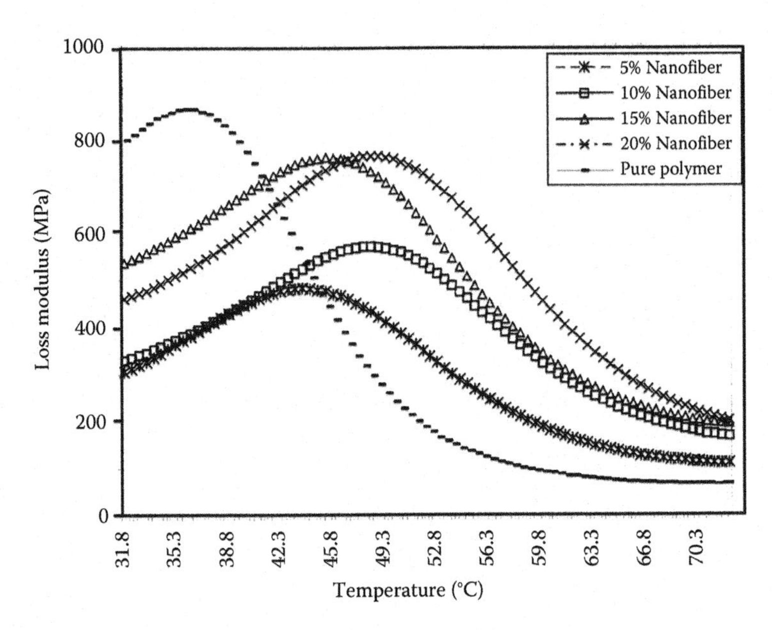

FIGURE 7.15
Effect of nanofiber reinforcement on loss modulus. (Reprinted from Manpreet, S., Thermal characterization of nanocomposite SMP for their mechanical and thermal properties, A thesis of graduate studies, Lamar University, Beaumont, TX, 2005. With permission.)

Loss modulus is an important parameter for damping properties for materials [39,40]. Figure 7.15 indicates the loss modulus of pure SMP and SMP composites. It exhibits that the loss modulus of pure SMP is about 800 MPa, while that of the SMP composites filled with 5% and 10% CNFs in weight is approximately 300 MPa. As the percentage of fiber increases to 15%, the increment in the loss modulus of the material is observed, which is contributed to the friction generated between the nanofibers. Pure SMP has the best damping characteristic compared with that of the nanofiber-filled SMP composite. As it is well known, the shape-memory effect of SMPs is seriously influenced by the friction force. Therefore, from the damping property determined through the loss modulus measurement, with an increase in CNF content, this filler will weaken the recovery behaviors.

7.3.3 Shape-Memory Polymer Filled with SiC Nanoparticles

Gall et al. [1,4] comprehensively investigated the thermomechanics of SMPs and their composites, including a SiC-particles-reinforced epoxy SMP. The SiC/SMP nanocomposite was found to have a higher elastic modulus, and was capable of generating higher recovery forces in comparison with a pure SMP specimen [4].

7.3.3.1 Thermal Transition and Stress–Strain Behavior

In the DMA test, dynamic thermal scan was performed from 0°C to 140°C at a constant heating rate of 5°C/min [1]. A static force of 50 mN, and a dynamic force of 40 mN were applied with a frequency of 1 Hz. As shown in Figure 7.16, the T_g of the pure SMP is marked out at the temperature 88°C. Above 118°C (T_g + 30°C), the storage modulus of the pure SMP, located at a rubbery state, is almost constant. The storage modulus of the pure SMP at 26°C is about two orders of magnitude larger than that at 118°C. In contrast, with the incorporation of 20 wt% SiC nanoparticles, T_g was increased by approximately 10°C. The experimental results revealed that at both 26°C and 118°C, the modulus of the SMP nanocomposite filled with 20 wt% of SiC is about twice as large as that of the pure SMP. Furthermore, the static tests results also demonstrated that SiC/SMP nanocomposites showed a higher elastic modulus than that of pure SMP (see Figure 7.17). Note that, at the same temperatures, the modulus determined by the static tests is 18%–28% lower than the storage modulus obtained from the dynamic tests. This difference may be attributed to the viscoelastic response of the SMP materials. In the static test, the force was applied at a rate of 10 mN/s, while in the dynamic test a dynamic force of 40 mN was applied with a frequency of 1 Hz.

As shown in Figures 7.16 and 7.17, it is obvious that the reinforced SMP presents a higher modulus than that of the pure SMP [1]. It should be firstly contributed to the higher elastic modulus of the SiC reinforcement in comparison with that of the SMP matrix. Note that the increased stiffness in SMP

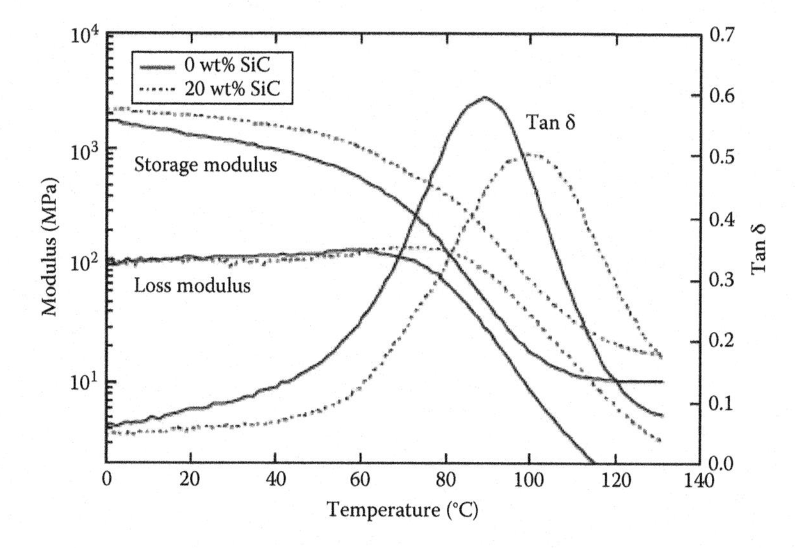

FIGURE 7.16
Storage modulus, loss modulus, and tangent δ of the SMP and SMP composite. (Reprinted from Manpreet, S., Thermal characterization of nanocomposite SMP for their mechanical and thermal properties, A thesis of graduate studies, Lamar University, Beaumont, TX, 2005. With permission.)

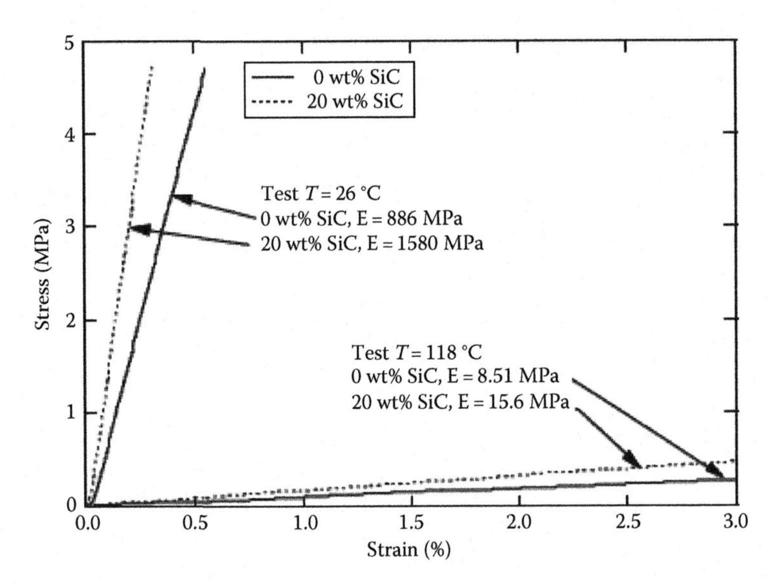

FIGURE 7.17
Elastic modulus of the SMP and the SMP composite at 26°C and 118°C. (Reprinted from Manpreet, S., Thermal characterization of nanocomposite SMP for their mechanical and thermal properties, A thesis of graduate studies, Lamar University, Beaumont, TX, 2005. With permission.)

nanocomposites is critical for the enhanced recoverable stress levels during shape recovery under constraint. The other reason is due to the chemical interaction between the polymer and particles at the polymer–particle interface. Meanwhile, this interactive action also influences the glass transition temperature of the nanocomposite. As shown in Figure 7.16, the T_g of SMP composites filled with nano reinforcement is higher than that of the pure SMP.

7.3.3.2 Recoverable Mechanical Property

During the evaluation of shape recovery, the experimental results are presented for the SMP and the SMP nanocomposite that had been pre-deformed at 118°C [1]. In Figure 7.18 the solid lines and dashed lines imply the response of the SMP and the SMP nanocomposite, respectively. As shown in Figure 7.18, the curves of the recovery stress as a function of temperature are initially steep and then flatten and converge to the pre-deformation stresses at 118°C. It is apparent that the recovery stress of the SMP/SiC nanocomposite is larger than that of the pure SMP at a certain fixed temperature. It demonstrates that the SiC reinforces the SMP and the recovery force consequently increases. For clarity, Figure 7.18 only presents recovery tests for the pre-deformation strain of 11%; tests at other strain levels yielded similar results.

In most cases, it is desirable to increase the actuation force in the progress of shape recovery with fixed strain. The recovery stresses at elevated

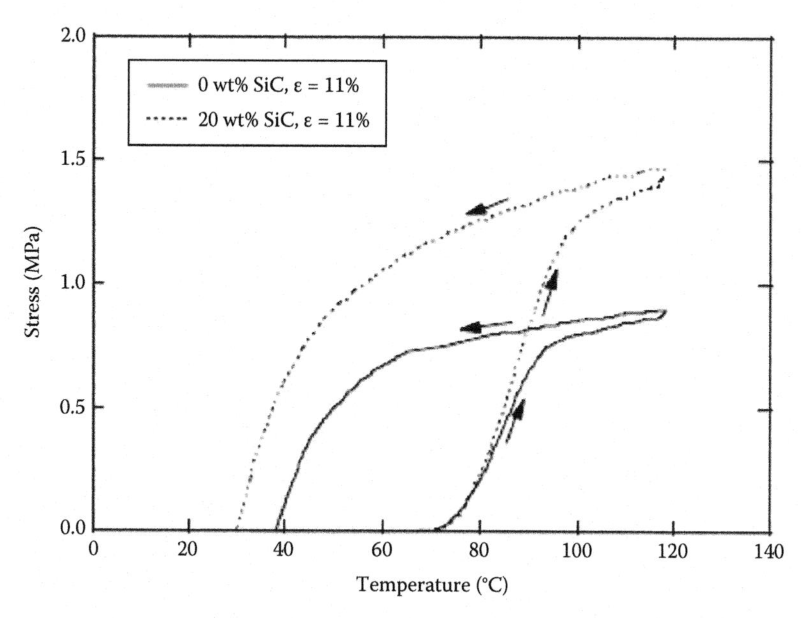

FIGURE 7.18
Stress recovery of the SMP and the SMP composite under strain constraint. (Reprinted from Manpreet, S., Thermal characterization of nanocomposite SMP for their mechanical and thermal properties, A thesis of graduate studies, Lamar University, Beaumont, TX, 2005. With permission.)

temperatures are mainly determined by the stress–strain relation at the high temperature range. The relatively high recovery stress shows potential for many applications that require good actuating performance. In addition, a burst peak stress can be obtained near T_g, which may be useful for SMP actuation. Since the initiation temperature of the shape recovery shifts to a lower temperature for pre-deformation at a temperature below T_g, it is can maintain the SMP device at a low temperature, e.g., the human body temperature, and obtain the peak recovery stress [41–43]. Based on the overall favorable properties of the SMP/SiC nanocomposite, it may receive more attention for researches and subsequent applications [44,45].

7.3.4 Shape-Memory Polymer Reinforced by Continuous Fibers

This section investigates the shape recovery behavior of a thermoset styrene-based shape-memory polymer composite (SMPC) reinforced by carbon-fiber fabrics [46]. The major advantages of SMPs are their extremely high recovery strain, low density, and low cost, etc. However, relatively low modulus and low strength are their intrinsic drawbacks. Fiber-reinforced SMPC which may overcome the above-mentioned disadvantages is studied by Leng et al. [46,47]. The investigation was conducted by three types of tests, namely, dynamic mechanical analysis (DMA), shape recovery test, and optical microscopic observation of the deformation mechanism for the SMPC specimen. Results reveal that the SMPC exhibits a higher storage modulus than that of the pure SMP. At and above T_g, the shape recovery ratio of the SMPC upon bending is above 90%. The shape recovery properties of the SMPC become relatively stable after some packaging/deployment cycles. Additionally, fiber microbucklings are a primary mechanism to approach a large strain in the bending of the SMPC.

To investigate the basic performance of pure SMP and SMPC at different temperatures, dynamic mechanical analyzer, DMA (NETZSCH, DMA242C), was used to determine the storage modulus and the value of tangent delta. A three-point bending mode was applied with a span of 40 mm. The dimensions of specimens were $50 \times 9 \times 3\,mm^3$. The scanning range of temperature was 0 ~ 120°C at a heating rate of 2°C/min and a frequency of 1 Hz. Moreover, in order to investigate the shape recovery performance of the SMPC, a shape recovery test upon bending of the SMPC specimen was performed in a water bath. In addition, to determine the microstructural mechanism of SMPC under the bending deformation, the optical microscope (ZEISS MC80DX) was used to observe the microstructural deformation of the SMPC specimens after shape recovery tests [46–48].

7.3.4.1 Dynamic Mechanical Analysis

The DMA test results in Figure 7.28 shows the storage modulus and tangent delta versus temperature for pure SMP and SMPC (Figure 7.19) [46]. According to the curves of storage modulus, it is clear that the SMPC holds a higher storage modulus than that of the pure SMP within the range of

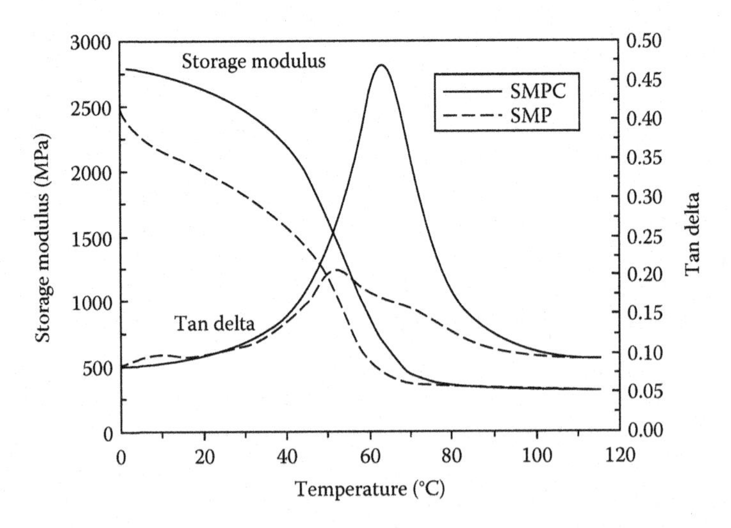

FIGURE 7.19
Storage modulus and tangent delta versus temperature for the pure SMP and the SMPC. (Reprinted from Lan, X. et al., *Smart Mater. Struct.*, 18, 024002, 2009.)

0~80°C. The storage modulus of the SMPC drops apparently within the glass transition region of about 40~80°C, namely $T_g - 20$°C to $T_g + 20$°C. The peak value of the tangent delta is defined as the T_g, and therefore the T_g of pure SMP and SMPC is found to be about 54°C and 64°C, respectively.

7.3.4.2 Shape Recovery Performance

A systematic shape recovery test of the SMPC specimen upon bending was performed. The typical procedure of the thermomechanical bending cycle for the SMPC includes the following steps (see Figure 7.20): (1) the specimen in its original shape is kept in a water bath for 5 min at $T_g + 20$°C; (2) the SMPC is bent to a storage angle θ_0 around a mandrel with the radius of 2 mm in the soft rubbery state, and then the SMPC is kept in a cool water with an external constraint to "freeze" the elastic deformation energy for 5 min (storage); and (3) the SMPC specimen fixed on the setup is immersed into another water bath at an elevated temperature, and then it recovers to an angle θ_N (recovery). The method used to quantify the precision of deployment is illustrated in Figure 7.20, where, r denotes the radius of mandrel, t represents the thickness of the SMPC specimen, θ_0 indicates an original storage angle of the specimen in storage state during the first bending cycle, $S(x_0, y_0)$ is a point selected to determine, θ_0. θ_N is the residual angle in the recovery state during the Nth thermomechanical bending cycle (N=1,2,3...). $R(x_N, y_N)$ is a testing point in order to calculate θ_N:

$$\theta_N = \text{ArcCot}\left(\frac{x_N}{y_N}\right) \quad (N = 1, 2, 3, \ldots\ldots, 0 \le \theta_N \le 180^\circ), \tag{7.12}$$

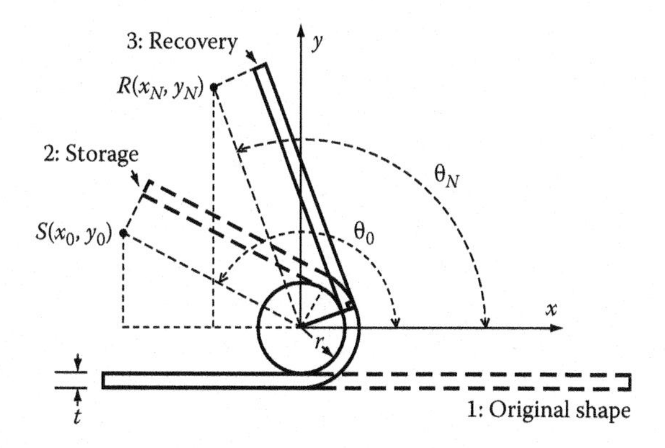

FIGURE 7.20
Schematic illustration for the test of shape recovery performances test. (Reprinted from Manpreet, S., Thermal characterization of nanocomposite SMP for their mechanical and thermal properties, A thesis of graduate studies, Lamar University, Beaumont, TX, 2005. With permission.)

The value of the shape recovery ratio is calculated by

$$R_N = \frac{\theta_0 - \theta_N}{\theta_0} \times 100\% \quad (N = 1, 2, 3......), \tag{7.13}$$

where

R_N denotes the shape recovery ratio of the Nth thermomechanical bending cycle.

$S(x_0, y_0)$ and $R(x_N, y_N)$ are measured by a vernier caliper with a resolution of 0.01 mm

Finally, θ_N and R_N are obtained through Equations 7.12 and 7.13. In the following tests, the radius r of mandrel and thickness t of the SMPC specimen are 2 mm and 3 mm, respectively.

To investigate the shape recovery velocity and shape recovery ratio R_1 of the SMPC at different temperatures, the test of recovery angle versus time was conducted at $T_g - 20°C$, $T_g - 10°C$, T_g, and $T_g + 10°C$. The predeformed temperature of the SMPC was set as the T_g of the SMPC (64°C). The original storage angle θ_0 was selected as 180°.

Figure 7.21 shows the relationship between recovery angle and time for four different SMPC specimens at various ambient temperatures. At and above the temperature of T_g, associated with a fast glass transition and strain energy dissipation of SMP, SMPC specimens deploy fast within the

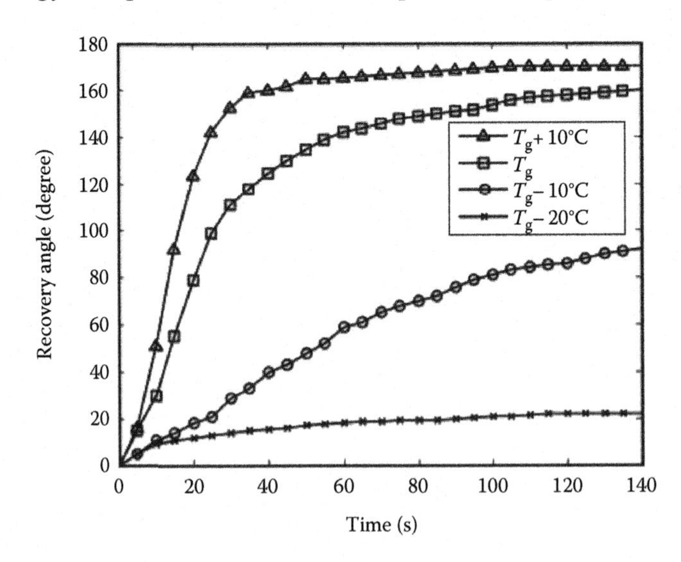

FIGURE 7.21
Recovery angle versus time during shape recovery process at different ambient temperatures ($T_g - 20°C$, $T_g - 10°C$, T_g, $T_g + 10°C$, where T_g is about 64°C). (Reprinted from Manpreet, S., Thermal characterization of nanocomposite SMP for their mechanical and thermal properties, A thesis of graduate studies, Lamar University, Beaumont, TX, 2005. With permission.)

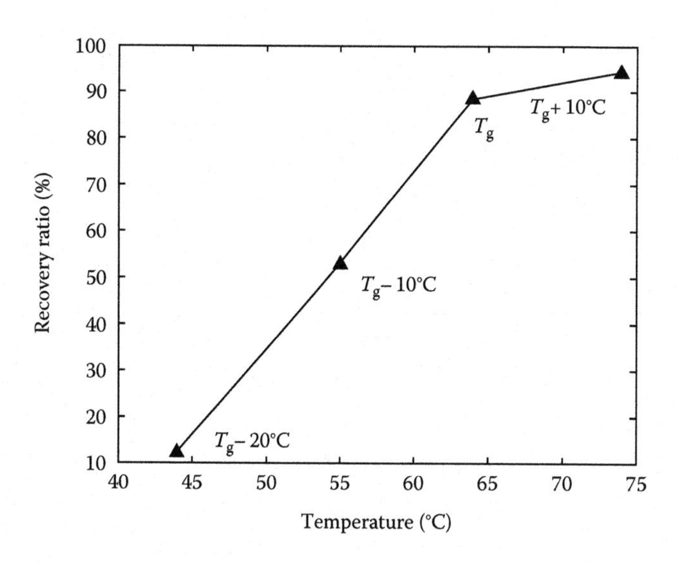

FIGURE 7.22
Shape recovery ratios at different ambient temperatures. Original storage angle θ_0 is 180° in storage state; Residual angle θ_1 in recovery state is selected at the 140th second during the first shape recovery cycle. (Reprinted from Manpreet, S., Thermal characterization of nanocomposite SMP for their mechanical and thermal properties, A thesis of graduate studies, Lamar University, Beaumont, TX, 2005. With permission.)

initial 30 s, and then the deployment velocity drops quickly owing to the termination of energy dissipation. Below T_g, the SMPC specimen deploys very slowly due to a partial glass transition and an energy dissipation of the SMP. In addition, based on Figure 7.21, a shape recovery ratio R_1 is defined (see Figure 7.22) where the residual angle θ_1 is selected at the 140th second in the first deformation cycle ($N=1$). As shown in Figure 7.22, at and above the temperature of T_g, the shape recovery ratio keeps relatively stable, and this corresponds to the full glass transition above T_g of the SMP.

In order to evaluate a degradation of the deployment, the shape recovery ratio R_N corresponding to the number of bending cycles N was tested at an original bending angle $\theta_0=180°$ at T_g +20°C (see Figure 7.23). It shows that the recovery ratio R_N decreases from 96% to 91% during the first 50 bending cycles. The shape recovery ratio becomes relatively stable after the 30th bending cycle, which keeps approximately at 90%.

Since the storage deformation of SPMC structures may be varied, the characterization of the shape recovery performance at different bending angles θ_0 was performed. The relationship between the residual angles θ_N ($N=1$, 2, 3, 4, and 5) and the original storage angles θ_0 ($\theta_0=45°$, 90°, 135°, and 180°) was investigated (Figure 7.24). Results indicate that the residual angle of the SMPC corresponding to the original bending angle of 45° is the lowest (3° ~ 5°). The SMPC laminate with the larger storage angle θ_0 results in a larger residual angle θ_N.

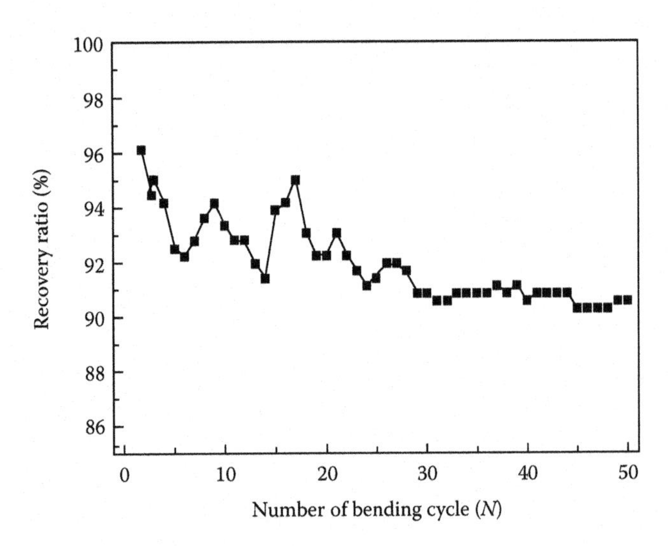

FIGURE 7.23
Shape recovery ratios versus the number of bending cycle at $T_g + 20°C$. (Reprinted from Manpreet, S., Thermal characterization of nanocomposite SMP for their mechanical and thermal properties, A thesis of graduate studies, Lamar University, Beaumont, TX, 2005. With permission.)

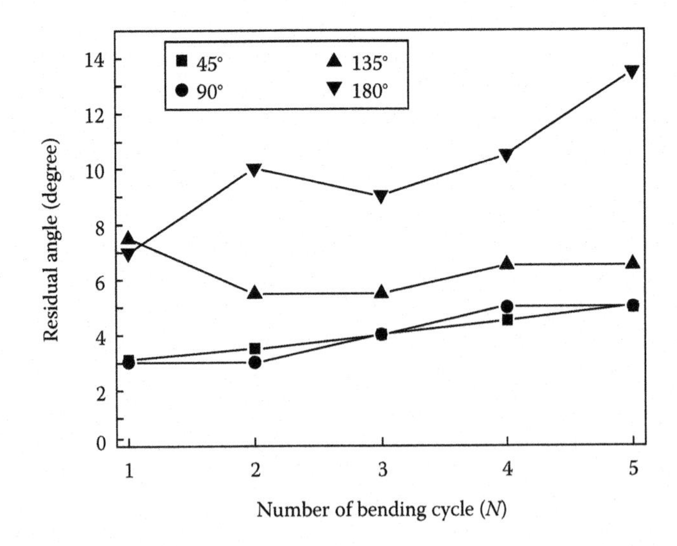

FIGURE 7.24
Residual angle θ_N ($N = 1, 2, 3, 4, 5$) versus the number of bending cycle at different original storage angles θ_0 (45°, 90°, 135°, 180°). (Reprinted from Liu, Y. et al., *Mech. Mater.*, 36, 929, 2004. With permission.)

7.3.4.3 Deformation Mechanism of Microstructure

The shape recovery performance of SMPCs is decided not only by the shape-memory effect of SMPs, but also by the microstructural deformation mechanism of fibers and SMPs [46]. Hence, the microscopic observation was conducted for the SMPC specimen during shape recovery process. As it is well known, the tensile stiffness of the carbon fibers is much higher than their compressive stiffness and the stiffness of the SMP matrix. Thus, we assume that, upon applying a bending force on SMPC specimen, the neutral axis of the bent specimen moves from the middle-plane towards the outer surface where the fibers are in a tensile stress state. As a result, all the other fibers except for those on the outer surface are in the compressive stress state [49]. Based on the assumption of the linear distribution of the compressive strains along the thickness of the specimen, we have

$$\frac{r}{t} = \frac{1}{\varepsilon} \tag{7.14}$$

where

r denotes the radius of the mandrel
t represents the thickness of specimen (see Figure 7.30)
ε is the maximum strain on the inner surface

The microstructure, particularly for the fiber microbuckling of the specimen, was observed by the optical microscope when 50 thermomechanical cycles are completed. We aim to investigate the microstructural mechanism of the SMPC specimen at large deformation upon bending in the thermomechanical cycles (see Figure 7.23). The microstructure of SMPC specimen was observed in three shapes after 50 thermomechanical cycles ($N = 50$), namely the original shape before bending (Figure 7.25a), the storage shape with the original storage angle $\theta_0 = 180°$ (Figure 7.25b), and the recovery shape $\theta_{50} \approx 16°$ (Figure 7.25c). In addition, for comparison, the microstructure of the SMPC specimen upon bending after the static three-point bending test at a room temperature is also investigated (Figure 7.25d).

Figure 7.25a indicates an optical microscopic image of the SMPC specimen in a side view right after fabrication without a deformation. Before applying the bending force, the bonding among transverse fibers, longitudinal fibers, and SMP matrix is pretty well, namely, no failure or delimitation in the composite. Figure 7.25b reveals a fiber microbuckling at the original storage angle $\theta_0 = 180°$. A large delimitation gap can be observed between transverse fiber tow and longitudinal fiber tow. The sine shape of the microbuckling of the transverse fiber tow can also be observed. Figure 7.25c presents the recovered configuration of the fiber microbuckling at the same location of Figure 7.25b. A small irreversible delaminate gap can also be observed between the transverse fiber tow and the longitudinal fiber tow. Figure 7.25d presents a

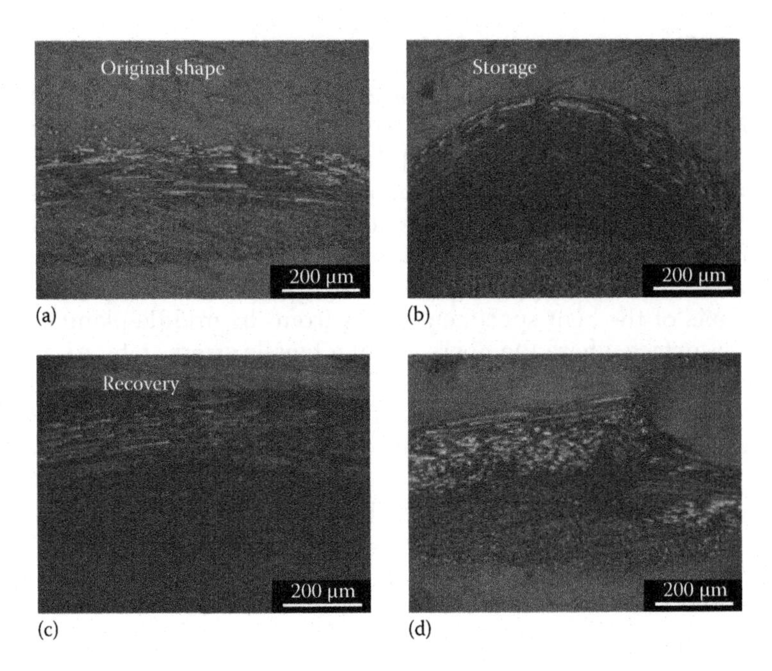

FIGURE 7.25
Optical microscopic images of microstructure of the SMPC specimen after 50 bending cycles in (a), (b), and (c) shape: (a) the original shape before bending, (b) the storage shape with an original bending angle $\theta_0 = 180°$, and (c) the recovery shape. (d) SMPC specimen upon bending after the static three-point bending test at a room temperature. (Reprinted from Liu, Y. et al., *Mech. Mater.*, 36, 929, 2004. With permission.)

buckled fracture of a transverse fiber tow after the static three-point bending test at the room temperature of ($T_g - 40°C$). It is clear that the buckling fracture of the transverse fiber tow occurs on the inner surface beyond the compressive strain limit of fibers.

In the thermomechanical cycle of predeformation and shape recovery stage ($T \geq T_g$), the stiffness of the SMP matrix was low enough to provide mobility for fibers to avoid the extremely high strain/stress in local areas and the permanent fiber buckling failures. Due to the microbucklings of the fiber and the SMP, only fiber microbucklings but no fiber fractures were observed during the thermomechanical cycles. It is obvious that the SMPC specimen approaches a much higher strain than the strain limit of carbon fiber associated with the microbuckling of the fiber and the SMP. In contrast, at $T \leq T_g - 40°C$, due to the relatively high stiffness of the SMP matrix, the mobility of the buckled fiber in high compressive strain state is limited by the strong constraint of the SMP matrix, which leads to the brittle fracture of the transverse fiber. Consequently, both fiber fractures and microbucklings were observed in the static three-point bending test at the room temperature.

SMPC materials are similar to traditional fiber-reinforced composites except for the use of a thermoset styrene-based shape-memory resin that enables much higher packaging strains than traditional composites without damage to the fibers or the resin. This high strain capacity can lead to SMPC component designs that can be packaged more compactly than designs made with other materials. In order to achieve high package strain and avoid fiber failure in the storage state, fiber microbuckling is required. With the microbuckling, SMPC materials are suited for use in deployable space structure components because of their high strain-to-failure capability [50,51].

7.4 Summary and Outlook

Several researches have been performed to improve the performance of SMPs, especially for the mechanical properties. Particle-filled and fiber-reinforced SMP composites have been employed to overcome the relatively poor mechanical property of pure SMPs, resulting in higher strength, stiffness, and recovery force, etc. These effective reinforcements include carbon or metallic particles, CNTs, CNFs, and chopped and continuous glass/carbon fibers. Based on the existing studies, it is prospected that particle fillers, particularly for nanosized particles will be widely used to improve the mechanical properties of polymer. The ongoing nanocomposites can be summarized as follows: introduce a novel developing filler into a polymer; improve the interaction and bonding between the nanoparticles and the molecular chains at a molecular level; and design the structure of the composite and provide approaches for fabrication. Although some achievements have been accomplished, fabrication technologies and characterization methods require a lot of researches. With these technologies and methods, it is declared that SMP materials will be more efficient and useful for their potential applications.

References

1. Liu, Y., Gall, K., Dunn, M. L., and McCluskey, P. 2004. Thermomechanics of shape memory polymer nanocomposites. *Mechanics of Materials* 36:929–940.
2. Li, F. and Larock, R. C. 2002. New soybean oil-styrene-divinylbenzene thermosetting copolymer. V. Shape memory effect. *Journal of Applied physics* 84:1533–1543.
3. Liang, C., Rogers, C. A., and Malafeew, E. 1997. Investigation of shape memory polymers and their hybrid composites. *Journal of Intelligent Material Systems and Structures* 4:380–386.

4. Gall, K., Mikulas, M., Munshi, N. A., Beavers, F., and Tupper, M. 2000. Carbon fiber reinforced shape memory polymer composites. *Journal of Intelligent Materials Systems and Structures* 11:877–86.
5. Ohki, T., Ni, Q. Q., Ohsako, N., and Iwamoto, M. 2004. Mechanical and shape memory behavior of composites with shape memory polymer. *Composites Part A: Applied Science and Manufacturing* 35:1065–1073.
6. Ohki, T., Ni, Q. Q., and Iwamoto, M. 2004. Creep and cyclic mechanical properties of composites based on shape memory polymer. *Science and Engineering of Composite Materials* 11:137–147.
7. Liang, B. H., Mott, L., Shaler, S. M., and Caneba, G. T. 1994. Properties of transfermolded wood-fiber polystyrene composite. *Wood and Fiber Science* 26:382–389.
8. Tobushi, H., Hashimoto, T., Hayashi, S., and Yamada, E. 1997. Thermomechanical constitutive modeling in shape memory polymer of polyurethane series. *Journal of Intelligent Material Systems and Structures* 8:711–718.
9. Tobushi, H., Okumura, K., Hayashi, S., and Ito, N. 2001. Thermomechanical constitutive model of shape memory polymer. *Mechanics of Materials* 33:545–554.
10. Liu, Y., Gall, K., Dunn, M. L., Greenberg, A. R., and Diani, J. 2006. Thermomechanics of shape memory polymers: Uniaxial experiments and constitutive modeling. *International Journal of Plasticity* 22:279–313.
11. Chen, Y. C. and Lagoudas, D. C. 2008. A constitutive theory for shape memory polymers. Part I: Large deformations. *Journal of the Mechanics and Physics of Solids* 56:1752–1765.
12. Chen, Y. C. and Lagoudas, D. C. 2008. A constitutive theory for shape memory polymers. Part II: A linearized model for small deformations. *Journal of the Mechanics and Physics of Solids* 56:1766–1778.
13. Zhou, B., Liu, Y. J., and Leng, J.S. 2009. Finite element analysis on thermomechanical behaviors of styrene-based shape memory polymer. *Acta Polymerica Sinica* 6:525–529.
14. Zhou, B., Liu, Y. J., Lan, X., Leng, J. S., and Yoon, S. 2009. A glass transition model for shape memory polymer and its composite. *International Journal of Modern Physics B* 23(6&7):1248–1253.
15. Zhou, B., Liu, Y. J., and Leng, J. S. 2009. Modeling the shape memory effect of shape memory polymer. *Proc. SPIE* (to be published).
16. Iijima, S. 1991. Helical microtubules of graphitic carbon. *Nature* (London) 354:56–58.
17. Subramoney, S. 1998. Novel nanocarbons—Structure, properties, and potential applications. *Advanced Materials* 15:1157–1171.
18. Wong, E. W., Sheehan, P. E., and Lieber, C. M. 1997. Nanobeam mechanics: Elasticity, strength, and toughness of nanorods and nanotubes. *Science* 277:1971–1975.
19. Rao, C. N. R., Satishkumar, B. C., Govindaraj, A., and Nath, M. 2001. Nanotubes. *Chemical Physics and Physical Chemistry* 2:78–105.
20. Baughman, R. H., Zakhidov, A. A., and Heer, W. A. D. 2002. Carbon nanotubes—The route toward applications. *Science* 297:787–792.
21. Xu, Y., Higgins, B., and Brittain, W. J. 2005. Bottom-up synthesis of PS–CNF nanocomposites. *Polymer* 46:799–810.

22. Lou, X., Detrembleur, C., Sciannamea, V., Pagnoulle, C., and Jerome, R. 2004. Grafting of alkoxyamine end-capped (co)polymers onto multiwalled carbon nanotubes. *Polymer* 45:6097–6102.
23. Wong, M., Paramsothy, M., Xud, X. J., Ren, Y., Li, S., and Liao, K. 2003. Physical interactions at carbon nanotube–polymer interface. *Polymer* 44:7757–7764.
24. Shaffer, M. S. P. and Windle, A. H. 1999. Fabrication and characterization of carbon nanotube/poly(vinyl alcohol) composites. *Advanced Materials* 11:937–941.
25. Seoul, C., Kim, Y. T., and Baek, C. K. 2003. Electrospinning of poly(vinylidene fluoride)/dimethylformamide solutions with carbon nanotubes. *Journal of Polymer Science Part B Polymer Physics* 41:1572–1577.
26. Seo, M. K. and Park, S. J. 2004. A kinetic study on the thermal degradation of multiwalled carbon nanotubes-reinforced poly(propylene) composites. *Macromolecular Materials Engineering* 289:368–374.
27. Bhattacharyya, A. R., Sreekumar, T. V., Liu, T., Kumar, S., Ericson, L. M., Hauge, R. H., and Smalley, R. E. 2003. Crystallization and orientation studies in polypropylene/single-wall carbon nanotube composite. *Polymer* 44:2373–2377.
28. Chang, T. E., Jensen, L. R., Kisliuk, A., Pipes, R. B., Pyrz, R., and Sokolov, A. P. 2005. Microscopic mechanism of reinforcement in single-wall carbon nanotube/polypropylene nanocomposite. *Polymer* 46:439–444.
29. Kashiwagi, T., Grulke, E., Hilding, J., Groth, K., Harris, R., Butler, K., Shields, J., Kharchenko, S, and Douglas, J. 2004. Thermal and flammability properties of polypropylene/carbon nanotube nanocomposites. *Polymer* 45:4227–4239.
30. Kumar, S., Doshi, H., Srinivasarao, M., Park, J. O., and Schiraldi, D. A. 2002. Fiber from polypropylene/nanocarbon fiber composites. *Polymer* 43:1701–1703.
31. Assouline, E., Lustiger, A., Barber, A. H., Cooper, C. A., Klein, E., Wachtel, E., and Wagner, H. D. 2003. Nucleation ability of multiwall carbon nanotubes in polypropylene composite. *Journal of Polymer Science Part B* 41:520–527.
32. Kashiwagi, T., Grulke, E., Hilding, J., Harris, R., Awad, W., and Douglas, J. 2002. Thermal degradation and flammability properties of poly(propylene)/carbon nanotube composites. *Macromolecular Rapid Communication* 23:761–765.
33. Liu, T., Phang, I. Y., Shen, L., Chow, S. Y., and Zhang, W. D. 2004. Morphology and mechanical properties of multiwalled carbon nanotubes reinforced nylon-6 composites. *Macromolecules* 37:7214–7222.
34. Kao, C. C. and Young, R. J. 2004. A Raman spectroscopic investigation of heating effects and the deformation behavior of epoxy/SWNT composites. *Composite Science Technology* 64:2291–2295.
35. Gojny, F. H. and Schulte, K. 2004. Functionalization effect on the thermomechanical behavior of multiwall carbon nanotube/epoxy-composites. *Composite Science Technology* 64:2303–2308.
36. Sandler, J. K. W., Kirk, J. E., Kinloch, I. A., Shaffer, M. S. P., and Windle, A. H. 2003. Ultra-low electrical percolation threshold in carbon-nanotube–epoxy composites. *Polymer* 44:5893–5899.
37. Tobushi, H., Hayahi, S., Ikai, A., and Hara, H. 1996. Basic deformation properties of a polyurethane-series shape memory polymer film. *The Japan Society Mechanical Engineering A* 62:576–582.
38. Ni, Q. Q., Zhang, C., Fu, Y. Dai, G., and Kimura, T. 2007. Shape memory effect and mechanical properties of carbon nanotube/shape memory polymer nanocomposites. *Composite Structures* 81:176–184.

39. Manpreet, S. 2005. Thermal characterization of nanocomposite shape memory polymer for their mechanical and thermal properties. A thesis presented to The Faculty of Graduate Studies Lamar University, requirement for the degree master of engineering science, Lamar University, Beaumont.
40. Ehrenstein, M. R., Rada, C., Jones, A. M., Milstein, C., and Neuberger, M. S. 2001. Switch junction sequences in PMS2-deficient mice reveal a microhomology-mediated mechanism of Ig class switch recombination. *Proceedings of the National Academy of Sciences of the United States of America* 98:14553–14558.
41. Zeng, Y. M., Hu, J. L., and Yan, H. J. 2002. Temperature dependency of water vapor permeability of shape memory polymer. *Journal of Dong Hua University* 19:52–57.
42. Benett, W. J., Krulevitch, P. A., Lee, A. P., Northrup, M. A., and Folta, J. A. 1997. Miniature plastic gripper and fabrication method, United States Patent 5,609,608.
43. Ferrera, D. A. 2001. Shape memory polymer intravascular delivery system with heat transfer medium, United States Patent 6,224,610.
44. Ash, B. J., Schadler, L. S., and Siegel, R. W. 2001. Thermal and mechanical properties of alumina/polymethylmethacrylate (PMMA) nanocomposites: Effects of strong and weak interfaces. Abstract of papers of the American Chemical Society 222:U286-U286.
45. Bhattacharya, S. K. and Tummala, R. R. 2002. Epoxy nanocomposite capacitors for application as MCM-L compatible integral passives. *Journal of Electronic Packaging* 124:1–6.
46. Lan, X., Wang, X. H., Liu, Y. J., Leng, J. S., and Du, S. Y. 2009. Fiber reinforced shape-memory polymer composite and its application in a deployable hinge. *Smart Materials and Structures* 18(024002):1–6.
47. Lan, X., Lv, H. B., Liu, Y. J., and Leng, J. S. Thermomechanical behavior of fiber reinforced shape memory polymer composite, SPIE International Conference on Smart Materials and Nanotechnology, SPIE 6423, July1–4, 2007, Harbin, China.
48. Lan, X., Lv, H. B., Leng, J. S., and Du, S. Y. Investigation of the mechanical behaviors for fiber reinforced shape memory polymer composite, 16th International Conference on Composite Materials, July, 2007, Kyoto, Japan.
49. Gall, K., Mikulas, M., Munshi, N. A., Beavers, F., and Tupper, M. 2000. Carbon fiber reinforced shape memory polymer composites. *Journal of Intelligent Material Systems and Structures* 11:877–886.
50. Abrahamson, E. R., Lake, M. S., Munshi, N. A., and Gall, K. 2003. Shape memory mechanics of an elastic memory composite resin. *Journal of Intelligent Material Systems and Structures* 14:623–632.
51. Schultz, M. R., Francis, W. H., Campbell, D., and Lake, S. M. 48th AIAA/ASME/ASCE/AHS/ASC Structures, Structural Dynamics, and Materials Conference 23–26 April, 2007. Honolulu, HI AIAA 2007-2401.

8

Applications of Shape-Memory Polymers in Aerospace

Yanju Liu
Department of Aerospace Science and Mechanics, Harbin Institute of Technology, Harbin, P.R. China

Jinsong Leng
Centre for Composite Materials and Structures, Harbin Institute of Technology, Harbin, P.R. China

CONTENTS

8.1 Introduction

Shape-memory materials are stimuli-responsive materials that can recover their original shape upon applying an external stimulus, such as heat, light, moisture. The use of this shape-memory material and its reinforced components have the potential to provide enhanced performance, improved efficiency of devices, and overcome certain drawbacks of the traditional mechanism.

As a novel kind of smart material, shape-memory polymers (SMPs) currently cover a broad application area ranging from outer space to subterranean regions, for example, smart fabrics and textiles [1,2], heat-shrinkable films for packaging or tubes for electronics [3], self-deployable sun sails used on spacecraft [4], self-disassembling mobile phones [5], intelligent medical devices [6], or implants for minimally invasive surgery. Among these SMPs, the thermoresponsive SMP is common. Recently, this new kind of smart material has been developed and qualified especially for deployable components and structures in aerospace. The applications include hinge, truss and boom, antenna, optical reflector, solar array, smart mandrel, morphing skin. For the traditional aerospace deployable structures, the change of structural configuration in-orbit is accomplished through the use of a mechanical hinge, stored energy devices or motor driven tools. In contrast, the deployment devices fabricated by SMPs and their composites can overcome certain inherent disadvantages. However, there are some intrinsic drawbacks for the traditional deployment devices, such as complex assembled processes, massive mechanisms, big volume, and undesired impacts during the deployment. In contrast, the novel deployable devices based on SMPs and their composite materials can overcome certain above-mentioned drawbacks.

In this chapter, the fundamental applications of SMPs in deployable devices are presented. Some applications demand that stored energy be used to exert force, enabling mechanical work, such as deployment of a packaged, load-bearing actuator. Composites in the deployment device are addressed especially in the applications of carbon fiber–reinforced SMPs.

8.2 Shape-Memory Polymer Composite Hinge

8.2.1 Finite Element Method Modeling and Analysis

Fiber-reinforced shape-memory polymer composites (SMPCs) can be used in actively deformable structures [7–12]. In these structures, the flexural deformation is the main mode for the deformation with thin shells, where the bending angle of flexure is almost larger than 90°, but the strain is often smaller than 10%. In this way, the shape recovery process of SMP composite structures concerns structural deployment dynamics. For the deployable hinge, the thermoset SMP composite structure can be simplified as a curved shell tape spring, as shown in Figure 8.1. A curved SMPCs shell is a thin-walled, straight strip of material with curved cross section. Referenced from the efficient analysis method of Prof. S. Pellegrino (Refs. [7,8]), the curved SMPC shell can be folded either in an equal sense, or in the opposite sense.

Figure 8.2 shows the principal strains on the surface $z = -t/2$ of the curved shell under the equal-sense bending. Figure 8.2a is the maximum strain,

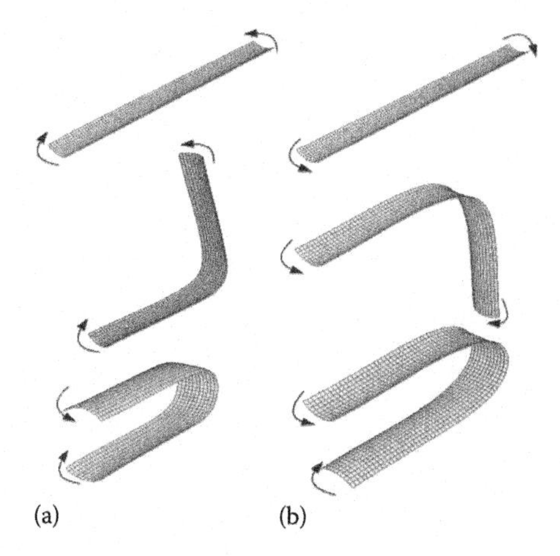

FIGURE 8.1
Two ways of folding a curved SMPC shell: (a) Equal-sense and (b) opposite-sense bending. (Reproduced from Lan, X. et al., *Proc. SPIE*, 7289, 728910, 2009.)

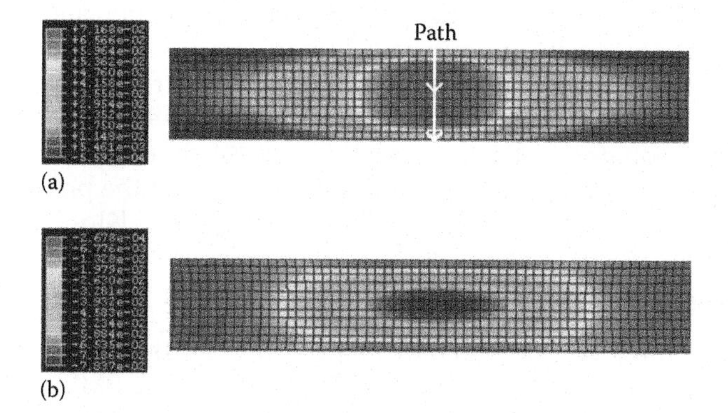

FIGURE 8.2
Principal strains on surface $z = -t/2$ curved shell under equal-sense bending; (a) maximum strain, ε_x; (b) minimum strain, ε_y. (Reproduced from Lan, X. et al., *Proc. SPIE*, 7289, 728910, 2009.)

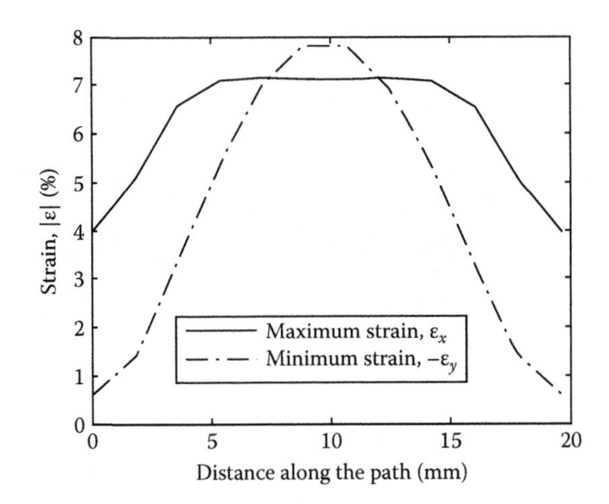

FIGURE 8.3
Principal strain of the point on the inner surface along the path. The solid curve: the maximum strain; the dash-dotted curve: the minimum strain. (Reproduced from Lan, X. et al., *Proc. SPIE*, 7289, 728910, 2009.)

ε_x, and Figure 8.2b indicates the minimum strain, ε_y. Only the top-surface strains are shown, as the distributions of strain on the opposite surface are essentially identical but with the sign reversed. It shows that the maximum strains are located in the central part and the distribution is somewhat uniform through the central part of the shell, corresponding to the fold region.

In order to investigate the regularity of the strain in the cross section of a curved shell, the strains of the nodes along the path, as shown in Figure 8.2a, are extracted at a bending angle of about 180°. The raw data are plotted in Figure 8.3. Figure 8.3 shows that the maximum strain is about 7% in the central part of the shell. This part corresponds to the fold region and it confirms that this region is somewhat uniformly curved.

To investigate the evaluation of the strain of the curved shell, the maximum strain nodes in the middle of the path at different steps during the deformation sequence is also extracted. Figure 8.4 shows the evaluation of the maximum strain in the deformation. It implies that the principal strain almost increases linearly during the deformation process [9].

8.2.2 Design and Deployment Demonstration

A deployable hinge was designed by using SMPC. Furthermore, a prototype of solar array was actuated using this SMPC hinge. The SMPC hinge consists of two curved circular SMPC shells in opposite directions, as shown in Figure 8.5.

The efficiency of Joule heating via the application of a voltage is important for the SMPC hinge. In order to investigate the temperature distribution of the hinge upon heating by an electrical current, an infrared camera (InfraTec,

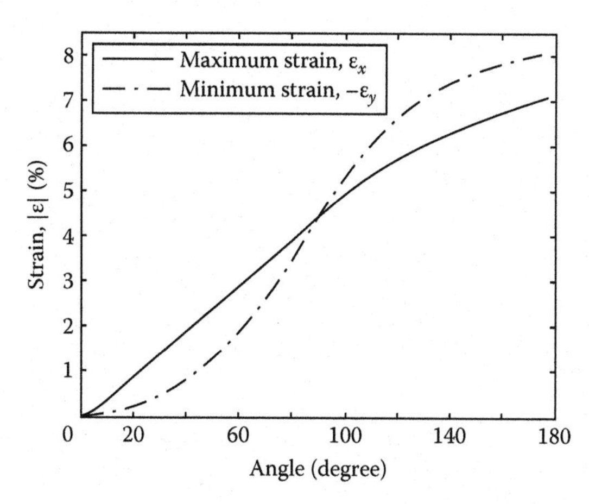

FIGURE 8.4
Evaluation of the principal strain of the point with maximum strain in the central region of shell. (Reproduced from Lan, X. et al., *Proc. SPIE*, 7289, 728910, 2009.)

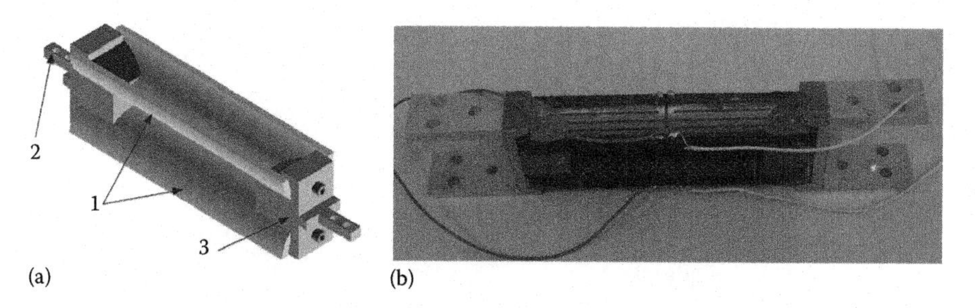

FIGURE 8.5
SMP composite hinge. (a) Illustration of the hinge (1, curved SMPC shell; 2 and 3, fixture of the hinge); (b) real scale hinge. (Reproduced from Lan, X. et al., *Proc. SPIE*, 7289, 728910, 2009.)

VarioCam) was used. Figure 8.6 shows the temperature distributions of the infrared image of the SMPC hinge during heating. A voltage of 20 V is applied on the embedded resistor heater in each curved SMPC laminate. The temperature of the SMPC hinge remains at about 100–130°C after heating for 100 s.

A prototype of solar array is actuated using the SMPC hinge. Figure 8.7 shows the deployment process of the SMPC hinge. A voltage of 20 V is applied on the embedded resistor heater in each circular laminate with a current of about 0.8 A. Hence, the total power of the hinge with two circular laminates is about 32 W. The temperature of the SMPC hinge remains at about 80°C after heating for 30 s. The original storage angle of the SMPC hinge in storage state is about 140°. The entire deployment process takes about 100 s. The

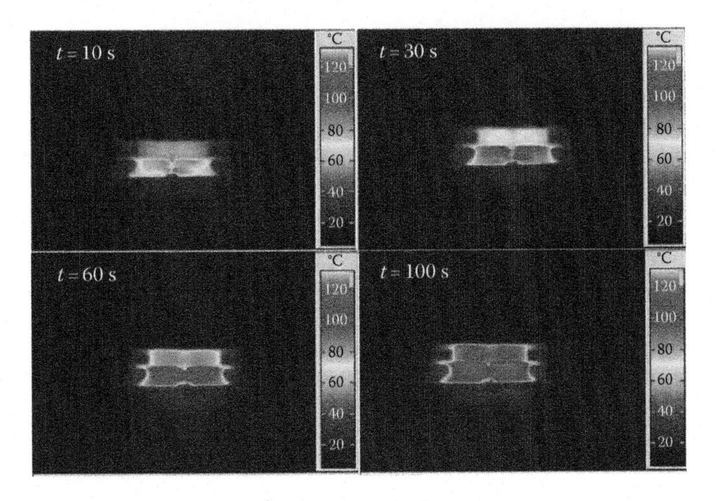

FIGURE 8.6
Temperature distributions of infrared image of SMPC hinge during heating by a 20 V voltage. (Reproduced from Lan, X. et al., *Proc. SPIE*, 7289, 728910, 2009.)

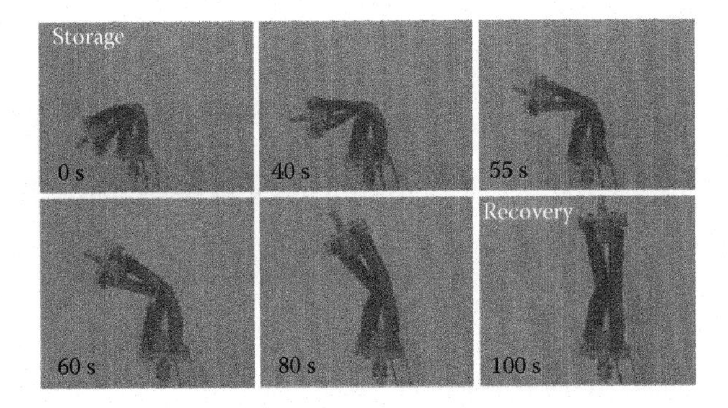

FIGURE 8.7
Shape recovery process of SMPC hinge. (Reproduced from Lan, X. et al., *Proc. SPIE*, 7289, 728910, 2009.)

deployment velocity of the hinge in the medium stage is relatively higher than those in the initial and final stages, as shown in Figure 8.8. The deployment ratio approaches approximately 100% [9].

Figure 8.9 shows a deployment process of a prototype of solar array that is actuated by an SMPC hinge. The prototype of solar arrays is suspended on a setup that simulates a zero-gravity environment. Heated by a voltage of 20 V, the SMPC hinge was bent to an original storage angle of 90° upon applying an external force. After fixing the storage shape at room temperature, the SMPC hinge was heated again by applying the same voltage. The prototype of-solar array deployed from 90° to ~0° for approximately 80 s.

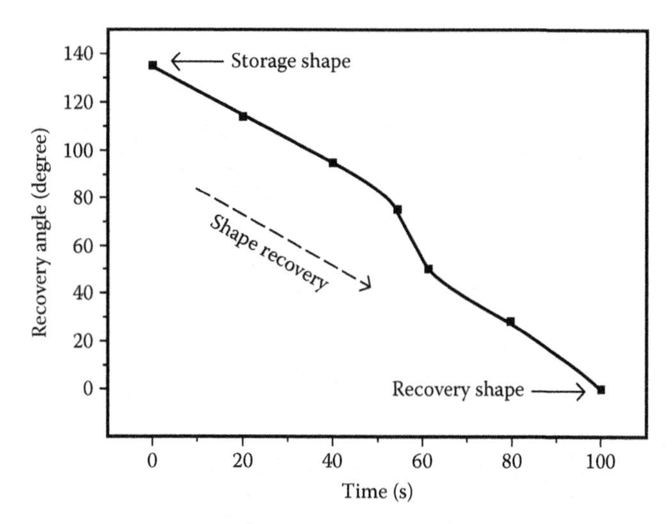

FIGURE 8.8
Relationship of recovery angle and time during the shape recovery process of SMPC hinge. (Reproduced from Lan, X. et al., *Smart Mater. Struct.*, 18, 024002, 2009.)

FIGURE 8.9
Shape recovery process of a prototype of solar array actuated by SMPC hinge. (Reproduced from Lan, X. et al., *Smart Mater. Struct.*, 18, 024002, 2009.)

8.2.3 Application in Solar Arrays

During the launching of the spacecraft, the area in the spacecraft is quite limited. Hence, the spacecraft needs lightweight, reliable, and cost-effective mechanisms for the deployment of radiators, solar arrays, and other devices. Composite Technology Development (CTD), Inc. has developed Epoxy SMP composites reinforced by carbon fiber (elastic memory composite, EMC) materials. The EMC materials show very high reversible strains, achieving high-deployed stiffness and strength-to-weight ratios. CTD has developed a deployable hinge fabricated by EMC materials (see Figure 8.10). Recently, CTD performed extensive ground testing on an EMC-deployable hinge that may be used for deployable spacecraft components (Figure 8.11) [10].

FIGURE 8.10
EMC flight hinge shown integrated onto ExpSA and packaged. (From Barrett, R. et al., Qualification of elastic memory composite hinges for spaceflight applications, in *47th AIAA/ASME/ASCE/AHS/ASC Structures, Structural Dynamics, and Materials Conference*, Newport, RI, May 1–4, 2006, AIAA-2006-2039.)

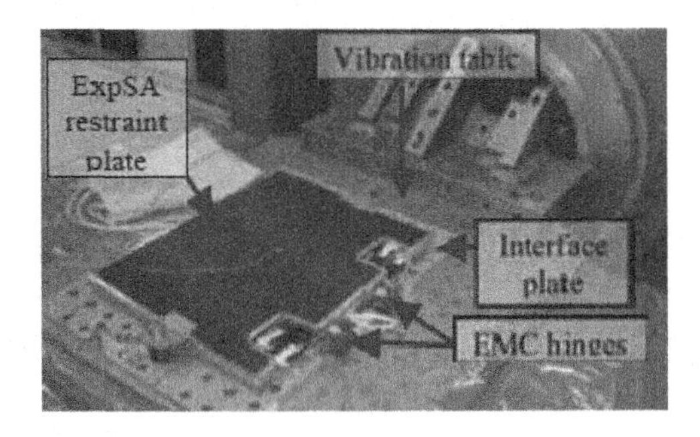

FIGURE 8.11
Random vibration test setup for ExpSA system with TEMBO® EMC hinges. (From Barrett, R. et al., Qualification of elastic memory composite hinges for spaceflight applications, in *47th AIAA/ASME/ASCE/AHS/ASC Structures, Structural Dynamics, and Materials Conference*, Newport, RI, May 1–4, 2006, AIAA-2006-2039.)

8.3 Shape-Memory Polymer Composite Boom

An EMC boom has been designed by CTD and used on a Micro-satellite. This extendable boom is lightweight and can support a variety of tip payloads (Figure 8.12). It has been identified to support a Micro-Propulsion Attitude Control System. The EMC is the central element of the boom. To stow the boom, the longerons are in a z-shape; thus, the EMC are flattened and bent in both the equal- and opposite-sense in a predeformed shape [11].

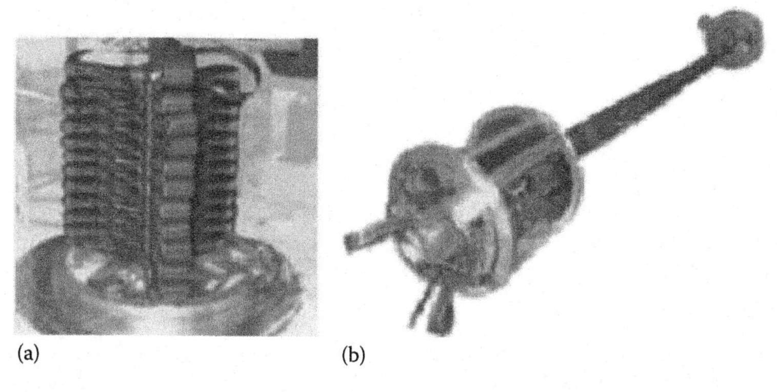

FIGURE 8.12
The proposed FalconSat-3 three-longeron EMC tubular boom in both a packaged (without the model MPACS unit, (a)) and deployed (b) configuration without the outer shroud. (Reproduced from Arzberger, S.C. et al., Elastic memory composites (EMC) for deployable industrial and commercial applications, in *SPIE Conference: Smart Structures and Materials 2005—Industrial and Commercial Applications of Smart Structures Technologies*, San Diego, CA, vol. 5762, pp. 35–47, 2005.)

To date, the deployable truss boom is considered in many ways to be the simplest and most efficient booms [12]. CTD proposed a three-longeron coilable truss boom that may be made by EMC, as shown in Figure 8.13a. EMC longerons could be fabricated by using a laboratory-scale pultrusion technique and evaluated in flexure and packaging ability. The three-longeron deployable booms could be packaged by twisting the boom into a helical shape when the EMC is above its transition temperature, as shown

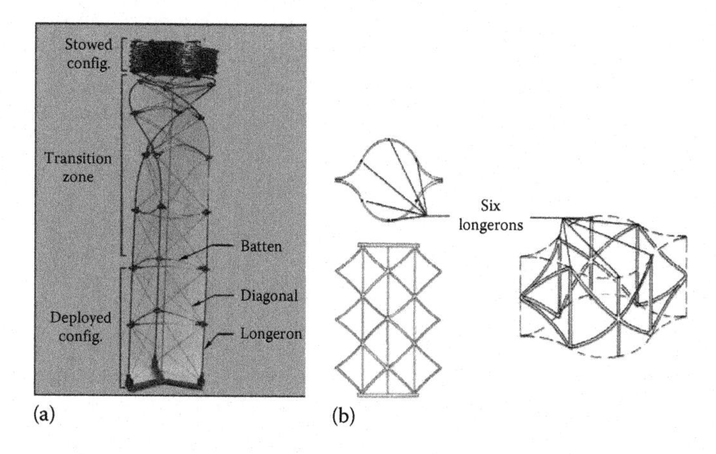

FIGURE 8.13
Examples of furlable truss booms: (a) Able Engineering's CoilABLETM boom. (b) Astro Aerospace's six-longeron collapsible/rollable boom. (Reproduced from Campbell, D. et al., Elastic memory composite material: An enabling technology for future furable space structures, in *46th AIAA/ASME/ASCE/AHS/ASC Structures, Structural Dynamics, and Materials Conference*, Austin, TX, vol. 10, pp. 6735–6743, 2005.)

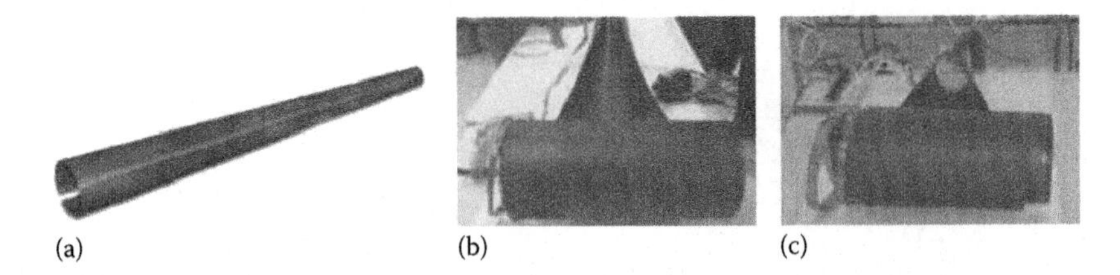

FIGURE 8.14
PowerSail 10-foot Storable Tubular Extendible Member (STEM) longeron with embedded heating elements; (a) fully deployed; (b) partially deployed; and (c) packaged. (Reproduced from Arzberger, S.C. et al., Elastic memory composites (EMC) for deployable industrial and commercial applications, in *SPIE Conference: Smart Structures and Materials 2005—Industrial and Commercial Applications of Smart Structures Technologies*, San Diego, CA, vol. 5762, pp. 35–47, 2005.)

in Figure 8.13. The boom needs high-strain capacity in the longeron material while maximizing its mechanical properties (e.g., strength, stiffness, and resistance against cycling forces). Thus, EMC reinforced by carbon fiber is suitable for this type of structure [11,12].

A new generation of deployable structure experiments was proposed to deploy large and lightweight solar arrays (Figure 8.14). This type of structure has very little deployed structural depth over a very large deployed area. In order to minimize the impacts to the spacecraft system, the structural mass and complexity should be minimized. Based on the above considerations, the structure fabricated by EMC is proposed for the deployed solar array and for the actuation to deploy the solar array from its packaged configuration. The longeron booms fabricated by the thin-film foldable EMC tubular are currently being fabricated to meet the needs for the deployable structure [11,12], as shown in Figure 8.14. Based on the experimental demonstration, a detailed theoretical study is being put forward. Geometric nonlinearity should be considered as the large deformation in the longeron. In addition, the presence of microbuckling of composite leading to material nonlinearity should also be considered.

8.4 Shape-Memory Polymer Composite for Deployable Optical Systems

The next generation space optical imaging system will consider using a deployable reflector to enlarge the aperture, which will provide a higher resolution or information translation capability. But the deployment precision and postdeployment stability of the optical reflector system are critical bottlenecks to achieve this goal. One of the key components is the deployable hinge connecting the main optical system and the reflectors. The hinges may result in a nonlinear micro-dynamic response, which may affect the dimensional stability and complicate the active alignment-control

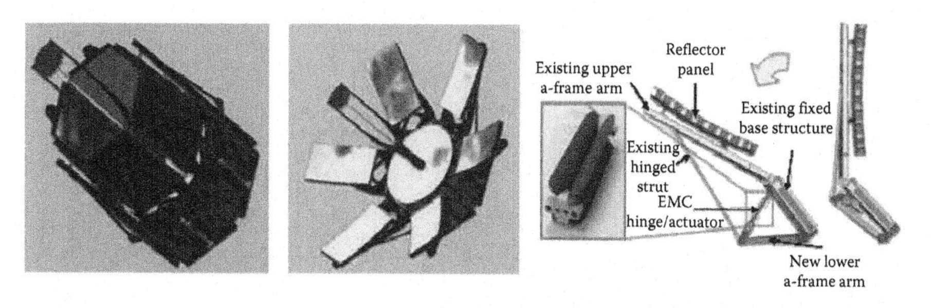

FIGURE 8.15
Graphical images of the packaged (left) and deployed (center) configurations for the LIDAR 3-m-class deployable reflector system and individual packaged and deployed configuration for each petal (right).

system. Hence, substantial work has been done to optimize design for mechanical hinges and latches that exhibit a higher precision and higher degree of micro-dynamic stability. A cooperative effort of CTD and the University of Colorado at Boulder has developed an EMC hinge to actuate the deployment of optical reflector system, which demanded a high micro-stability and precision (see Figure 8.15). The EMC hinge shows a potential to fix the optical reflector during the launch, accurately deploy the reflector on orbit with a low shock to system, and finally self-lock and maintain the deployment shape when the system is in the working state. This deployment system, mainly composed by EMC hinge, eliminates the need for a self-lock latch. That is, the EMC combines the deployment function during the deployment process and a structural supporting function when the system works. In addition, the EMC hinge is relatively lightweight; has a low cost; is easy to use; and has high reliability and a low coefficient of thermal expansion [13].

8.5 Shape-Memory Polymer Composite for Ground-Based Deployable Mirrors

In this section, a new kind of shape-memory composite material is used for the fabrication of thin, lightweight deployable mirrors. With its unique properties, SMPs are suitable for this application. The well-known shape-memory process contributes to the deployment of the novel ground-based mirrors. The mirror consists of shape-memory composite substrate coated on the reflective side of the composite reflector (see Figure 8.16). The reflective surfaces are mainly composed of electroplated nickel to provide high-quality reflectance. The electroplated nickel metal surface, which is less than 30 μm thick, is adhered on the surface of the SMP-composite mirror. The

FIGURE 8.16
A representative composite, replica optic produced using EMC resin TEMBO™ BG-1.3(v.2). (Reproduced from Arzberger, S.C. et al., *Proc. SPIE*, 5179, 143, 2003.)

substrate has the ability to be deformed for packaging and then perform a good shape recovery upon Joule heating from the external sources. With certain reinforcements (i.e., fibers, particulates, or nano-reinforcements), the shape-memory composite substrates have better mechanical and electrical properties.

8.6 Shape-Memory Polymer Composite Reflector

8.6.1 Stiffeners for a Flexible Reflector

The "SpringBack"-flexible reflector is a novel kind of moderate-precision reflector. The surface of this kind of reflector is fabricated by a thin open-weave carbon fiber/epoxy material, as shown in Figure 8.17. Before launching, the reflector can be stowed into a package of cylindrical shape, and the external restrain will store the elastic strain energy. Once on orbit, the reflector can be released by the system stimulus and recovered to the working state [14]. However, there are still many limitations before this kind of reflector can be wildly applied. It includes low packaging efficiency compared to open-mesh reflector systems of comparable size, strain limitations of the composite laminates [14].

Recently, a new kind of solid surface deployable reflector was proposed by Harris Corporation, which is similar to the springback reflector. By enabling large-aperture antennas to be stowed within an existing launch vehicle, the reflector is envisioned to significantly promote the ability of the satellite communication systems. In addition, Harris Corporation is now considering the use of EMC materials for the system (Figure 8.18) to realize the full

FIGURE 8.17
TDRS-H "Taco Shell" reflector. (Courtesy of The Boeing Company, Chicago, Illinois.) (Reproduced from Keller, P.N. et al., Development of elastic memory composite stiffeners for a flexible precision reflector, in *AIAA/ASME/ASCE/AHS/ASC Structures, Structural Dynamics and Materials Conference*, Newport, RI, vol. 10, pp. 6984–6994, 2006.)

FIGURE 8.18
SMP composite reflector. (Reproduced from Keller, P.N. et al., Development of elastic memory composite stiffeners for a flexible precision reflector, in *AIAA/ASME/ASCE/AHS/ASC Structures, Structural Dynamics and Materials Conference*, Newport, RI, vol. 10, pp. 6984–6994, 2006.)

potential of the FPR design [14]. The stiffener around the edge of the antenna can be made by EMC materials. With this stiffener which has shape-memory effect, the antenna can be stowed in the spacecraft and then actuated to deploy the antenna, and finally hold the surface of the antenna in a precision micro-system.

8.6.2 Truss Structure

Recently, ILC Dover developed a new kind of inflatable truss structure assembled from cylindrical booms. The SMP was used to actuate the inflatable truss from high packaging state to a large deployment state. In order to

FIGURE 8.19
SMP composite truss in packed and deployed configurations. (Reproduced from Lin, J.K.H. et al., Shape memory rigidizable inflatable (RI) structures for large space systems applications, in *Collection of Technical Papers—AIAA/ASME/ASCE/AHS/ASC Structures, Structural Dynamics and Materials Conference*, Newport, RI, vol. 5, pp. 3695–3704, 2006.)

show the deployment process, the electrical resistance heaters were adhered onto the surface of the booms to heat the SMP. The truss is demonstrated in both the deployed and underployed positions, as shown in Figure 8.19. This SMP truss can be designed to approach a compaction ratio of about 100:1.

8.6.3 Parabolic Dish Antenna Reflector

The large-aperture inflatable antenna is one important kind of deployment reflector. However, the deformation of central part of the inflatable antenna is mostly very large, which often makes it very difficult to realize inflatable deployment. Recently, SMP composite materials have been envisioned to be used to fabricate the central part of the antenna to realize a large deformation (Figures 8.20 and 8.21). In the design, both the dish and the supporting structure will be made from the SMP composite. To better fit this application, further research and development work are on-going to improve the properties of the SMP composite.

8.6.4 Singly Curved Reflector

A novel kind of singly curved parabolic antenna scale model has also been designed and fabricated from the SMP materials and its composite. In order to improve reflectivity, a coated polyimide film was attached to the inner surface of the antenna, as shown in Figure 8.22. The reflector could be heated above T_g and tightly rolled. After the antenna was packed and cooled to room temperature, it was placed in an oven and heated to $T_g+20°C$ for the deployment of the reflector. The feasibility test of this application was successful and further work is designed in this area, aiming to applying this technology to larger-sized antennas [15].

FIGURE 8.20
SMP reflector (2 m diameter) for the JHU/APL hybrid inflatable antenna. (Reproduced from Lin, J.K.H. et al., Shape memory rigidizable inflatable (RI) structures for large space systems applications, in *Collection of Technical Papers—AIAA/ASME/ASCE/AHS/ASC Structures, Structural Dynamics and Materials Conference*, Newport, RI, vol. 5, pp. 3695–3704, 2006.)

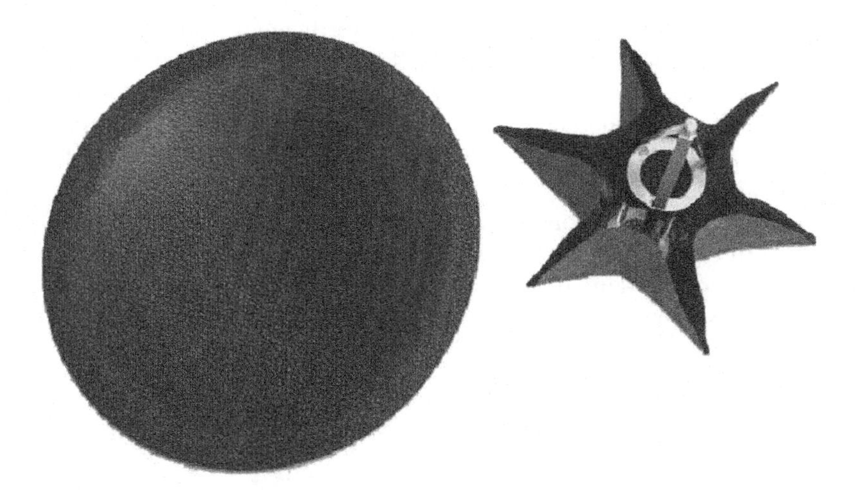

FIGURE 8.21
SMP reflector (0.5 m diameter) in both deployed and packed configurations. (Reproduced from Lin, J.K.H. et al., Shape memory rigidizable inflatable (RI) structures for large space systems applications, in *Collection of Technical Papers—AIAA/ASME/ASCE/AHS/ASC Structures, Structural Dynamics and Materials Conference*, Newport, RI, vol. 5, pp. 3695–3704, 2006.)

8.7 Shape-Memory Polymer for Morphing Structures

Flight vehicles are often envisioned to be multifunctional so that they can perform more missions during a single flight, such as an efficient cruising and a high maneuverability mode. However, when the airplane moves

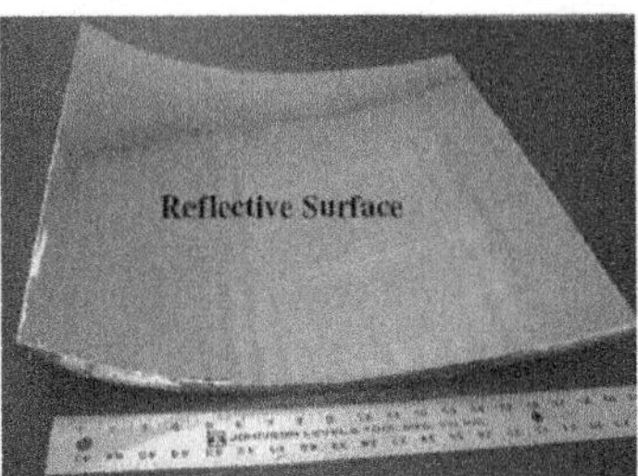

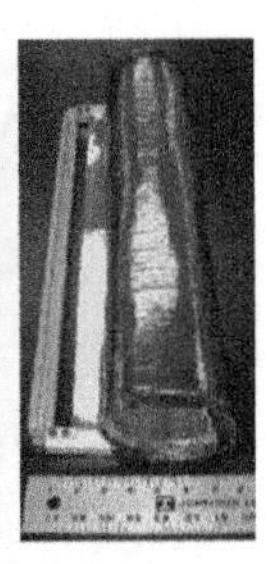

FIGURE 8.22
Singly curved parabolic reflector (0.5 m SMP) (deployed/packed). (Reproduced from Lin, J.K.H. et al., Shape memory rigidizable inflatable (RI) structures for large space systems applications, in *Collection of Technical Papers—AIAA/ASME/ASCE/AHS/ASC Structures, Structural Dynamics and Materials Conference*, Newport, RI, vol. 5, pp. 3695–3704, 2006.)

towards other portions of the flight envelope, its performance and efficiency may deteriorate rapidly. In order to solve this problem, researchers have proposed to radically change the shape of the aircraft during flight. By applying this kind of technology, both the efficiency and flight envelope can be improved. This is because that different shapes correspond to different trade-offs between beneficial characteristics, such as speed, low energy consumption, and maneuverability, as shown in Figure 8.23 [16]. Figure 8.24 is a spider plot that attempts to graphically capture the advantage of morphing an aircraft [17].

Recently, NextGen Aeronautics was developing morphing technologies, as shown in Figure 8.25. Current effort focuses on the wind tunnel test of a full-scale morphing wing under transonic Mach numbers [18]. The main aim of NextGen Aeronautics is to show the morphing process during the flight

FIGURE 8.23
Different shapes of a bird's wing. (Reproduced from Cui, E.J. et al., *Aeronauti. Manuf. Technol.*, 08, 38, 2007.)

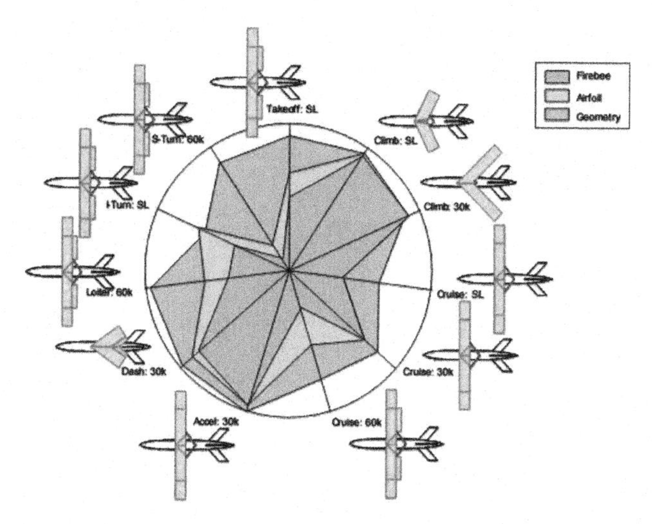

FIGURE 8.24
Qualitative spider plot. (Reproduced from Joshi, S.P. et al., Comparison of morphing wing Strategies based upon aircraft performance impacts, in *45th AIAA/ASME/ASCE/AHS/ASC Structures, Structural Dynamics, and Materials Conference,* April 19–22, 2004, Palm Springs, CA, AIAA 2004-1722, 2004.)

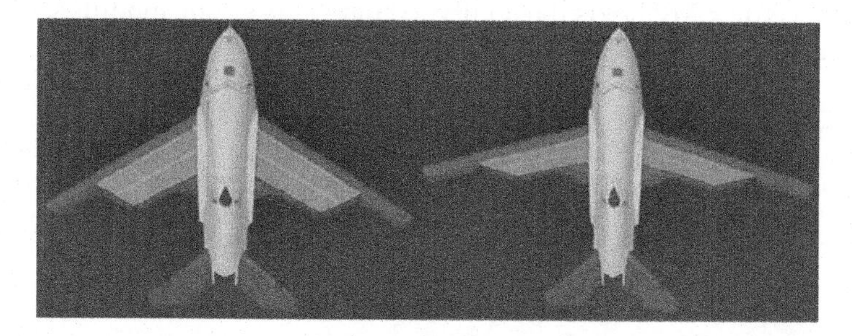

FIGURE 8.25
Morphed configurations of the MFX-1. (Reproduced from Flanagan, J.S. et al., Development and flight testing of a morphing Aircraft, the NextGen MFX-1, in *48th AIAA/ASME/ASCE/AHS/ ASC Structures, Structural Dynamics, and Materials Conference,* Honolulu, HI, AIAA 2007-1707, 2007.)

and establish the ground theory for the future technology development, ultimately leading to aerodynamically efficient, shape-changing, morphing wings on operational aircraft that would provide optimal, uncompromised performance during complex military missions.

The Defense Advanced Research Projects Agency (DARPA) is also developing the morphing technology to demonstrate such radical shape changes. As illustrated in Figure 8.26, Lockheed Martin is addressing technologies to achieve a z-shaped morphing change under the DARPA's program fund [19].

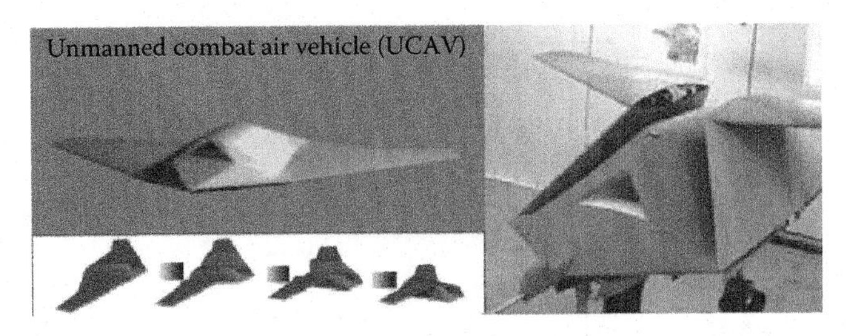

FIGURE 8.26
z-Shaped morphing wing produced by Lockheed Martin. (Reproduced from Love, M.H. et al., Demonstration of morphing technology through ground and wind tunnel tests, in *48th AIAA/ASME/ASCE/AHS/ASC Structures, Structural Dynamics, and Materials Conference*, Honolulu, HI, AIAA 2007-1729, 2007.)

However, finding a proper skin under certain criteria is crucial to develop a morphing aircraft. Generally, a wing skin is necessary, especially for the wing of a morphing aircraft. Researchers focus their works on investigating proper types of materials that are currently available to be used as a skin material for a morphing wing. In this case, the SMP shows more advantages for this application. It becomes flexible when heated above the transition temperature, and then it returns to a solid state when the stimulus is terminated. Since the SMP holds the ability to change its elastic modulus, it could potentially be used in these concept designs.

8.7.1 Folding Wing

Cornerstone Research Group (CRG) developed an improved SMP, which appears to be a prime candidate for seamless skin at the wing fold (see

FIGURE 8.27
Initial SMP skin prototype. (Reproduced from Bye, D.R. and McClure, P.D., Design of a morphing vehicle, in *48th AIAA/ASME/ASCE/AHS/ASC Structures, Structural Dynamics, and Materials Conference*, Honolulu, HI, 2007-1728, 2007.)

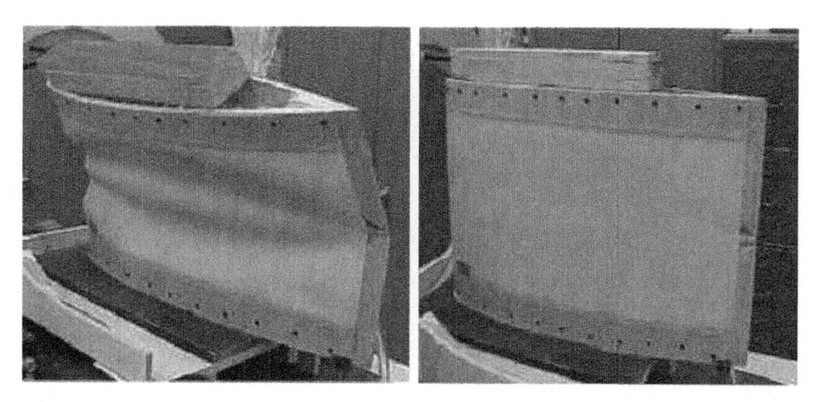

FIGURE 8.28
SMP on a two-dimensional test jig. (Reproduced from Bye, D.R. and McClure, P.D., Design of a morphing vehicle, in *48th AIAA/ASME/ASCE/AHS/ASC Structures, Structural Dynamics, and Materials Conference*, Honolulu, HI, 2007-1728, 2007.)

Figure 8.27) [20]. It has demonstrated the feasibility of morphing SMP skin concepts in a two-dimensional test jig, as shown in Figure 8.28.

Several methods for triggering the shape-memory effect of the SMPs have been developed. The most common and simpliest triggering method to actuate the SMP is by using direct Joule heating. In order to thermally activate the SMP skin, electrically conductive wire springs were embedded in the SMP matrix and an electrical current was passed through the electrically conductive wire springs to heat the SMP above its transition temperature. The parallel wire springs are embedded into the SMP, as shown in Figure 8.29. The advantage is that the inherent separation does not occur between the heating elements and the SMP upon elongation.

SMP embedded with electrically conductive wire springs is fabricated, as shown in Figure 8.30. During the heating process within 60 s, the SMP

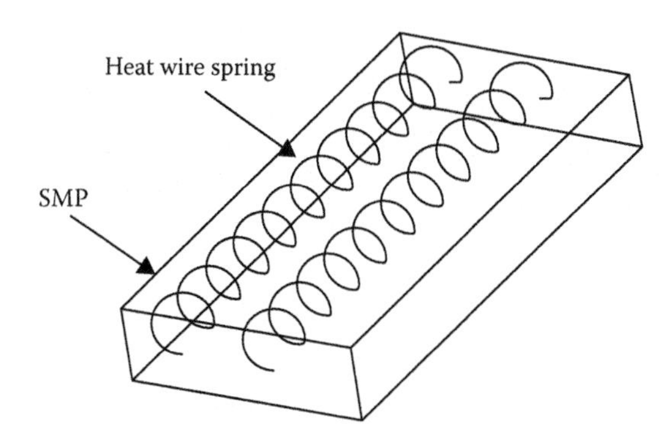

FIGURE 8.29
Illustration of SMP embedding heating wire springs. (Reproduced from Yin, W. et al., *Proc. SPIE*, 7292, 72921H-1, 2009.)

FIGURE 8.30
SMP prototype embedding heating wire springs. (Reproduced from Yin, W. et al., *Proc. SPIE*, 7292, 72921H-1, 2009.)

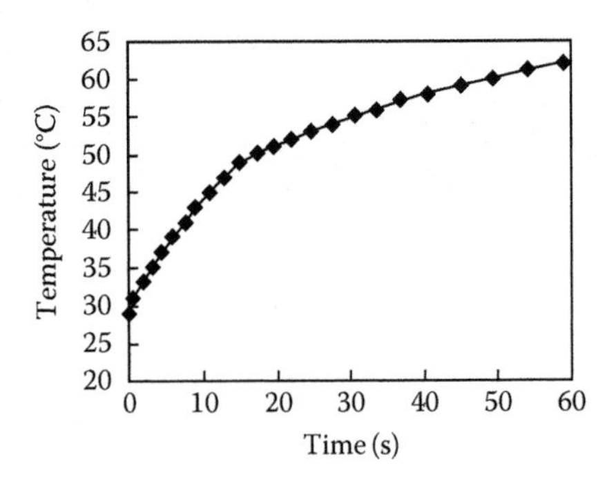

FIGURE 8.31
Temperature increase of SMP embedding wire springs when heated. (Reproduced from Yin, W. et al., *Proc. SPIE*, 7292, 72921H-1, 2009.)

is heated to its glass transition temperature of 53°C at about the 20th s, as shown in Figure 8.31 [19].

Another type of an electrically conductive wire, NiCr wire, is also taken into account, as shown in Figure 8.32. Several finite element models were created to determine duration for which the SMP plate can be heated up to its transition temperature by the resistive wires (Figure 8.33).

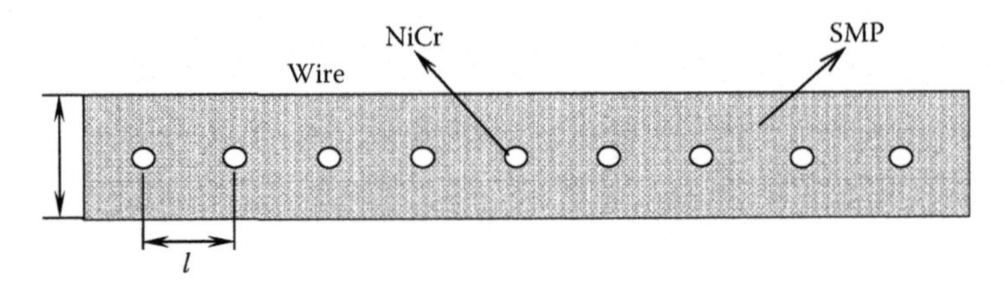

FIGURE 8.32
Cross section of SMP-NiCr composite.

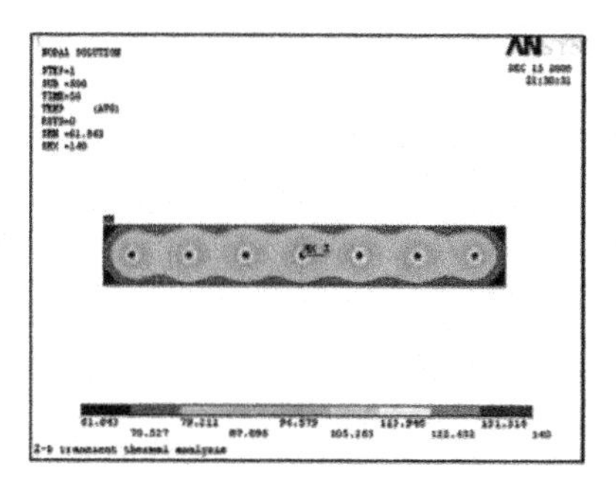

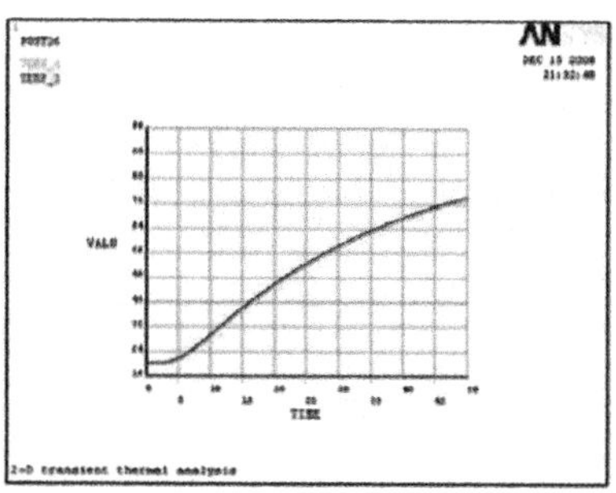

FIGURE 8.33
Thermal analysis for SMP.

8.7.2 Variable Camber Wing

The morphing concept of a variable camber wing is shown in Figure 8.34 [19]. It comprises a flexible SMP skin, a thin metallic sheet, and a honeycomb structure. The metallic sheet is used to replace the traditional hinges to keep the surface smooth during the camber changing. Honeycomb, which is high-strain capable in one direction without dimensional change in the perpendicular in-plane axis, provides distributed support to the flexible skin. Flexible skin is covered to create the smooth aerodynamic surface. FBG sensors are bonded on the upper surface of the metal sheet to sense its deflection.

The elastic modulus of the SMP reduces with temperature increasing. The elastic modulus of SMP will drop one or two orders of magnitude when the SMP is heated above its transition temperature. SMP skin can be elongated or sheared when SMP is in the state of low modulus. The problem that may

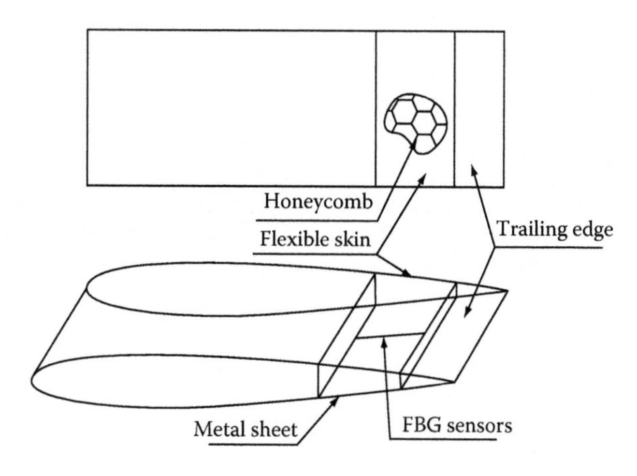

FIGURE 8.34
Schematic of the variable camber wing. (Reproduced from Yin, W. et al., *Proc. SPIE*, 7292, 72921H-1, 2009.)

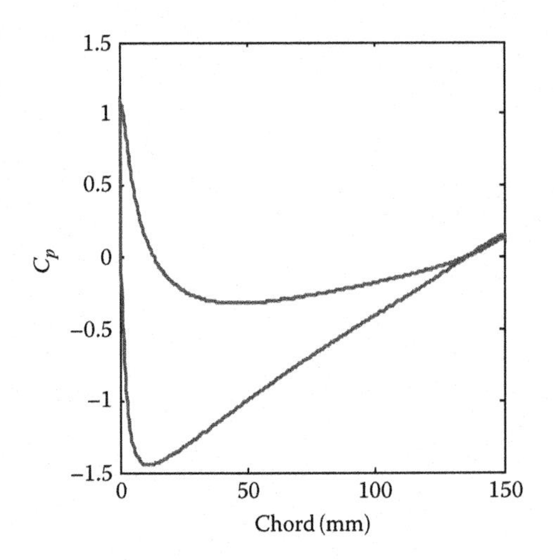

FIGURE 8.35
Pressure distribution of NACA0020 ($M=0.2$, $\alpha=5°$). (Reproduced from Yin, W. et al., *Proc. SPIE*, 7292, 72921H-1, 2009.)

be faced is that whether SMP skin has high enough out-of-plane stiffness to withstand aerodynamic loads.

The baseline airfoil is assumed to have an NACA 0020 profile and a chord of 150 mm, as shown in Figure 8.34. Figure 8.35 shows the pressure distribution of NACA0020 with the free stream velocity of 0.2 M, pressure of 1 bar, and air density of 1.225 kg/m^3 at 5° angle of attack. The pressure coefficient on the upper surface is negative, and a bubble may occur in the upper flexible skin under aerodynamic pressures.

The out-of-plane deformation of SMP skin at different temperatures is shown in Figure 8.36. It can be seen that the out-of-plane deformation increases with the temperature increasing. When the temperature is as high as 53°C, the maximum displacement of the SMP skin is about 7 mm. Figure 8.37 exhibits an aerodynamic pressure distribution of the NACA 0020 with SMP skin at different temperatures.

The out-of-plane deformation of the SMP skin is powerful and affects the aerodynamic pressure distribution. The lift decreases by about 34.8% and the drag increases by about 35.4% when SMP skin is heated to 53°C, as shown in Table 8.1.

One possible way of reducing or eliminating the out-of-plane deformation of the flexible skin is to use the prestrain method. Figure 8.38 shows the variation in the out-of-plane displacement of the SMP skin at 53°C as the prestrain increases. The out-of-plane displacement decreases due to the increase in a prestrain, and the maximum displacement decreases by about 72% when the prestrain is equal to 0.1, and therefore the prestrain is set as 0.1.

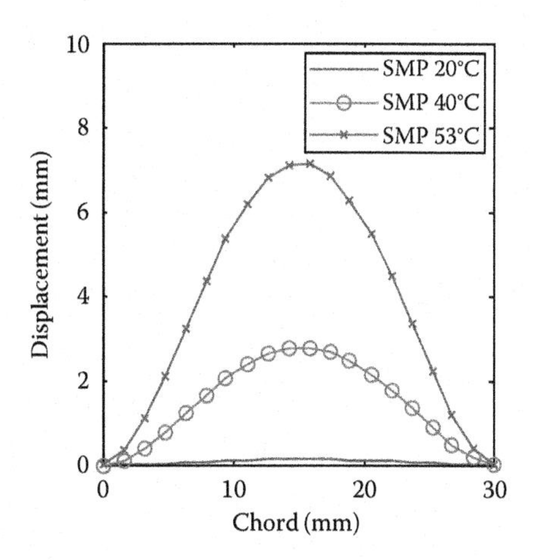

FIGURE 8.36
Deformation of SMP skin at different temperatures. (Reproduced from Yin, W. et al., *Proc. SPIE*, 7292, 72921H-1, 2009.)

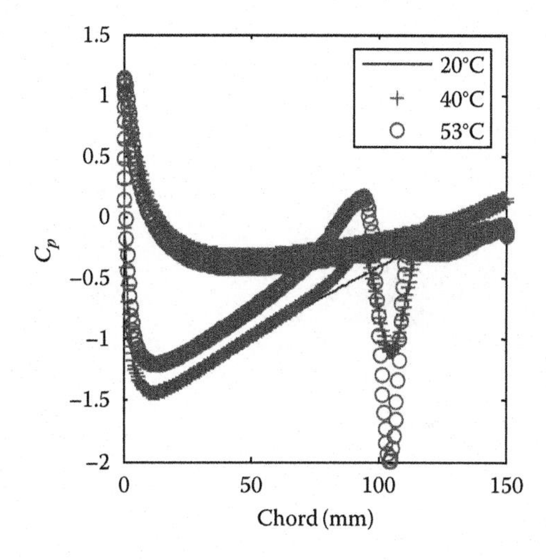

FIGURE 8.37
Aerodynamic pressure coefficient of the NACA 0020 with SMP skin at different temperatures. (Reproduced from Yin, W. et al., *Proc. SPIE*, 7292, 72921H-1, 2009.)

TABLE 8.1

Aerodynamic Characteristics of Variable Camber Wing with SMP Skin at Different Temperatures

Temperature (°C)	Lift (N)	Drag (N)	Lift-to-Drag Ratio
20	238.195	25.658	9.283
40	239.515	26.526	9.029
53	155.278	34.746	4.469

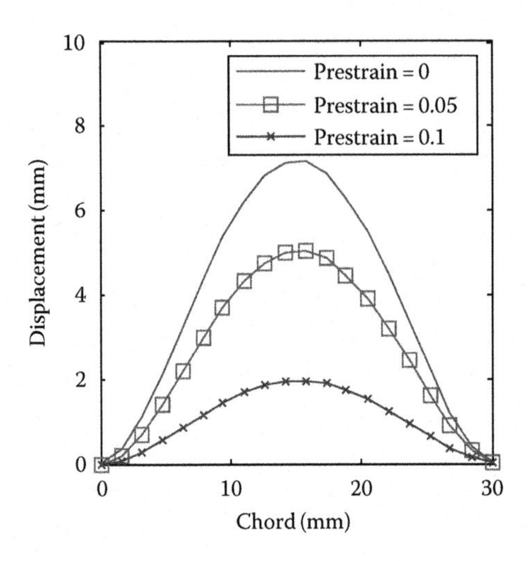

FIGURE 8.38
Deformation of the SMP skin (53°C) under different prestrains. (Reproduced from Yin, W. et al., *Proc. SPIE*, 7292, 72921H-1, 2009.)

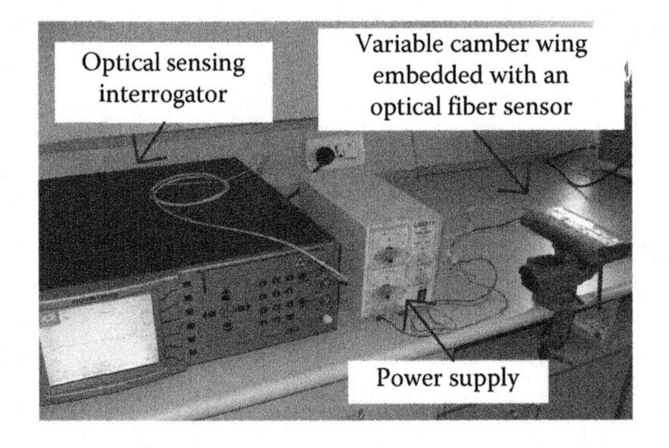

FIGURE 8.39
Experimental setup to test the reconstruct the shape of variable camber wing. (Reproduced from Yin, W. et al., *Proc. SPIE*, 7292, 72921H-1, 2009.)

Experimental setup of the variable camber wing is shown in Figure 8.39. Micro Optics si425 Optical Sensing Interrogator is used to demodulate the FBG sensor.

The optical fiber Bragg grating (FBG) sensors for the measurement of strain have been developed for many years, and a lot of sensing techniques have been established. In this study, the FBG sensors are bonded on the upper surface of the metal sheet to measure the strains, and the shape of the bending metallic sheet can be reconstructed by using a FBG sensor. Suppose that the deflection equation corresponding to the polynomial is expressed as follows:

$$w = a_3x^3 + a_2x^2 + a_1x + a_0 \tag{8.1}$$

Since the boundary of the variable camber wing is clamped on the left edge, w denotes the deflection, x represents the span. The parameters a_0 and a_1 can be set as zero. And the parameter a_2 is proportional to the parameter a_3 with the proportionality constant 90 calculating through simulating analysis by finite element method. Then the unknown parameter of the polynomial is reduced to one parameter: a_2 or a_3. Therefore, the relationship between the wavelength and a_2 or a_3 is:

$$6a_3x + 2a_2 = \frac{2}{t(1-p_e)} \times \frac{\lambda_1 - \lambda_0}{\lambda_0} \tag{8.2}$$

where
- λ_1 is the wavelength after shape change
- λ_0 is the wavelength before shape change
- p_e is the photoelastic coefficient
- t is the thickness of the variable wing prototype on the left end

Determined by a single FBG sensor, the original configuration (Figure 8.40a) and deformed configuration (Figure 8.40b) of the variable camber wing are reconstructed as shown in Figure 8.41a and Figure 8.41b, respectively. Experimental results (Figures 8.40 and 8.41) show that the shapes of the metal sheet can be reconstructed using a single FBG sensor.

Figure 8.42 shows the model used for the wind tunnel testing. The heating area of the SMP skin (1 mm thickness) is $150 \times 20\,\text{mm}^2$. The temperature sensor is embedded between the inner surface of the SMP skin and the honeycomb. A power supply is used to heat the springs embedded in the SMP skin.

The temperature of the SMP skin heated by the springs embedded in the SMP matrix is determined in two conditions. The first case is in the still air. The second case is in the air at 12 m/s.

(a)

(b)

FIGURE 8.40
Photograph of the original and morphing configurations of the variable camber wing. (a) Original configuration (0°), (b) morphing configuration (15°). (Reproduced from Yin, W. et al., *Proc. SPIE*, 7292, 72921H-1, 2009.)

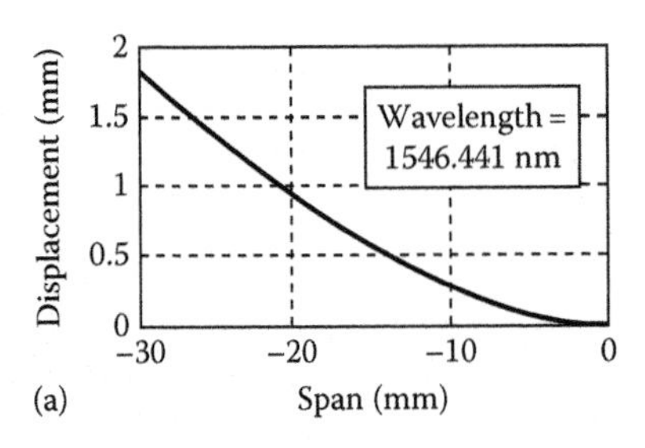

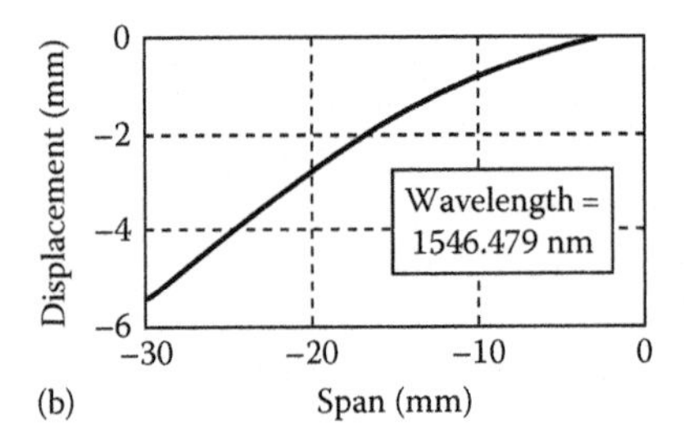

FIGURE 8.41
Reconstructed sheet shapes of the original and morphing configurations of the variable camber wing by using FBG sensor. (a) Original configuration (0°), (b) morphing configuration (15°). (Reproduced from Yin, W. et al., *Proc. SPIE*, 7292, 72921H-1, 2009.)

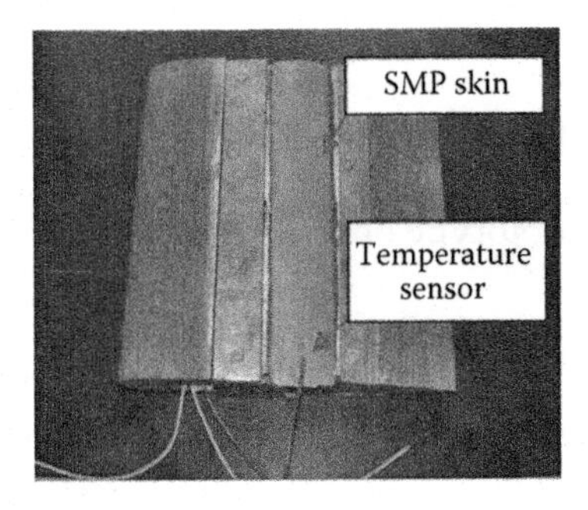

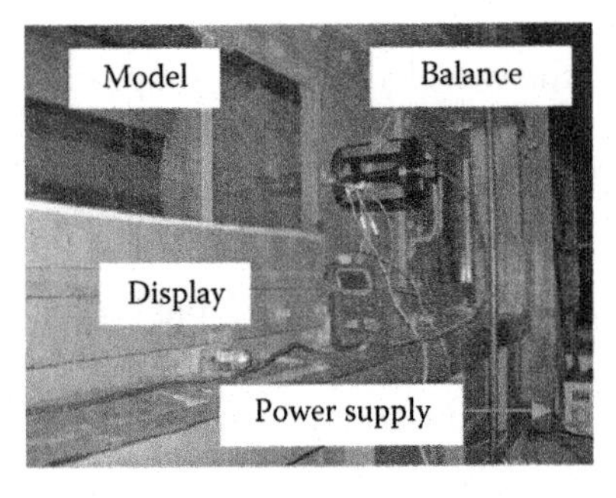

FIGURE 8.42
Wind tunnel test of the variable camber wing.

Figure 8.43 shows the temperature versus time relationship for the SMP skin under different wind velocities (0, 12 m/s). The transition temperature of SMP is 53°C; the skin reached the very same temperature at a time of about 24.5 s when the wind velocity was set as 0 m/s; whereas under the same power supply, it took at least 70 s to reach the same temperature when the wind velocity was 12 m/s.

8.7.3 Deployable Morphing Wing

The SMP-based morphing wing is illustrated in Figure 8.44 [22]. The wing, with an airfoil of NACA 0020, consists of carbon fiber–reinforced SMP (CF-SMP) skin and SMP filler with nine through holes. Before the realistic flight test of a morphing aircraft, the SMP morphing wing is compressed and curled on the aircraft. After the SMP composite is heated to a temperature higher than T_g by the inner heating system, the morphing wing is unlocked and deployed. During the deformation, the CF-SMP skin provides most recovery stress for the wing. However, because the modulus of

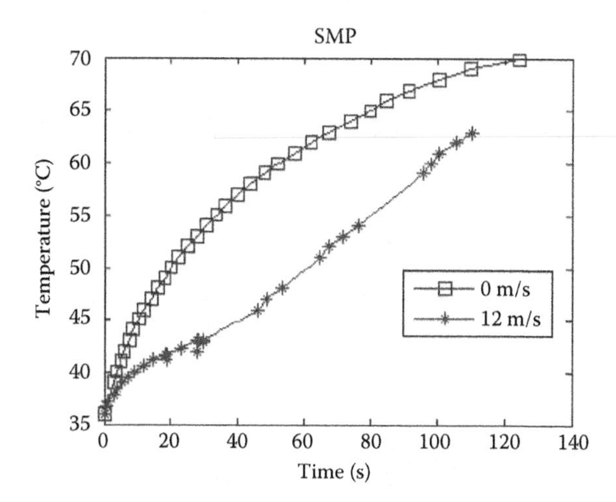

FIGURE 8.43
Temperature–time relationship at different wind velocity.

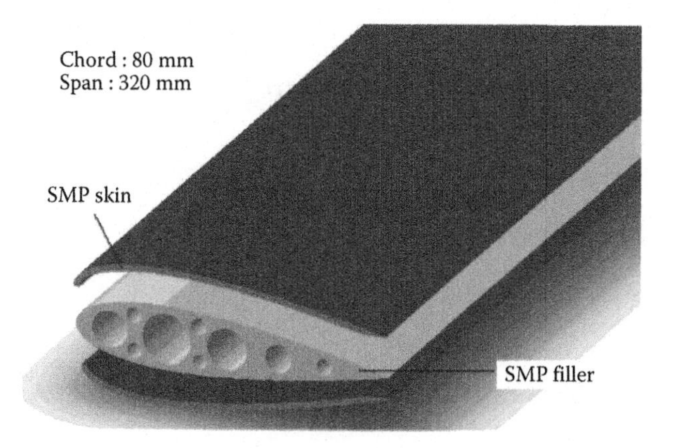

FIGURE 8.44
Illustration of the designed SMP-based deployable wing. (Reproduced from Yu, K. et al., *Proc. SPIE*, 7375, 737560, 2008.)

SMP decreases significantly after being heated, the SMP filler is necessary to sustain the whole skin under the air loads generated by the wing.

The aerodynamic characteristic of the deployed SMP wing is calculated by the CFD. The initial condition of the simulation is set as a steady flow, with a velocity of 100 m/s. For the two-dimensional airfoil, the static pressure and the temperature on the surfaces are shown in Figure 8.45.

The air loads and the temperature fields obtained from the aerodynamic analysis are used as the applied loads on the SMP skin in ANSYS. Figure 8.46

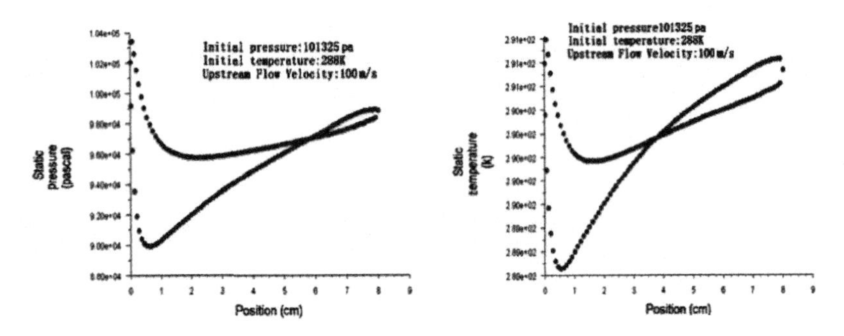

FIGURE 8.45
Static pressure and static temperature on the surface of airfoil. (Reproduced from Yu, K. et al., *Proc. SPIE*, 7375, 737560, 2008.)

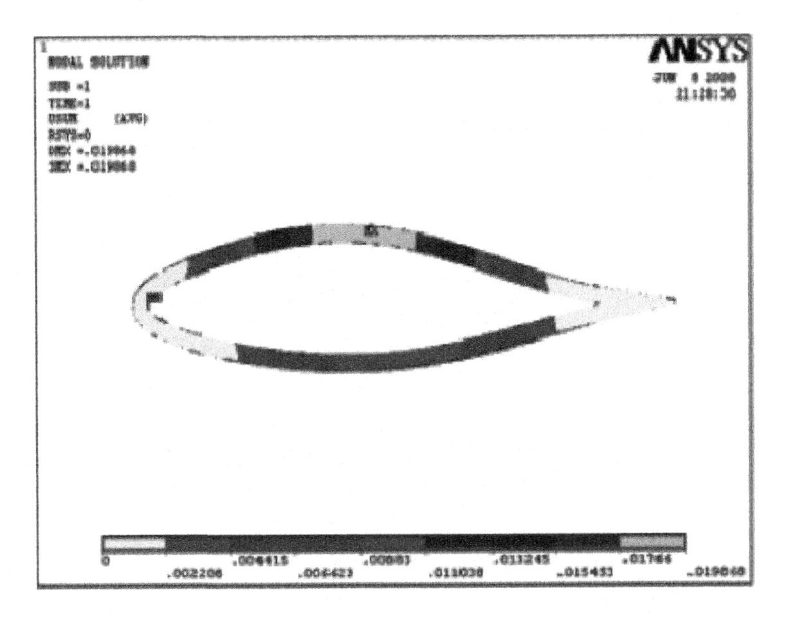

FIGURE 8.46
Static deformation of the SMP skin. (Reproduced from Yu, K. et al., *Proc. SPIE*, 7375, 737560, 2008.)

shows the static deformation of the SMP skin calculated by finite element method. Primary tests are conducted to investigate the stability and feasibility of the designed morphing wing. Figure 8.47 shows the deployment process of the prototype of SMP skin.

During the test, the CF-SMP composite skin prototype deployed steadily and precisely in the air. However, even though the carbon fiber is embedded into SMP as the reinforcing material, due to the significant decrease in the modulus of the SMP, the speed of deployment and the recovery force are still much lower than expected. This indicates that the deployment precision will

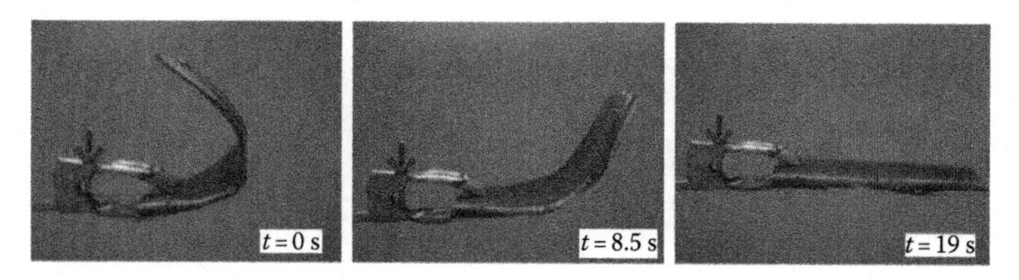

FIGURE 8.47
Snapshot series of deployment of CF-SMP composite. (Reproduced from Yu, K. et al., *Proc. SPIE*, 7375, 737560, 2008.)

decrease under the air loads. In view of this drawback, other reinforcing materials with better performance are appreciated. In our test, the following two improvements are proposed.

Firstly, the modulus of SMA wire rises dramatically with the increase of temperature. The modulus of the austenite is 3–4 times higher than that of the martensite. This characteristic helps to increase the modulus of the wing skin, and upgrade the spreading speed indirectly.

Secondly, although the elastic steel slice does not have the shape-memory capability, the duration of the deformation is also decreased owing to its great flexibility, especially for the first stage of the deployment. The stiffness and the strength of the composite skin are increased for the using reinforcement phase. Moreover, the deployment process also shows excellent stability and precision.

Figure 8.48 shows the comparison of the spreading speed between the three composite materials. The morphing skin reinforced by SMA wires and elastic steel slice has a higher modulus during the deformation. So the

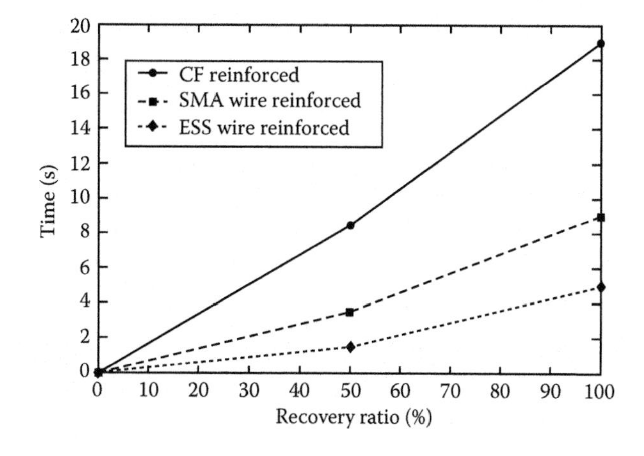

FIGURE 8.48
Comparison of deployment speed between the three kinds of composite skin. (Reproduced from Yu, K. et al., *Proc. SPIE*, 7375, 737560, 2008.)

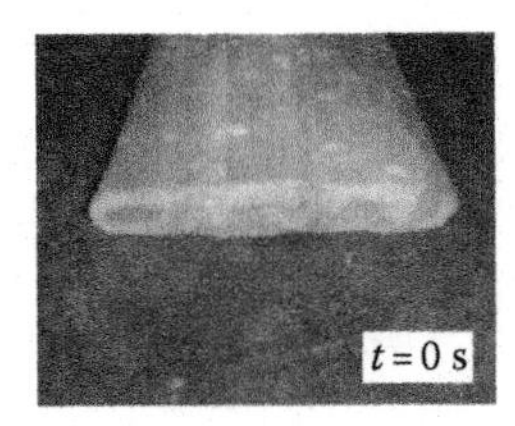

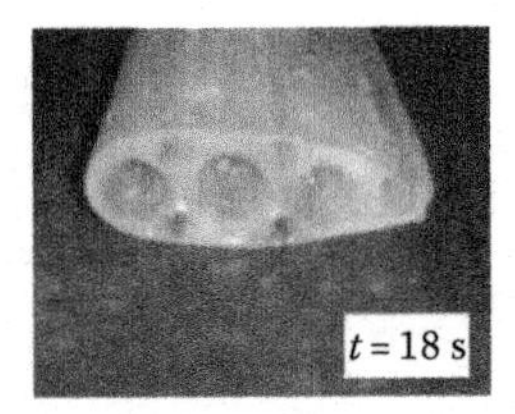

FIGURE 8.49
Self-recovered SMP filler under an IR heat lamp. (Reproduced from Yu, K. et al., *Proc. SPIE*, 7375, 737560, 2008.)

spreading speed is dramatically increased. In this way, the deployment precision of the wing is improved.

During the deployment process of the designed missile wing, the skin provides main recovery force. However, to increase the deployment precision of the airfoil, the filler is necessary to sustain the air loads, as shown in Figure 8.49.

Nine through holes of different sizes increase the allowable deformation of the SMP filler, which makes the compression of the wing more convenient. The test of its deformation ability shows that the filler recovers its shape precisely in the air, and the shape recovery ratio approaches is up to 100%.

Styrene-based SMPs have become popular because they have the ability to be injection-molded and the ability for the SMPs to be colored since the original color is transparent. Based on its unique properties the glass transition temperature can be designed for a specific application [23]. The fundamental investigation may lay a foundation for the SMP to be used for morphing wing.

8.8 Reusable Shape-Memory Polymer Mandrel

Cornerstone Research Group, Inc. (CRG) has successfully used thermosetting SMP mandrels to fabricated filament winding composite with complex-curved shapes. Using the SMP mandrel, it will be very easy to extract the filament winding composite when the SMPs return to its original shape, as shown in Figures 8.50 and 8.51. SMP mandrels for filament winding allow for a reusable, quick, and low-cost mandrel system [24].

CRG has successfully demonstrated the manufacturing process, using both the bottle-shape and the air duct SMP mandrels, as shown in Figure 8.52. During the composite processing, an SMP mandrel is used on the filament winder. Using fiberglass, the machine filament is wound to the bottle-shape SMP mandrel. After the part was cured, the mandrel was heated above its transition temperature and then removed from the composite part [24]. Recently, CRG was proposed to improve the SMP mandrel technology

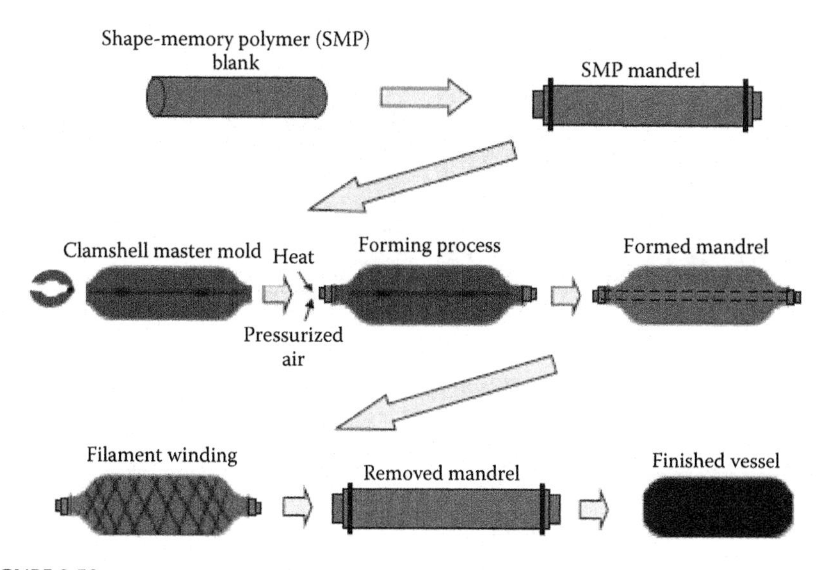

FIGURE 8.50
Illustration of the working process of SMP mandrel. (Reproduced from Everhart, M.C. et al., *Proc. SPIE*, 5388, 87, 2004.)

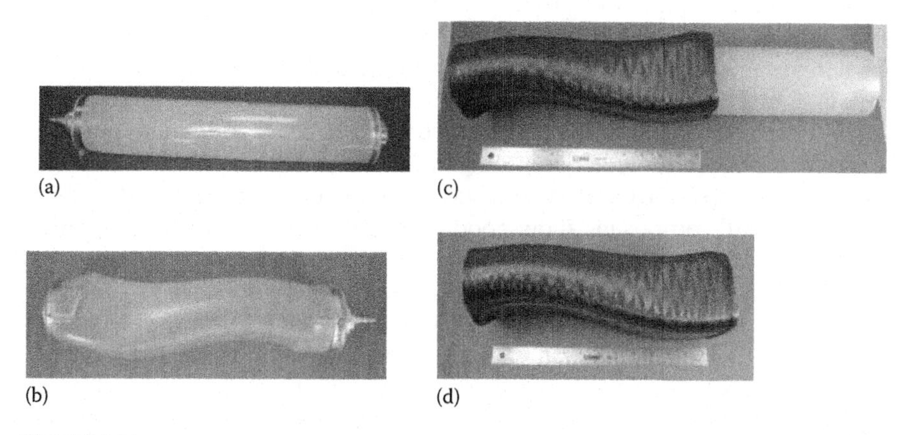

FIGURE 8.51
SMP composite mandrel; (a) Mandrel in memory shape with pressure fittings on ends; (b) deformed air duct mandrel; (c) and (d) air duct shape carbon fiber-wound part. (Reproduced from Everhart, M.C. et al., *Proc. SPIE*, 5762, 27, 2005.)

by adding high-strain fiber reinforcement (HSFR), which both raises the toughness of the SMP and allows the SMP to elongate at a large deformation. The resulting SMP composite is able to produce mandrels durable enough to withstand multiple uses to produce complex-shaped composite in high production rate manufacturing.

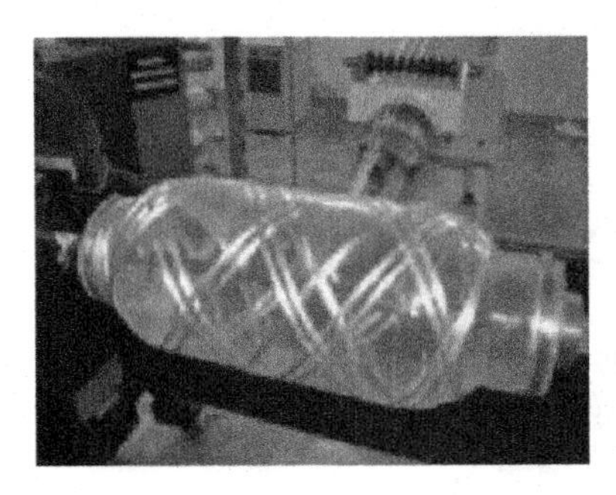
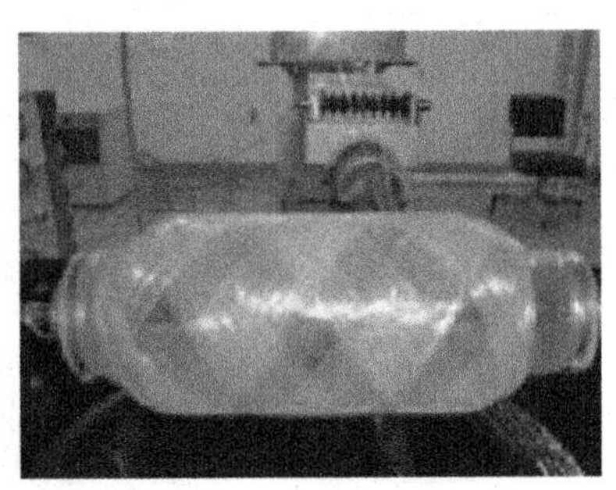

FIGURE 8.52
Filament wound part with mandrel. (Reproduced from Everhart, M.C. et al., *Proc. SPIE*, 5762, 27, 2005.)

8.9 Summary and Outlook

With the rapid development of the SMPs and SMP composites in the related areas, they show great potential for many aerospace devices. SMP materials are low-cost, have easy control over the recovery temperature, are lightweight, space-qualified, structural composites with the ability to be mechanically deformed, store strain, and provide reliable and predictable shape recovery upon exposure to a specific thermomechanical cycle. As well known, pure SMPs are not suitable for many practical applications that require particular functions (e.g., high strength, high recovery force, and good electrical conductivity). Recent developments have expanded the useful range for SMP materials. The continuous-fiber-reinforced SMP composites represents excellent mechanical properties. As both a functional and structural material, the continuous-fiber-reinforced SMP composites show good potential in many advanced applications. When fiber-reinforced SMP are used as actuator materials, there are no moving parts. The novel class of the reinforced SMP composite is currently being developed to meet the future space mission requirements. These applications include deployable structures (i.e., booms, trusses and reflectors) or morphing structures (i.e., folding wing and morphing skins). The production of the materials with improved performances is anticipated to provide enabling technologies for the future industrial and commercial needs.

References

1. Zeng, Y. M., Hu, J. L., and Yan, H. J. 2002. Temperature dependency of water vapor permeability of shape memory polymer. *Journal of Dong Hua University* 19: 52–57.
2. Mondal, S. and Hu, J. L. 2006. Temperature stimulating shape memory. Polyurethane for smart clothing. *Indian Journal of Fibre & Textile Research* 31: 66–71.

3. Charlesby, A. 1960. *Atomic Radiation and Polymers*. Pergamon Press, New York.
4. Campbell, D., Lake, M. S., Scherbarth, M. R., Nelson, E., and Six, R. W. 2005. Elastic memory composite material: An enabling technology for future furable space structures. In *46th AIAA/ASME/ASCE/AHS/ASC Structures, Structural Dynamics, and Materials Conference*, Austin, TX, AIAA-2005-2362.
5. Hussein, H. and Harrison, D. 2004. Investigation into the use of engineering polymers as actuators to produce 'automatic disassembly' of electronic products. In *Design and Manufacture for Sustainable Development*, Bhamra, T. and Hon, B., (eds.), Wiley-VCH, Weinheim, Germany.
6. Yee, J. C. H., Soykasap, O., and Pellegrino, S. 2004. Carbon fibre reinforced plastic tape springs. In *45th AIAA/ASME/ASCE/AHS/ASC Structures, Structural Dynamics & Materials Conference*, Palm Springs, CA, April 19–22, 2004.
7. Seffen, K. A. and Pellegrino, S. 1999. Deployment dynamics of tape spring. *Proceedings of the Royal Society of London A* 455: 1003–1048.
8. Metcalfe, A., Desfaits, A. C., Salazkin, I., Yahia, L. H., Sokolowski, W. M., and Raymond, J. 2003. Cold hibernated elastic memory foams for endovascular interventions. *Biomaterials* 24: 491–497. AIAA 2004-1819.
9. Lan, X., Wang, X., Liu, Y., and Leng, J. S. 2009. Fibre reinforced shape-memory polymer composite and its application in a deployable hinge. *Smart Materials and Structures* 18: 024002.
10. Barrett, R., Francis, W., Abrahamson, E., and Lake, M. S. 2006. Qualification of elastic memory composite hinges for spaceflight applications. In *47th AIAA/ASME/ASCE/AHS/ASC Structures, Structural Dynamics, and Materials Conference*, Newport, RI, May 1–4, 2006, AIAA-2006-2039.
11. Campbell, D. and Lake, M. S. 2005. Elastic memory composite material: An enabling technology for future furlable space structures. In *AIAA/ASME/ASCE/AHS/ASC Structures, Structural Dynamics and Materials Conference*, Austin, TX, vol. 10: 6735–6743.
12. Arzberger, S. C., Tupper, M. L., Lake, M. S. et al. 2005. Elastic memory composites (EMC) for deployable industrial and commercial applications. In *SPIE Conference: Smart Structures and Materials 2005—Industrial and Commercial Applications of Smart Structures Technologies*, San Diego, CA, vol. 5762: pp. 35–47.
13. Arzberger, S. C., Munshi, N. A., and Lake, M. S. Elastic memory composites for deployable space structures. www.CTD-materials.com
14. Keller, P. N., Lake, M. S., Codell, D., Barrett, R., Taylor, R., and Schultz, M. R. 2006. Development of elastic memory composite stiffeners for a flexible precision reflector. In *AIAA/ASME/ASCE/AHS/ASC Structures, Structural Dynamics and Materials Conference*, Newport, RI, vol. 10: 6984–6994.
15. Lin, J. K. H., Knoll, C. F., and Willey, C. E. 2006. Shape memory rigidizable inflatable (RI) structures for large space systems applications. In *Collection of Technical Papers—AIAA/ASME/ASCE/AHS/ASC Structures, Structural Dynamics and Materials Conference*, Newport, RI, vol. 5: 3695–3704.
16. Cui, E. J., Bai, P., and Yang, J. M. 2007. The development way of smart morphing aircraft. *Aeronautical Manufacturing Technology*, (Chinese) 08: 38–41.
17. Keihl, M. M., Bortolin, R. S., and Sanders, B. et al. 2005. Mechanical properties of shape memory polymers for morphing aircraft applications. In *SPIE Conference: Smart Structures and Materials 2005—Industrial and Commercial Applications of Smart Structures Technologies*, San Diego, CA, vol. 5762: 143–151.

18. Flanagan, J. S., Strutzenberg, R. C., Myers, R. B., and Rodrian J. E. 2007. Development and flight testing of a morphing Aircraft, the NextGen MFX-1. In *48th AIAA/ASME/ASCE/AHS/ASC Structures, Structural Dynamics, and Materials Conference*, Honolulu, HI, AIAA 2007-1707.
19. Love, M. H., Zink, P. S., Stroud, R. L., Bye, D. R., Rizk, S., and White, D. 2007. Demonstration of morphing technology through ground and wind tunnel tests. In *48th AIAA/ASME/ASCE/AHS/ASC Structures, Structural Dynamics, and Materials Conference*, Honolulu, HI, AIAA 2007-1729.
20. Bye, D. R. and McClure, P. D. 2007. Design of a morphing vehicle. In *AIAA Structures, Structural Dynamics, and Materials Conference*, Honolulu, HI, 2007-1728.
21. Yin, W., Fu, T., Liu, J., and Leng, J. 2009. Structural shape sensing for variable camber wing using FBG sensors. In *Proceeding SPIE*, vol. 7292 72921H-1, *SPIE International Conference on Smart Structures/NDE*, San Diego, CA, March 8–12, 2009.
22. Yu, K., Yin, W., Sun, S. et al. 2009. Design and analysis of morphing wing based on SMP composite. In *Proceeding SPIE*, vol. 7290, 72900S (2009), *SPIE International Conference on Smart Structures/NDE*, San Diego, CA, March 8–12, 2009.
23. Yu, K., Yin, W., Liu, Y., and Leng, J. 2008. Application of SMP composite in designing a morphing wing. In *International Conference on Experimental Mechanics 2008 (ICEM 2008)*, Nanjing, China, November 8–11, 2008.
24. Everhart, M. C. and Stahl, J. Reusable shape memory polymer mandrels. www.CTD-materials.com

9

Shape-Memory Polymer Foam and Applications

Witold M. Sokolowski
Jet Propulsion Laboratory, California Institute of Technology, Pasadena, California

CONTENTS

9.1 Introduction

Currently existing approaches for the deployment of large structures including structures deployed in space such as solar arrays, solar sails, sunshields, or radar antennas typically rely on electromechanical mechanisms and mechanically expandable booms for deployment and to maintain them in the fully deployed, operational configuration. These support structures and their associated deployment mechanisms, launch restraints, and controls comprise sometimes more than 90% of the total mass budget for a deployed assembly.[1] In addition, they significantly increase the stowage volume, cost, complexity, and modes of failure. Therefore, one of the efforts at the National Aeronautic and Space Administration (NASA) and the Department of Defense (DoD) has been to develop expandable structures with relatively low mass and small launch volume to be used in low-cost missions.

As a result, space-inflatable structures have emerged in the last 12 years.[2,3] Inflatable technology is very attractive for space applications because inflatable structures are lightweight and have a small packing volume. However, some complete space-inflatable systems are not simple, since besides an inflatable structure they must include an inflation system (a gas container(s), the plumbing, a launch restrainer, a controlled deployment device, etc.) that increases the total weight, stowage volume, and complexity.[4,5] In addition, inflatables are vulnerable in space due to potential debris and micrometeorite strikes that may damage these structures.[6]

The development of structures made of cold hibernated elastic memory (CHEM) technology is one of the most recent results of the quest for simple, reliable, and low-cost self-deployable structures. The CHEM technology utilizes shape-memory polymers (SMPs) in open-cell foam structures or sandwich structures made of SMP foam cores and polymeric laminated-composite skins.[7] These lightweight foam structures are deployed via shape-memory and the foam's elastic recovery. The key to this technology is the use of SMP material systems. These materials behave very differently, depending upon whether they are above or below the glass transition temperature (T_g). Above T_g these materials are flexible and rubbery; below T_g, they are glassy and rigid. Most important of all, structures formed initially below T_g "remember" their shapes and sizes through successive warm/cold cycles and, if unconstrained, return to their original shapes when warm. Thus, a structure can be formed below T_g, warmed above T_g to make it flexible, folded or rolled for stowage, cooled below T_g so it can be stored in the compressed state without external forces, transported to space, warmed above T_g to allow it to self-deploy back to its original shape, and cooled below T_g to rigidize it for use. This approach provides a simple end-to-end process for stowing, deployment, and rigidization that has benefits of low mass, low stowage volume, low cost, and great simplicity. In addition to other appealing properties, shape-memory foam structures have debris impact energy absorption and dynamic damping capabilities.[8]

There are other SMP systems being developed in industry. The Elastic Memory Composites (EMC) have been developed recently by Composite Technology Development Co., Lafayette, CO.[9] EMC materials are similar to traditional fiber-reinforced composites except for the use of a thermoset shape-memory resin that enables EMC materials to achieve higher failure strains and provide higher packing capability than traditional composites through a specific thermomechanical load cycle.[10] Cornerstone Research Group Inc., Dayton, OH. has been developing dynamic polymer composite (DPC) systems. DPC materials are like other high-performance composites except that they use polystyrene-based SMPs in the matrix. Fabrication with these resins allows flexibility above its T_g and a high strength and stiffness at lower temperatures.[11]

This chapter describes the CHEM foam technology, provides some basic property data, discusses its advantages over other deployable structures, and identifies potential space, commercial, and biomedical applications. Some of these applications have been experimentally and analytically investigated with encouraging results. Present and future improvements in design, manufacturing, and processing of CHEM materials that will broaden potential applications are revealed here as well. Section 9.3.3, dedicated to potential solar sail structure applications, describes some advanced concepts including an ultralightweight self-deployable porous CHEM membrane. Although the space community is the original major beneficiary, a number of potential applications are also anticipated for the "earth environment." CHEM developers strongly believe that this technology has great promise for a host of commercial and biomedical applications. Some of these potential and already investigated CHEM applications are described in this chapter.

9.2 Overview of CHEM Foam Deployable Structure

9.2.1 Description

The CHEM technology utilizes SMP in open-cell foam structures or sandwich structures with a core made of SMP foam and polymeric laminated-composite skins.[12] These materials are polyurethane-based thermoplastic polymers with wide T_g ranges. They are unique because they exhibit large changes in elastic modulus E above and below the T_g. A large amount of inelastic strain (up to 400%) may be recovered by heating.[13,14] The reversible change in the elastic modulus between the glassy and rubbery states of the polymers can be as high as 500 times, and is shown in Figure 9.1. In addition, these materials also have high damping properties in their transition temperature range and large temperature dependence on gas permeability. Mechanical and chemical properties, durability, and moldability are practically the same as in conventional polyurethanes. The material's shape-memory function allows repeated shape changes and shape retention without material degradation. This phenomenon is explained on the basis of molecular structure and molecular movements, and is described elsewhere.[15,16]

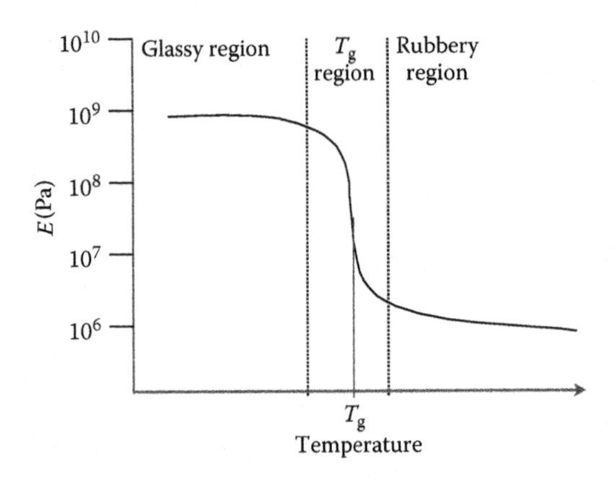

FIGURE 9.1
Elastic modulus E vs. temperature of polyurethane-based SMP.

In CHEM foam technology, the T_g can be tailored to rigidize the structure in the fully deployed configuration. The stages involved in the utilization of a CHEM structure are illustrated in Figure 9.2 and are as follows.[17] The original structure is fabricated and assembled in a room held below T_g. Later, the structure is warmed above T_g to make it flexible and is rolled or folded up for stowing. Then, the packaged structure is cooled below T_g so that it becomes firm in the compressed state. As long as the temperature is maintained below T_g, no external forces are needed to keep the structure compressed. Next, the packaged structure is warmed above T_g in an unconstrained configuration. Memory forces and the foam's elastic recovery cause the structure to naturally deploy back to its original shape and size without external actuation.

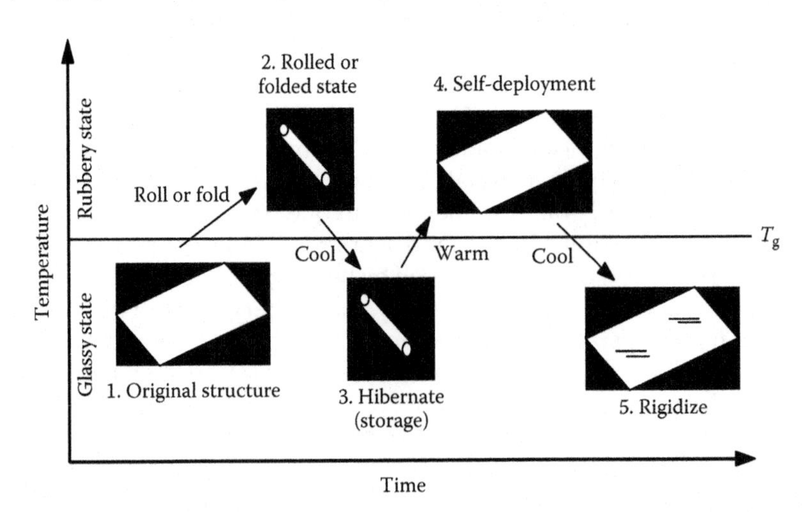

FIGURE 9.2
CHEM processing cycle.

Finally, the deployed structure is cooled below T_g to rigidize it, whereupon it is put into service.

An attractive aspect is the wide range of T_g that can be selected for deployment and rigidization. The T_g of the polyurethane-based SMPs ranges from −75°C to +100°C, thus allowing a wide variety of potential space and commercial applications for different environments. In these applications, the T_g of a CHEM structure should be slightly higher than the maximum ambient temperature. Heat would only be applied briefly for deployment, followed by radiative cooling to initiate rigidization. Very high ratios of the elastic modulus E below the T_g to the modulus above T_g (up to 500 for solid SMP) enable users to keep its original shape in a stowed, hibernated condition without external compaction forces for an unlimited time below the T_g. Furthermore, a narrow transition-temperature range for full transformation from a glassy to a rubbery state reduces the heat consumption during deployment (shape restoration).

CHEM structures are under development by the Jet Propulsion Laboratory (JPL) and industry. They are based on polyurethane SMPs that have been developed by Mitsubishi Heavy Industries in the last 18 years. Experimental results that have been obtained so far are very encouraging; the accumulated data indicate that the CHEM technology performs robustly in the Earth and space environments. Furthermore, the test and evaluation results, and preliminary analyses show that the CHEM technology is a viable way to provide a lightweight, compressible structure that can recover its original shape after long-term compressed storage.[18]

9.2.2 Properties of Baseline SMP Foam Material

A baseline SMP foam, with $T_g = 63°C$, was developed for convenience and simplicity of demonstration and testing in the Earth environment. The basic mechanical and thermal properties as well as the behavior under cyclic stress–strain–temperature loading were investigated at the University of Cambridge.[19] Here are the results.

9.2.2.1 Mechanical Properties

The basic mechanical properties were examined and compared with the foam developer's (Mitsubishi Heavy Industries—MHI).

The data indicated

- Good agreement existed between the results of these tests and developer's data.
- High ratio of E below T_g or E above T_g.

9.2.2.2 Stress–Strain–Temperature Cycles

The testing simulated the use of the shape-memory foam function in a CHEM processing cycle: heating above T_g and compaction → cooling below T_g and

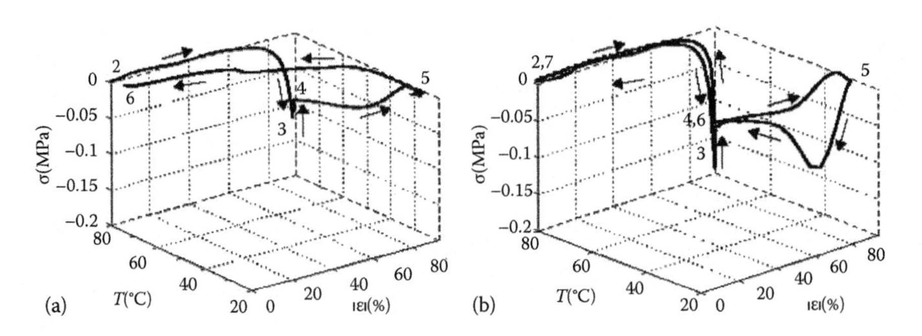

FIGURE 9.3
Stress–strain–temperature. Heating-sample heated to T_g+20ÆC. (1); Compression-sample compressed to about 95% strain. (2–3); Relaxation-sample left at T_g+20 until no further stress changes were noted. (3–4); Packing-sample cooled whilst still being hold at high strain. (4–5); Recovery-stress reduced to zero and sample reheated to T_g+20ÆC. (5–6).

hibernation → heating above T_g and deployment → cooling below T_g and rigidization (Figure 9.3)

Results indicated

- Complete shape recovery after the CHEM processing cycle.
- Recovery forces (stresses) nearly the same as compaction forces.
- The foam behaves in a typical elastic–plastic fashion at both above and below the T_g.

9.2.2.3 Thermal Properties

Specific heat was measured in the glassy state, during glass transition phase and in soft state, shown in Figure 9.4.

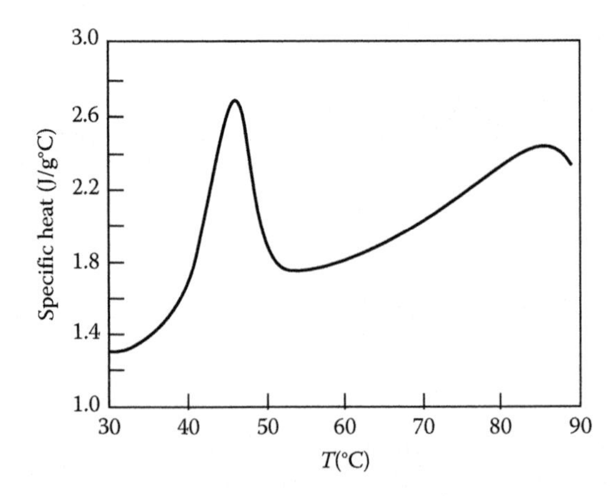

FIGURE 9.4
Specific heat vs. temperature.

Results demonstrated

- Specific heat increases with temperature with a pronounced peak during glass transformation phase.
- Thermal conductivity λ was predicted from theory for the uncompressed foam (0.027 W/m K) and compressed foam (0.12 W/m K).

The basic properties of the foam, designated as MF5520 herein, are given in Table 9.1.

During CHEM technology development, evaluation and test results were very encouraging.[12] All structural models including rods, tubes, wheels, chassis, boards, and tanks demonstrated the basics of CHEM concept such as:

- High full/stowed volume ratios above T_g
- Long-term stowage unconstrained in cold hibernated condition: over 1 year and continued
- Deployment when heating above T_g
- Precision original shape restoration after long stowage
- Rigidization of original shape when cooling below T_g

An additional SMP foam, designated as M-18G with T_g = 4°C was developed specifically for Mars applications.[12] The elastic modulus of M-18G was increased 3 times by adding chopped fiberglass reinforcement. Typically,

TABLE 9.1

Properties of CHEM Foam

Properties	MF 5520
Density (g/cm³)	0.032
T_g (°C)	63
Compressive strength (MPa)	0.09–0.102
Tensile strength (MPa)	0.2
E (compression) below T_g (MPa)	2.57–2.69
E (tension) below T_g (MPa)	11.4
E (compression) above T_g (MPa)	0.042–0.064
Coefficient of thermal expansion (glassy state) (ppm/°C)	27.5
Thermal conductivity (W/m K)	0.027
Thermal conductivity (95% compressed) (W/mK)	0.12
Specific heat (30°C) (J/kg K)	1320
Outgassing (Wt. loss–WVR) %	1.17

Note: WVR, water vapor recovered.

conventionally made CHEM foams have relatively low strength and structural rigidity. However, CHEM foam cores can be used in high-load carrying applications when combined with laminated-composite skins to form sandwich structures.

9.2.3 Characteristics of CHEM Foam Structure

The overall simplicity of the CHEM process is one of its greatest assets. In other approaches to expandable structures specifically in space-deployable subsystems, stowage and deployment are difficult and challenging, and introduce a significant risk, heavy mass, and high cost. Simple procedures provided by CHEM technology greatly simplify the overall end-to-end process for designing, fabricating, deploying, and rigidizing space gossamer structures. The CHEM technology avoids the complexities associated with other methods for deploying and rigidizing structures by eliminating deployable booms, deployment mechanisms, and inflation and control systems that can use up the majority of the mass budget. A long line of CHEM structure's major advantages are listed below.[8]

- *Low mass and stowage volume*

Polymer foam structure assures lightweight: almost two orders of magnitude lighter than aluminum. Incorporation of SMPs in an open cellular structure affirms high compressibility and full/stowed volume ratios.

- *High reliability and low cost*

No deployment mechanisms, controls, or inflation systems, etc. Already developed solid SMPs are inexpensive. Short time for technology development is anticipated.

- *Self-deployability and simplicity*

Precision deployment by elastic recovery and shape-memory of SMP foam. Simple deployment and rigidization. A structural and thermal isotropy behavior results in predictable thermal and temporal dimensional stability.

- *High dynamic damping and clean deployment and rigidization*

Foam acts like a structure composed of thousands of interconnected springs. Deployment by elastic recovery and shape-memory effects and rigidization by transition from rubbery to glassy state assure a clean, contamination-free environment.

- *No long-term stowage effects and ease of fabrication*

CHEM structures can be stowed in glassy state for an unlimited time without any compression set. They offer indefinite storage/shelf life in rubbery state compared with restricted storage or refrigeration of other polymers. Good

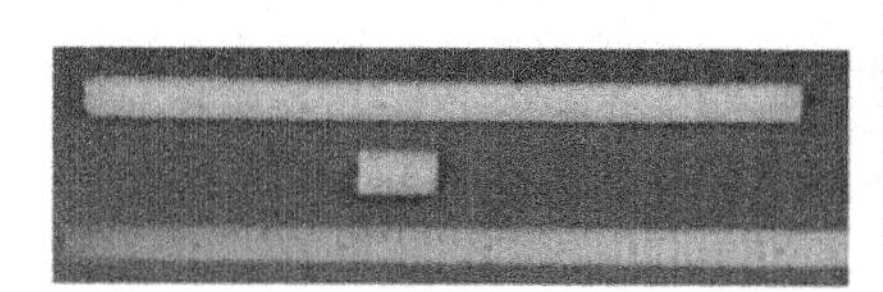

FIGURE 9.5
Stowed and deployed CHEM structures.

machinability in glassy and rubbery states. Cutting and shaping possible by conventional and computer numerically controlled (CNC) machining.

- *Impact* and *radiation resistant and thermal* and *electrical insulators*

Polyurethane-based CHEM foams belong to the preferred class of space radiation resistance materials. They can effectively absorb the energy of impact or of forces generated by deceleration without creating high damaging stresses. Very low thermal and electrical conductivity.

The stowed and deployed CHEM structures are shown in Figure 9.5. The disadvantage of a CHEM structure is the heat energy needed for deployment. However, the solar heating deployment approach appears to be feasible. Previously conducted studies and analyses indicate that solar radiation could be utilized as the heat energy for deployment in Mars and Earth environments.[20,21] Briefly, in this concept a shape-memory structure is compacted, and in the hibernated state can be covered by a thermal control blanket that has a high ratio of solar absorptivity-to-infrared emissivity. Prior to heating, the assembly must be kept out of direct solar radiation environment. When exposed to solar radiation, heat is generated inside the package and the original structure is deployed. After full deployment, the thermal blanket is removed and the structure is rigidized by the ambient (space) environment.

9.3 CHEM Structure Space Applications

9.3.1 Investigated Space Applications

CHEM structure technology provides NASA a robust, innovative self-deployable structure with significantly higher reliability, lower cost, and simplicity over other expandable/deployable structures to be used on many future space missions in Earth and space science programs. A myriad of CHEM applications are anticipated for space robotics and other support structures for telecommunication, power, sensing, thermal control, impact

TABLE 9.2

Investigated Space CHEM Applications

Applications	TRL	Comments
Nanorover wheels	4	Integrated with a nanorover and demonstrated in a laboratory experiment
Precision soft lander	2–3	Safe and stick-at-the-impact-site landing Small model proof-of-concept
Sensors delivery systems	2–3	CHEM-based integrated sensors are dropped and deployed in different planetary locations
Horn antenna	2–3	Deployable conical corrugated horn antenna Small model proof-of-concept
Radar antenna	2–3	Three-layer membrane design Small model proof-of-concept
Thermal-meteoroid shield	2	Lightweight deployable thermal and meteoroid protecting system
Habitats structures	2–3	Shelters, hangars, crew cabins, trans habs Small model proof-of-concept
In situ propellant production tanks	2–3	Small model proof-of-concept

Note: CCSL, CFRP/CHEM Spring Lock; CFRP, carbon fiber reinforced polymer.

and radiation protection subsystems, as well as for space habitats. Therefore, various feasibility studies and preliminary investigations have been conducted on potential CHEM space applications under various programs at JPL.[22–26]

Investigated CHEM space applications and their present technology readiness level (TRL) are shown in Table 9.2. TRLs range from TRL-1, for which basic principles have been observed and reported, to TRL-9, in which an actual flight system has been proven. Some of these studies are described in the following subsections.

9.3.1.1 Advanced Self-Deployable Wheels for Mobility Systems

Ultralightweight self-deployable and rigidizable wheels were developed and demonstrated at JPL utilizing CHEM structure process technology.[8,22] During this investigation, several different wheel designs were developed and evaluated for a prototype nanorover. The structural models of different designs were fabricated from the CHEM foam and assessed using a CHEM thermomechanical processing cycle. All wheels recovered completely after several cycles and a wheel design with the fastest recovery (deployment) was selected for a nanorover. Full-scale structural wheels were fabricated and assembled on a two-wheeled prototype nanorover, shown in Figure 9.6. Finally, the compacted wheels were successfully deployed at ~ 80°C and subsequently rigidized at room temperature in an atmospheric as well as in a simulated low pressure (6 millibar) Mars environment.

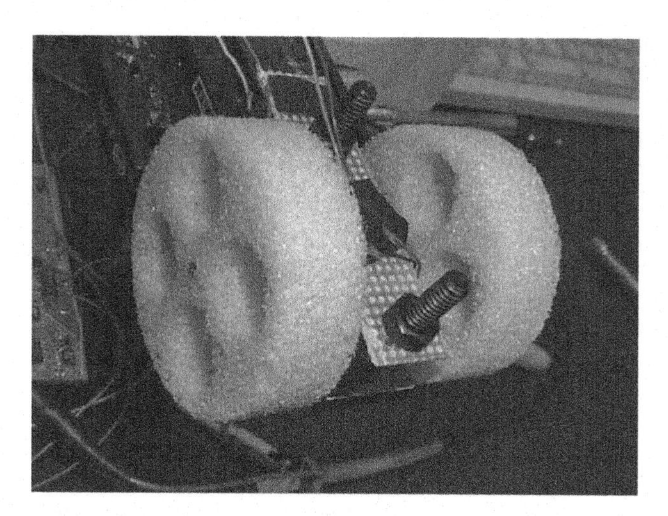

FIGURE 9.6
CHEM nanorover wheels.

Demonstrated complete wheel recovery after a CHEM processing showed that this structural concept is a viable way to provide ultralightweight, compressible, self-deployable wheels that can recover (deploy) their original size after cold hibernation storage. The recovery forces were able to fully deploy compacted CHEM wheels without any mechanical or inflatable systems. A high ratio of elastic modulus E below T_g or E above T_g in CHEM foams indicated very effective rigidization and eliminated support structures to maintain CHEM wheels in fully deployed, operational configuration.

The present mechanically deployed rover wheels are relatively heavy, complex, not quite reliable, and not autonomous with small full/stowed volume ratios. In addition, the MEMS and microelectronics embedded in CHEM foam structures will be self-protected from harmful effects of debris/micrometeoroid impact, or thermal environment. Autonomous, lightweight, self-deployable wheels technology for future space vehicles is critical and enables further robotics and future human exploration of space.

9.3.1.2 CHEM Horn Antenna

Development of ultralight deployable systems is one of the critical needs for many missions including recently proposed JASSI (Juniper Deep Atmospheric Sounder and Synchrotron Imager) deployable horn antenna. Preliminary investigation and analysis results indicated that CHEM self-deployable and rigidizable foam structure technology is one of the promising methods for this application.[8,23]

During these studies, several different designs of baseline 3.5 m long conical corrugated horn antenna were developed and structural/dynamic analyses were performed for each design configurations. A small CHEM structural antenna model was fabricated and a thin conductive Al layer was successfully deposited on the inside surface of the model (Figure 9.7). This structural model went through the CHEM processing cycles demonstrating the basics of CHEM concepts such as high full/stowed volume ratio, cold hibernated stowage, deployment when heating above T_g, original shape restoration and rigidization when cooling below T_g.

These studies indicated the feasibility of using CHEM foam structure technology for self-deployable horn antennas. Present mechanically deployable antennas are heavy, complex, not reliable, and packaging volume inefficient. CHEM foam antennas will provide a novel, self-deployable antenna structure technology with significant higher reliability, low mass, low cost, and simplicity.

9.3.1.3 Precision Soft Lander

There is a demand and need for precise, reliable, safe, low cost landing systems for small, single landers as well as for large multiprobe missions for planetary and small body exploration. Current spacecraft use complex systems such as aeroshells coupled to parachutes, solid rockets, and ultimately airbags to minimize the impact of landing, or use all propulsive soft landing approaches. Airbags are being used for intermediate-sized landers but they are too complicated and expensive for small (1–50 kg class) landers. As lander mass increases airbag systems become too heavy. Airbags have problems as well. On first impact, they produce lander bounce making it difficult

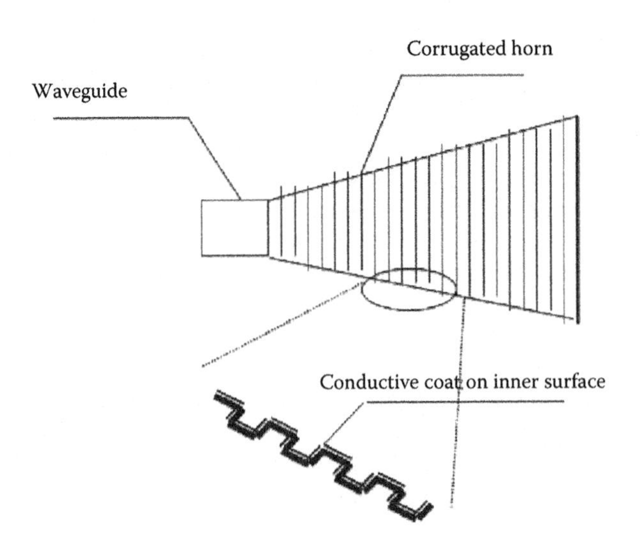

FIGURE 9.7
CHEM horn antenna.

to land within a small, scientifically interesting target area. Also, past mission experience indicates that the lander platform is not sufficiently stable to perform some precision operations even after airbags deflation. In addition, airbags do not provide sufficient thermal insulation against heat loss from the lander body to the cold ground.

Preliminary investigation indicated that problems of achieving a soft, stable, precision landing can be solved with the Precision Soft Lander (PSL) concept.[8,24] This concept employs a CHEM foam technology as the main thermal and energy absorbing element of the system. The PSL concept is described as follows.

A CHEM foam pad has a T_g well above ambient temperature. It will be compacted, at the temperature above T_g to about a tenth or less of its original volume, then cooled below T_g and later installed on a spacecraft without compacting restraints. Upon entry of the spacecraft into a planetary atmosphere, the temperature will rise above T_g causing the pad to expand to its original volume and shape. As the spacecraft decelerates and cools, the temperature will fall below T_g rigidizing the foam structure. The structure will absorb kinetic energy during ground impact by inelastic crushing, thus protecting the payload from damaging shocks and providing a safe stick-at-the-impact-site landing. Thereafter, this pad will serve as a mechanically stable, thermally insulating platform for the landed spacecraft.

When developed, the PSL system will offer a near-term technical solution for access to scientifically interesting sites including difficult and hard-to-reach areas. In addition, it has the potential to be highly reliable: no moving part, no actuators, and no subsystems that have to be deployed by other mechanical mechanisms.

9.3.1.4 Radar Antenna

Preliminary studies indicated the CHEM foam structure could be used for self-deployable, lightweight radar antennas. A novel CHEM structure-based radar antenna is described in Refs. [8,25]. The radar antenna is a flat array consisting of microstrip patches with microstrip transmission lines as power dividers. To achieve the required dual polarizations with 80 MHz of bandwidth, the antenna is designed with a three-layer membrane. The top layer has 18×6 radiating square patches, the middle layer is the ground plane, and the bottom layer has the power dividing transmission lines. Each layer is a 2-mil thick polyimide material with 5 μm copper deposited on it. The top layer has a spacing from the middle layer, while the bottom layer is spaced from the middle ground plane. The CHEM foam structure, which is to be made into flat sheets, is placed between the three membrane layers to not only serve as spacers but also to be used as an antenna deployment mechanism. In other words, the three membranes and the two sheets of CHEM foam material are pressed together to form a very thin structure and then is rolled up for stowage. During deployment, the CHEM foam returns

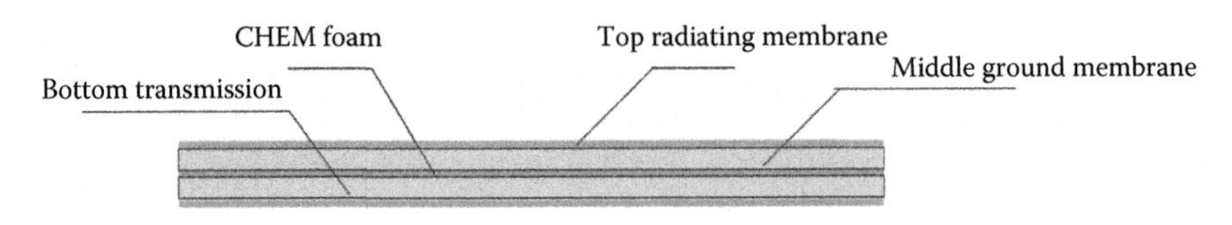

FIGURE 9.8
CHEM structure-based radar antenna.

to its original shape and size and unrolls the structure to form a flat antenna. The CHEM structure-based radar antenna is shown in Figure 9.8 above.

Currently, existing approaches for producing large deployable antenna structures in space typically rely upon mechanical mechanisms or inflatable booms to deploy structures and maintain them in the fully deployed, operational configuration. These support structures are heavy, expensive, and not reliable. The CHEM structure-based radar antenna when successfully developed, will have application for Earth mapping L-band synthetic aperture radars (SAR) to provide measurement of changes in ice water, soil moisture, global ecosystem, as well as Earth surface deformation. In addition, this structure technology could be used for other applications such as space deployable solar arrays.

9.3.1.5 Sensor Delivery System

Future planetary exploration missions to Mars and other planets/small bodies are aimed at understanding the global geology and climate history. While orbital platforms provide a detailed understanding of a surface at the order of 10 m resolution, detailed in situ exploration at the submeter level has been limited to landed missions such as Mars Pathfinder in 1997. The in situ exploration of science sites has been envisioned using mobile robots or rovers, however, due to the assumed rugged terrain, rovers may have difficulties accessing these sites due to mobility and power availability limitations.

Recently, another approach for widespread planetary exploration is the deployment of a network of sensors located in diverse and hard-to-reach locations on the planet surface. However, the major technological challenge to overcome is how we can deploy and scatter them across a planetary surface in an effective, reliable, and simple way.

The proposed development of CHEM-based sensor delivery system (SDS) is the answer to that technical challenge. Recently conducted preliminary studies confirmed a feasibility of low mass, low packing volume, self-deployable SDS technology to future planetary missions, specifically to planet Mars.[26] Integrated SDS system utilizes a CHEM open cellular structure as the main structural element. The CHEM-based integrated sensors are in hibernated, compacted condition during the launch and flight stages and then are dropped in different locations from a planetary lander or aerial vehicles. They are deployed and, if necessary, rigidized during the falling or on planetary surface. The CHEM foam packaging structure absorbs a large amount of impact energy without generating high damaging stresses, and the SDS system will be able to robustly survive the surface impact. Once deployed, the CHEM structure will expand to deploy the elements of the sensor system. These sensors would be scattered across scientifically interesting but hard-to-reach surface sites to form a network of sensors for in situ detection of life.

The development of a simple, low cost CHEM-based sensor system that leads to the deployment of a large number of sensors to a planet surface will benefit NASA by providing a means for robust planetary exploration over a wide surface area. In situ exploration of scientifically interesting sites by such a system will certainly improve NASA's ability to reach difficult terrain settings without the need for mobility as is required for rovers.

9.3.2 CHEM Improvements for Large Structures

One of the major efforts in CHEM technology in the last several years has been to improve and optimize the quality and physical properties of the CHEM foams. Wright Material Research Co. (WMR), working with JPL under the Small Business Innovation Research (SBIR) Program, has developed new CHEM microfoam materials.[27] WMR utilized its proprietary foam processing to develop the microfoams from the SMP raw materials. These CHEM microfoams exhibit micron-size cells that are uniformly and evenly distributed within the cellular structure. The cell sizes can be controlled during the foam forming process. When compared with conventionally made CHEM foams, the CHEM microfoams have enhanced physical and mechanical properties, improved isotropy of properties, increased compressive and tensile strength, tear strength, and fracture toughness. Also, because of the very small size of the cells, the material is well suited for ultralightweight porous membrane or thin film space applications.

Other activities have been started to develop CHEM sandwich structures intended for high-load support-structure applications. WMR under the SBIR contract has worked on CHEM sandwich structures that involve fiber-reinforced SMP composite face sheets and CHEM foam cores.[28] The same memory-polymer was used for the face sheets and for the foam cores. This approach maximized the packaging ratio and reduced stowage volume of the sandwich structure.

In the future, it is planned to develop new CHEM microfoams that are reinforced with carbon nanotubes. This material system will increase the mechanical strength as well as the thermal conductivity. The improvement in thermal conductivity is important because it will cut down the time required to deploy a CHEM structure in space.

Preliminary investigations and analyses have indicated that CHEM foams by themselves cannot be used for high-load-carrying support structures. However, it appears that the sandwich CHEM structures or a hybrid design of CHEM foams and polymer composites could be used to develop support structures for large deployable solar sails, antennas, telescopes, sunshields, and solar arrays.[29] For these applications, the CHEM foams are combined with laminated-composite skins to augment strength or stiffness, or to serve as a deployment mechanism.

In one of these applications, Composite Optics Inc. (COI), working with JPL, developed a spring-lock truss-element concept for large (>50 m) boom structures that involved a unique hybrid design of CHEM foams and polymer composites.[30,31] This truss element, referred to herein as the CCSL truss element, is essentially two carbon-fiber-reinforced polymers (CFRP) that are separated by blocks of CHEM foams, as shown in Figure 9.9.

The two material systems are used in a manner that allows the truss element to be stowed in a small volume and then deployed without the use of complex mechanisms. The CCSL concept uses the CHEM foam to lock the truss element in the stowed and deployed states to control the deployment, and to enhance the buckling resistance. The CFRP tape-like skins are used to provide high axial stiffness to the truss members, to reduce weight, to provide elastic spring energy for deployment, and to increase the overall stability of the boom. The baseline truss-boom design was a three-legged CCSL-longeron truss-element configuration supported with diagonals and horizontals, all connected at joints. All other trust elements were made from CFRP material to maintain ultralow mass, high buckling resistance, and high durability. The results of a preliminary investigation on the CCSL truss element were encouraging and, as a result, COI has successfully demonstrated a proof-of-concept for the CCSL structure.

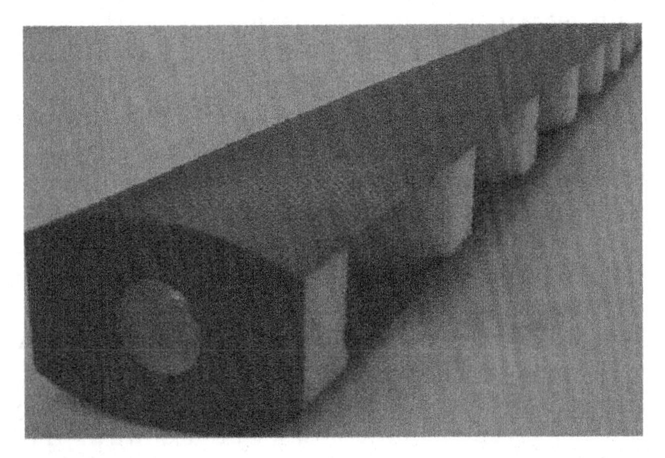

FIGURE 9.9
CCSL truss element.

9.3.3 Potential Advanced CHEM Structures

The potential use of CHEM microfoams reinforced with carbon nanotubes that may be developed in the future are being considered for thin-membrane applications, specifically for solar sails.[32,33] In particular, a CHEM membrane without support booms that is deployed by using shape-memory and elastic recovery is envisioned. In this advanced structural concept, the CHEM membrane structure is warmed up to allow packaging and stowing prior to launch, and then cooled to induce hibernation of the internal restoring forces. In space, the membrane "remembers" its original shape and size when warmed up. After the internal restoring forces deploy the structure, it is then cooled to achieve rigidization. For this type of structure, the solar radiation could be utilized as the heat energy used for deployment. This solar-sail concept, that uses an ultralightweight microporous membrane that is deployed by shape-memory, could advance solar-sail technology and enable the development of sail materials with areal densities less than $2\,g/m^2$. In addition, highly integrated multifunctional CHEM membranes with embedded thin-film electronics, sensors, actuators, and power sources could be used to perform other spacecraft functions such as communication, navigation, science gathering, and power generation (Figure 9.10).

Certainly, more research, experiments, and analyses must be done to fully realize the potential of the self-deployable CHEM membranes.[33] Future research is needed to determine if the shape-memory elastic recovery forces in thin CHEM membrane will be substantial enough to provide a viable structural concept. Likewise, research is needed to determined if the stiffness of rigidized membrane that is reinforced with carbon nanotubes is high enough to support deployed structure in space. Structural models need to be

FIGURE 9.10
Potential boomless CHEM membrane in solar sail.

developed to demonstrate, characterize, and improve the understanding of self-deployment and rigidization mechanisms. All these activities will help to assess the applicability of future structural concepts.

The pay off of this research could be high. This advanced membrane concept represents the introduction of a new generation of self-deployable structures. If developed, this innovative technology will introduce a new paradigm for defining configurations for space-based structures and for defining future mission architectures. It will provide new standards for fabricating, stowing, deploying, and rigidizing large deployable structures in a simple, straightforward process. This new technology could be one of the precursors and fundamental technologies for future next generation, thin-film, self-deployable structures, in general, and for solar sails specifically, without booms and other support structures.

9.3.4 Comparison with Other Deployable Structures

Currently existing approaches for deployment of large, ultralightweight gossamer structures in space typically rely upon electromechanical mechanisms and mechanically expandable or inflatable booms for deployment and to maintain them in a fully deployed, operational configuration. These support structures with the associated deployment mechanisms, launch restraints, inflation systems, and controls can comprise more than 90% of the total mass budget. In addition, they significantly increase the stowage volume, cost, and complexity.

In a preliminary investigation, a generic elastic-memory, self-deployable structure was compared with other deployable structures. The results of this comparison are summarized in Table 9.3.[34] Similarly, the mass budgets for the

TABLE 9.3

Comparison of Mechanical, Inflatable, and Elastic Memory Deployable Structures

Characteristics	Mechanical Deployment	Inflatable	Shape-Memory Deployment
Mass	Heavy	Up to 3 × lighter	>10× lighter
Stowage volume	Bulky	2–30 × smaller	50 × smaller
Reliability	Good	Good	Potentially highest
Cost	High	Lower	Lowest
Simplicity	No	Better	Best
Deployment/inflation subsystem	Yes	Yes	No
Impact resistance	Not good	Not good	Good
Dynamic damping	Good	Very good	Very good
Clean deploy/rigidization	Clean	Potential gas leak	Clean[a]
Stowage effects	Yes	Yes	No
Fabrication	Difficult	Easier	Easiest

[a] Meets NASA outgassing requirements.

TABLE 9.4

Mass Budgets for ISIS Inflatable Sunshield and Elastic Memory Concept

ISIS		Shape-Memory Deployment Mass (kg)
Sunshield Subsystem[a]	**Inflatable Mass (kg)**	
Membrane	2.58	7.1[b]
Support structures	7.87	~1.0[c]
Inflation system	15.43	0
Launch restraint	30.22	0
Container	34.02	0
Controlled deployment	11.87	0
Thermal insulation	1.38	1.4[d]
High absorptivity blanket	0	0.7[e]
Subtotal	104.35	~10.2

[a] Sunshield area = 28.19 m^2.
[b] 0.25 kg/m^2 areal density or lower.
[c] Attachment to the spacecraft.
[d] Uses the same insulation as ISIS.
[e] Absorbs solar heat for deployment.

inflatable sunshield used for the Inflatable Sunshield in Space (ISIS) experiment and for a corresponding hypothetical elastic-memory-deployable sunshield are presented in Table 9.4. ISIS was a one-third-scale Next Generation Space Telescope (NGST) sunshield flight demonstration experiment. The total weight of the ISIS four-membrane sunshield design was 104.35 kg. The dominant items contributing to the mass budget were the support tubes (7.87 kg), the inflation system (16.43 kg), a launch restrainer (30.22 kg), a controlled-deployment device (11.87 kg), and a container (34.02 kg). These items created an areal density greater than 3 kg/m^2. Most of these items are eliminated by using CHEM technology, thus drastically reducing the areal density, stowage volume, and cost. Also, the CHEM processing cycle of stowing, deployment, and rigidization, results in significant reductions in complexity and improvements in reliability, compared to existing alternative deployment methods.

9.4 Potential Commercial Applications of CHEM Structures

Although the space community is the major beneficiary, a lot of potential CHEM commercial applications are also foreseen for the earth environment.[35] Such applications could be made of CHEM foam with a T_g slightly above the highest ambient temperature. The CHEM products will be compacted, stowed unconstrained in small volumes at room temperature, transported if needed, and deployed at required locations. They also can be dropped from aerial vehicles and deployed in hard-to-reach locations. Sensors, circuitry, and automated components will be easily integrated to enable CHEM-based systems to act as smart structures which operate autonomously. In addition, CHEM structures are potentially self-repairable by temporary heating and recooling.

Depending upon the application, CHEM technology may utilize just a CHEM foam or a sandwich structure in case of high-load carrying applications. The sandwich CHEM structures involve fiber reinforced polymer composite skins and CHEM foam cores.

CHEM structures can be compacted and packaged in different ways. Some of these techniques are shown in Figure 9.11. The structures can be compacted by compression, rolling, or a combination of these two or other methods. Also, the packaging could be in forms such as accordion folding, roll up, combination of these two, or other appropriate packaging for selected CHEM applications.

The CHEM products can be deployed by using different heat sources. They can be heated for instance in ovens, or by portable heaters such as a hair dryer. In some applications, conductive embedded fibers can provide the resistive heating for deployment as well as reinforcement for higher load carrying products. Recently conducted studies and analyses indicate that solar radiation could be utilized as the heat energy for deployment in Mars and Earth environments.[20,21] The CHEM product in a compacted, hibernated state can be covered by a thermal control blanket that has a high ratio of solar absorptivity-to-infrared emissivity. When exposed to solar radiation, heat is radiated inside the package and the original structure is deployed. After full deployment, the thermal blanket is removed and the structure is rigidized by ambient temperature.

Intrinsic properties of CHEM foams such as impact/shock resistance, thermal and electrical insulation, structural or sound/electromagnetic wave attenuation, make them a potential technology for numerous self-deployable commercial products. Here is a long list of interesting potential CHEM applications[35]:

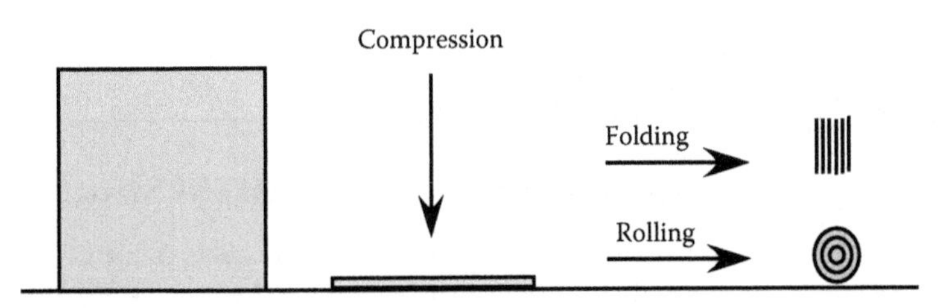

FIGURE 9.11
Compaction/packaging of CHEM products.

- Deployable thermal insulation
  - Insulation for building and transport systems such as refrigerated trucks, railway cars, ships designed to carry liquid natural gas
  - Coolers, thermoses, liquefied gas tanks
  - Hot/cold-refrigerated storage
  - Thermal insulation shields/barriers
- Deployable packaging/impact absorption
  - Packaging of impact and thermal sensitive products
  - Sensors/drug/food air delivery/landing systems
- Deployable structures
  - Thermally insulated deployable shelters, storage places, hubs
  - Temperature-controlled rooms
  - Prefabricated walls/slabs
  - Access denial barriers/walls
- Deployable recreation/sport products
  - Tents and camping equipment
  - Boats, kayaks, rafts
  - Life jacket, floating wheels
  - Water and snow skis
  - Surf and snow boards
- Deployable automotive applications
  - Car safety and security subsystems
  - Human protection products
- Deployable food equipment
  - Dishes and meal containers
  - Plates and coffee cups
  - Hot/cold storage for food
- Deployable toys
  - Toys that take the advantage of simple reversible compaction, deployment, and rigidization CHEM cycle.
  - Decoys with high fidelity features
- Sound/electromagnetic attenuation
  - High damping sound and electromagnetic wave shielding
  - Sound and electromagnetic wave-controlled rooms

However, this long list is by no means meant to be exhaustive. CHEM developers strongly believe that this technology has great promise for a host of commercial applications when fully developed. They have been already contacted regarding potential CHEM applications in recreation, automotive, packaging, construction, and biomedical sectors. Some development work is under way in automotive and packaging areas.

9.5 Medical Applications of CHEM Structures

Polyurethane-based CHEM foam materials have very unique and appealing properties for the design and manufacturing of self-deployable medical devices. Similar to solid SMP, they have an excellent biocompatibility, and the T_g can be tailored for shape restoration/self-deployment when inserted in the human body.[35]

A combination of these properties plus natural CHEM foam porosity, lightweight, high full/stowed volume ratio, and precision of original shape restoration suggest that it has potential to be utilized in functional and deployable elements of different clinical devices. They can be miniaturized, deformed, and inserted in the body through small catheters. Then, under the body heat, they can precisely deploy/recover a much larger predetermined required shape in satisfactory position.[36] The CHEM part will self-expand when inserted in the body as it reaches a temperature of 37°C. Also, a CHEM foam porosity can be adjusted to the needs of the application. In addition, the CHEM technology provides a simple end-to-end process for stowing and deployment, and avoids the complexities associated with other methods for deployment of medical devices.

Consequently, a lot of potential CHEM medical applications are foreseen for vascular and coronary grafts, orthopedic braces and splints, medical prosthetics and implants, just to name a few.[36] The CHEM could be used to design artificial grafts for replacing diseased arteries as well as a scaffold for tissue engineering. It may possibly be used as a three-dimensional matrix to support bone growth in vitro and in vivo. CHEM could be appropriated for both soft tissue and hard tissue engineering, because of the wide range of T_gs (between −70°C and +70°C) and large reversible changes of elastic modulus. It may perhaps be utilized for breast implants and others. Polyurethane is already used to cover silicone or saline implants.

Presently, several important CHEM applications are being considered for self-deployable vascular and coronary devices. One of these potential applications is the removal of a blood clot (thrombus) from the arterial network. Feasibility studies and preliminary in vitro demonstrations on SMP conducted by M. Metzger et al. were optimistic.[37] The formation or lodging of thrombus in the arterial network supplying the brain, typically causes the ischemic strokes. The stroke is the third leading cause of death and the principal cause of long-term disability in North America. The strokes can by caused also by the intracranial aneurysm. However, present endovascular interventions on aneurysms have important drawbacks such as a significant incidence of residual lesion, deficient healing at the neck, recanalization, or recurrences. Therefore, the search for new and more effective methods has been continued.[38] The CHEM foam materials have appealing characteristics for the design of endovascular devices. Their unique properties suggest that they have a large potential to be used as an embolic agent and filling material to occlude aneurysms.

The CHEM foams were experimentally investigated for endovascular treatment of aneurysm at CHUM Research Center, Notre-Dame Hospital, Montreal, in collaboration with École Polytechnique, Montreal with encouraging results.[39] Lateral wall venous pouch aneurysms were constructed on both caroid arteries of eight dogs. The aneurysms were occluded per-operatively with CHEM blocks. Internal maxillary arteries were occluded via a 6F transcatheter technique using compressed CHEM blocks. Angiography and pathology were used to study the evolution of the occlusion and neointimal formation at the neck of experimental aneurysms after 3 and 12 weeks. The CHEM extract demonstrated no evidence of cell lysis or cytoxicity and no mutagenicity. The efficient vascular embolization was confirmed in the aneurysms and good neointimal formation over the neck of treated aneurysms was demonstrated at the CHEM interface. Maxillary arteries embolized with CHEM foam remained occluded during this experiment. The major conclusion of the investigation was that the foamy nature of this new embolic agent favors the ingrowth of cells involved in neointima formation, and new embolic devices for endovascular interventions could be designed using CHEM's unique physical properties.

References

1. Carey, J., Goldstein, E., Cadogan, D., Pacini, L., and Lou, M., Inflatable sunshield in space (ISIS) versus next generation space telescope (NGST) sunshield—A mass properties comparison, *AIAA Structures, Structural Dynamics and Materials Conference*, April 3–6, 2000, Atlanta, GA.
2. Dornheim, M., Inflatables structures taking to flight, *Aviation Week & Space Technology*, 150(4), January 25, 1999, 60–62.
3. Dornheim, M. and Anselmo, J., Complex antenna is star of Mission 77, *Aviation Week & SpaceTechnology*, 144(22), May 27, 1996, 58–59.
4. Huang, J., The development of inflatable arrays antennas, *IEEE Antennas & Propagation Magazine*, 43(4), August 2001, 44–50.
5. Lou, M., Fang, H., and Hsia, L., Self-rigidizable inflatable boom, *Journal of Spacecraft & Rockets*, 39(5), September–October 2002, 682–690.
6. Njoku, E., Sercel, J., Wilson, W., Moghaddam, M., and Rahmat-Samii, Y., Evaluation of inflatable antenna concept for microwave sensing of soil moisture and ocean salinity, *IEEE Transactions and Remote Sensing*, 37(1), January 1999, 63–78.

7. Sokolowski, W., Cold hibernated elastic memory self-deployable and rigidizable structure and method therefore, Patent no: US 6,702,976 B2, March 9, 2004.
8. Sokolowski, W. and Hayashi, S., Applications of cold hibernated elastic memory (CHEM) structures, *Proceeding of SPIE 10th International Symposium on Smart Structures and Materials*, March 2–6, 2003, San Diego, CA.
9. Lake, M., Hazelton, C., Murphey, T., and Murphy, D., Development of coilable longerons using elastic memory composite material, AIAA Paper 2002-1453, April 2002.
10. Lake, M., Munshi, N., Tupper, M., and Meik, T., Applications of elastic memory composite materials to deployable space structures, AIAA Paper 2001-4602, August 2001.
11. Cullen, S., Roberts, T., and Tong, T., Studies of shape memory behavior of styrene-based network copolymers, *Proceedings of the First World Congress on Biomimetics*, December 9–11, 2002, Albuquerque, NM.
12. Sokolowski, W., Chmielewski, A., Hayashi, S., and Yamada, T., Cold hibernated elastic memory (CHEM) self-deployable structures, *Proceeding of SPIE'99 International Symposium on Smart Structures and Materials*, March 1–5, 1999, Newport Beach, CA.
13. Hayashi, S. and Shirai, Y., Development of polymeric shape memory material, *Mitsubishi Technical Bulletin*, No. 184, December 1988.
14. Hayashi, S., Tobushi, H., and Kojima, S., Mechanical properties of shape memory polymer of polyurethane series, *JSME International Journal*, Series I, 35, July 1992, 206–302.
15. Hayashi, S., Properties and applications of polyurethane-series shape memory polymer, *International Progress in Urethanes*, 6, 1993, 90–115.
16. Hayashi, S., Ishikawa, N., and Giordano, C., High moisture permeability for textile applications, *Polyurethane World Congress*, Vancouver, British Columbia, Canada, October 1993, pp. 400–404.
17. Sokolowski, W., Tan, S., and Pryor, M., Lightweight shape memory self-deployable structures for Gossamer applications, AIAA Paper 2004-1660, April 2004.
18. Sokolowski, W. and Ghaffarian, R., Surface control of cold hibernated elastic memory self-deployable structure, *Proceeding of SPIE International Symposium on Smart Structures and Materials*, February 26–March 2, 2006, San Diego, CA.
19. Watt, A., Pellegrino, S., and Sokolowski, W., Thermo-mechanical properties of a shape-memory polymer foam, Collaboration between University of Cambridge and JPL, www2.eng.cam.ac.uk/~amw33/foams.pdf, April 17, 2000.
20. Sokolowski, W., Awaya, H., and Chmielewski, A., Solar heating for deployment of foam structures, NASA Novel Technology Report (NTR), NPO-20961, *NASA Tech Briefs*, 25(10), October 2001, 36–37.
21. Kirkpatrick, E. and Sokolowski, W., Heating methods for deployment of CHEM foam structures, *Proceeding of International Conference on Environmental Systems (ICES)*, July 7–10, 2003, Vancouver, British Columbia, Canada.
22. Sokolowski, W. and Rand, P., Advanced lightweight self-deployable wheels for mobility systems, NASA Novel Technology Report (NTR), NPO-21225, *NASA Tech Briefs*, 27(2), February 2003, 10a–12a.
23. Sokolowski, W., Levin, S., and Rand, P., Lightweight self-deployable foam antenna structures, NASA Novel Technology Report (NTR), NPO-30272, *NASA Tech Briefs*, 28(7), July 2004, 38–39.

24. Sokolowski, W. and Adams, M., Soft landing of spacecraft on energy-absorbing cushions, NASA Novel Technology Report (NTR), NPO-30435, *NASA Tech Briefs*, 27(1), January 2003, 69.
25. Sokolowski, W. and Huang, J., Novel self-deployable radar structure, NASA Novel Technology Report (NTR), NPO-30742, *NASA Tech Briefs*, 28(10), October 2004, 63–64.
26. Sokolowski, W. and Baumgartner, E., New sensor delivery system, NASA Novel Technology Report (NTR), NPO-30654, *NASA Tech Briefs*, 27(11), November 2003, 58.
27. Tan, S., Space rigidizable deployable ultra-lightweight microcellular CHEM foams, NASA SBIR Phase II Final Report, August 31, 2004.
28. Tan, S., Self-deployable ultra-lightweight modular unit for habitat structural applications, SBIR Phase I Final Report, July 17, 2004.
29. Sokolowski, W. and Tan, S., Advanced self-deployable structures for space applications, *Journal of Spacecraft and Rockets*, 44(4), July–August 2007, 750–754.
30. Pryor, M. and Sokolowski, W., Deployable truss elements for space based structures, COI/JPL NRA 99-05 Proposal, February 2000.
31. Sokolowski, W., Tan, S., and Pryor, M., Lightweight shape memory self-deployable structures for Gossamer applications, AIAA Paper 2004-1660, Palm Springs, CA, April 2004.
32. Sokolowski, W. and Tan, S., Self-deployable membrane structure, NASA New Technology Report, NTR-41759, January 2006.
33. Sokolowski, W., Tan, S., Willis, P., and Pryor, M., Shape memory self-deployable structures for solar sails, Submitted to *SPIE'08 International Symposium on Smart Materials, Nano+Micro-Smart Systems*, December 9–12, 2008, Melbourne, Australia.
34. Sokolowski, W., Ultra-lightweight shape memory selfdeployable structures for solar sails NRA Proposal, 01-OBPR-08-G, December 21, 2001.
35. Sokolowski, W., Potential bio-medical and commercial applications of cold hibernated elastic memory self-deployable foam structures, *Proceedings of SPIE International Symposium on Smart Materials, Nano-and Micro-Smart Systems*, December 12–15, 2004, Sydney, Australia.
36. Sokolowski, W., Metcalte, A., Hayashi, S., Yahia, L., and Raymond, J., Medical application of shape memory polymers, *Institute of Physics Publishing, Biomedical Materials*, 2007, 2.
37. Metzger, M.F., Wilson, T.S., Schumann, D., Mathews , D.L., and Maitland, D.J., Mechanical properties of mechanical actuator for treating Ischemic stroke. *Biomedical Microdevices*, 4(2), 2002, 89–96.
38. Raymond, J., Metcalfe, A., Salazkin, I., and Schwarz, A., Temporary vascular occlusion with poloxamer 407, *Biomaterials*, 25(18), August 2004, 3983–3989.
39. Metcalfe, A., Desfaits, A.C., Salazkin, I., Yahia, L., Sokolowski, W.M., and Raymond, J., Cold hibernated elastic memory foams for endovascular interventions. *Biomaterials*, 24(3), February 2003, 491–7; Erratum in: *Biomaterials*, 24(9), April 2003, 1681.

10

Shape-Memory Polymer Textile

Jinlian Hu
Institute of Textiles and Clothing, Hong Kong Polytechnic University, Hong Kong, China

CONTENTS

10.1 Introduction

Shape-memory polymers (SMPs), a type of shape-memory material, are defined as polymeric materials with the ability to sense and respond to external stimuli in a predetermined shape [1,2]. Compared with shape-memory metals and ceramics, SMPs possess such characteristics as light weight, corrosion resistance (compared with metals), formability, workability, and low cost [3,4]. Such polymers as polynorbornylene, transpolyisoprene, styrenebutadiene copolymer, polyethylene, polyester copolymerized with others, and block polyurethanes, etc., have been found to have shape-memory effect [4,5].

10.1.1 Different Shape-Memory Polymers

Since SMPs were first developed in France and commercialized in Japan in 1984, research activities have continued in Japan, Europe, Korea, China including Taiwan, and the United States. A variety of polymers were developed with shape-memory effects. In this section, we briefly introduce several typical SMPs.

10.1.1.1 Early Developed SMPs

The first SMPs were polynorborene based with a glass transition temperature (hereinafter referred to as T_g) range of 35°C–40°C. Developed by the French CdF Chimie Company, they were commercialized by Nippon Zeon Co. in Japan [4,5]. They are suitable for developing apparel textiles, but their T_g is very difficult to adjust. Because of their high molecular weight (100 times of ordinary plastics), their process ability is limited. It has recently been reported that the reinforced polynorbornene has a great advantage for higher temperature applications [5,6].

Kuraray Company in Japan developed *trans*-polyisoprene-based SMPs with a T_g of 67°C in 1987. The third kind of SMPs, introduced by Asahi Company, was styrene butadiene based and had a T_g ranging from 60°C to 90°C. Both of these polymers have a high T_g, which limits their applications to apparel areas, and their process ability is also reportedly poor [4,5,7].

10.1.1.2 Shape-Memory Polyurethanes (SMPUs)

Mitsubishi Heavy Industry (MHI) later succeeded in developing the polyurethane-based thermoplastic polymers (SMPUs) [4,7]. The T_g values of MHI SMPUs lie in a broad temperature range from −30°C to 65°C, and the materials are more processible than the earlier SMP materials [4,7]. Because SMPUs have demonstrated improved process abilities, these types of SMPs receive the widest attention and have diverse applications. Hayashi and Tobushi et al. [3,8,9,10] investigated the mechanical properties, and systematically examined the structure–property relationships and the shape-memory behaviors of SMPUs. Li et al. also reported their studies on shape-memory effect of segmented polyurethane with different molecular weights and different lengths of hard and soft segments, and put forward some theories on their relations with shape-memory effects [1,11,12,15]. There are a number of papers that applied MHI SMPUs to many areas [16,17–20,23]. It is reported that a large amount of elastic strain (up to 400%) can be recovered by heating the SMPUs, and the reversible change in the elastic modulus between the glassy state and the rubbery state of the materials (elasticity memory) can be as high as 500 times [3,4,7].

10.1.1.3 Others

Shape-memory effect of ethylene vinyl acetate (EVA) copolymers was also investigated by Li et al. (1999) [37]. Their results indicated that it is possible to prepare good SMPs with a large recoverable strain and a high final recovery rate by introducing sufficient numbers of cross-links and a gel content higher than about 30% into EVA samples with a two-step method. The response temperature is lower than 100°C, and it can be controlled continuously in principle by using EVA copolymers of different compositions. Luo et al. synthesized a series of ethylene oxideethylene terephthalate segmented copolymers (EOET) with long soft segments and studied the thermally stimulated shape-memory behavior of the materials. They discovered that the length of the soft segment, the hard segment content, and the processing conditions can have an important influence on shape-memory effects [24]. In addition, Li et al. reported their investigation of the production of SMP with polyethylene/nylon six graft copolymers [25]. The nylon domains, which serve as physical cross-links, play a predominant role in the formation of a stable network for graft copolymers.

The study shows that shape-memory materials with thermoplastic characteristics can be prepared not only by block copolymerization but also by graft copolymerization, and the reaction processing method might be a promising and powerful technique for this purpose. They concluded that polyethylene/nylon six blend specimens are able to show good shape-memory effect under normal experimental conditions [25].

In a European patent, another series of SMP resins were reported, consisting essentially of a block copolymer having an A-B-A block structure in the polymer chain [26].

This polymer has an average molecular weight within the range of 10,000–1,000,000. Block A is a polymer block comprising a homopolymer of a vinyl aromatic compound, a copolymer of a vinyl aromatic compound and another vinyl aromatic compound, a copolymer of a vinyl aromatic compound and a conjugated diene compound and/or a hydrogenated product thereof. Block B is a polymer comprising a homopolymer of butadiene, a copolymer of butadiene with another conjugated diene compound, a copolymer of butadiene with a vinyl aromatic compound and/or a hydrogenated product thereof.

Another polymer material was reported to have shape-memory characteristics. The material is prepared from a polymer substance obtained by dissolving an amorphous fluorine containing polymer in an acrylic monomer, and polymerizing the monomer with or without permitting the resulting polymer to form an interpenetrating polymer network [27].

10.1.2 Characteristics of SMPs

10.1.2.1 Structure of SMPs

Basically, SMPs should be made of copolymers. Within such polymers, there are hard and soft segments, and the soft segments must be large enough to allow considerable free rotation when temperature changes and result in the recovery of the deformations previously exerted on the structure. The formation of a network structure with hard point/segments (fixed phase) and amorphous/soft segments/regions (reversible phase) are the two necessary conditions for their good shape-memory effect.

We take SMPUs as an example to explain the structure of such polymers. In polyurethanes, hard segments such as bifunctional diisocyanates (OCN-R-NCO) form the fixed phase in which there are physical cross-linking points through polar interaction, hydrogen bonding, and crystallization. Such physical cross-linking points cannot be broken at temperatures below 120°C [4]. Meanwhile, the soft segment such as polyol (HO-R′-OH) domains form the reversible phase, and the observed shape-memory effect is due to the rotation of soft segments. Depending on the weight and length of soft segments, the molar ratio of the soft segment to hard segment and the manufacturing procedure of the resins, the static and dynamic properties of the polymers are easily controlled [1,13,14,15]. The shape recovery temperature can be freely tailored in the range of room temperature from −30°C to 65°C so as to meet the requirements of specific applications.

10.1.2.2 Shape-Memory Behavior of SMPs with Thermosensitivity

Shape recovery due to shape-memory of a polymer may be triggered by heat, light, electricity, and other stimuli. The SMPs to be discussed in this chapter are thermosensitive SMPs. In this case, the shape-memory behavior refers to the phenomenon that a material recovers its original shape if it is heated above the phase transition temperature of the reversible phase without external forces because the molecular movement becomes active. In this type of SMPs, the shape-memory process can be described as follows.

The phase transition temperature of the reversible phase (T″) of SMPs is lower than that of the fixed phase (T′). When the reversible phase is heated to the softened state (T4) and treated under the external force, the shape of the polymers would change. Then the deformed specimen is refrigerated to the hardened state (T5). Because the strength of the hardened reversible phase is larger than the recovery force of the fixed phase, the deformation is "frozen" even if the external load is released. The deformation would not recover forever if the temperature were lower than the phase transition temperature of the reversible phase (T″). Once the operating temperature (T6) is heated higher than T″, the factor of resisting the elastic recovery of the fixed phase disappeared, that is, the reversible phase was softened and the strength of it is reduced. Then the specimen would recover the original shape because of the elastic recovery that is contributed by the fixed phase. Consequently, the SMPs show the mechanical behavior that included fixing the deformation as the plastic and also recovering the deformation as the rubber [1,7,11,15,21,22].

10.1.2.3 Other Properties of SMPs

In addition to the above shape-memory behavior with respect to thermosensitivity, SMPs have other intrinsic properties. Because there are extensive investigations into SMPUs and their products, we summarize these good properties based on the research into SMPUs as follows [2,4,8,10,28,29].

First, with the temperature rising especially near or higher than T_g, it is found that the moisture permeability of the thin film of SMPUs becomes extremely high. The reason is that the molecular movement of the amorphous polymeric soft segments becomes active at temperature near and higher than the T_g and form "free space" in the soft segment domains, and the "free space" allows water vapor molecules with an average diameter of 3.5Å be easily transmitted through the polymer thin film. On the other hand, the volume of SMPs will expand when the temperature is heated to the temperature above T_g while the refractive index decreases sharp at temperature higher than T_g. It is evidenced that the materials have large changes in dielectric constant at temperature below and above the T_g. Moreover, they have excellent chemical properties and do not dissolve in any acid or base. DMF (*N,N*-dimethylformamide) is the only one that is capable of dissolving SMPs.

10.2 Overview of SMPs Used for Textiles

There are more ways of applying the SMPs than other shape-memory materials to textiles, clothing, and related products. The main forms of SMPs used in textiles includes shape-memory fibers, shape-memory yarns, shape-memory fabrics, and shape-memory chemicals. However, patents and literature mainly focused on woven fabric or, nonwovens where shape-memory fibers are bonded with an adhesive. A patent (U.S. patent no. 5,128,197 7/1992) [30] on woven fabric of shape-memory yarns has different functions depending on the T_g of the SMP in the fabric. The woven fabric having a low T_g does not wrinkle and deform and can be applied to the crease of slacks and the pleats of skirts. The woven fabric having a high T_g can be applied to the collars, cuffs, and shoulder pads of utility shirts. The woven fabric made of SMP with the T_g higher than normal temperature gives a hard-hand feel at normal temperature. There has been proposed a nonwoven fabric, which is composed of fibers of a resin having shape-memory property and adhesive of a resin having shape-memory property. It tends to be uneven in thickness and it is difficult to distribute the adhesive uniformly, and is high in cost owing to the expensive adhesive.

Several kinds of SMPs can be employed to prepare SMP fibers. They include SMPUs (including modified SMPUs by incorporating of ionic or mesogenic components into the hard segment phase), polyethylene terephthalate–polyethylene oxide copolymer, polystyrene–poly (1,4-butadiene) copolymer, polyethylene/nylone-6-graft copolymer, triblock copolymer made from poly(tetrahydrofuran) and poly (2-methyl-oxazoline), thermoplastic polynorbornene, and other polymers that show shape-memory effect by cross-linking after spinning, such as polyethylene, poly(vinyl chloride), and polyethylene–poly(vinyl chloride) copolymer.

The used SMPU is synthesized by solution polymerization or bulk polymerization. In both the synthesis processes, prepolymerization technology is preferred to get higher polymer molecular weight. All the chemicals are demoistured prior to use. The synthesis equipments are also cleaned to get rid of any organic or moisture residual.

The fibers made of SMPs are spun by wet spinning, dry spinning, melt spinning, reaction spinning, and electric spinning. In wet spinning the polyurethane polymer is dissolved in an appropriate solvent, such as DMF, at a suitable solid concentration of 15%–35% with a viscosity of 30–150 Pa·s, or directly used after synthesis by solution polymerization. This solution is extruded through orifices into a coagulating bath. The winded up speed is relatively low compared with the other three methods. In drying spinning a highly viscous solution is put through a spinneret, and simultaneously hot air is supplied to evaporate the solvent. If diamines are used as extenders urea-urethane groups are formed and highly elastic fibers with good heat stability are obtained. The spinning speed is from 200–1000 m/min which is much higher than that of wet spinning. Dry spinning is an expensive and environmentally unfriendly process. In chemical spinning, in general, a prepolymer is spun by including a glycerin in coagulation bath of aliphatic diamines with portions of triamines. To the melt spinning, thermoplastic SMPU with high thermal stability and mechanical properties was spun using single screw extruder. Because no DMF and water bath are required, it is more clean and safe.

Shape-memory yarns include the shape-memory fibers alone; blended yarn of SMP fibers; and natural, regenerated, or synthetic fibers. The blended yarns can be core spun yarn, friction yarn, and fancy yarn. Textile articles also include sewing threads of SMP fibers and ordinary natural or synthetic fibers.

Shape-memory fabrics include woven, knitted, braided, and nonwoven fabric. The above-mentioned fiber, yarn, and fabric, easily returns to its original shape, which it remembers when it is heated above the shape-memory switch temperature, even if it wrinkles or deforms. Therefore, in this case, the fabric can be applied to the wrinkle free, reform recovery, and many kinds of design and fancy in art.

The fabric can be applied to textiles depending on the properties and uses that are desired such as collars, cuffs of shirts, and any other apparel which needs shape fixity; elbow, knee of apparels, and any other cloths which need the recovery of bagging; and shape fixity of denim, velvet, cord, knitting fabric, and any other fabrics.

Shape-memory finishing fabric can be acquired by coating with shape-memory emulsion or combining shape-memory film. A Japanese patent [32,33] published a nonwoven fabric combined with SMP and shape-memory adhesive. A U.S. patent [31] published a shape-memory fabric adhesives ordinary fabric and SMP powder.

Hong Kong Polytechnic University has studied shape-memory finishing chemicals and technologies for cotton fabrics, wool fabrics, and garment finishing. This method is more environmental friendly, simpler finishing technology than other traditional finishing methods, and the fabrics have good temperature sensitivity.

10.3 Importance of Textile

The important and promising uses of shape-memory materials mainly focus on biomedicine, textiles and apparels, toys, packaging, national defense, and industrial fields. Shape-memory materials used in apparels, bandages, and so on, are mainly shape-memory alloys. A smart shirt with automatic crimple sleeves for summer, reported in Italy, was fabricated from a core yarn of nickel titanium alloy which gives a hard-hand feel, and is expensive to manufacture due to the cost of nickel titanium alloy. Many patents and literature also report on the use of shape-memory alloys weaved in the fabric. But the fabric or garments made with shape-memory alloy have the following disadvantages:

1. They have a hard-hand feel
2. Processing is different
3. Switch temperature limits
4. The fibers and fabrics are limited in the structure and in the use of apparel

But SMP fibers are more suitable for textile applications than shape-memory alloys. SMP fibers can be spun by common spinning methods such as wet spinning or dry spinning, and the switch temperature can be designed and controlled. We can adjust the shape-memory properties and switch temperature according to the product's demands. The appearance and weave ability is similar to the common fiber.

Shape-memory textiles with SMP fibers have better temperature responsive ability; have better capabilities for developing 3-D textiles than spandex and polyester; better elasticity than fabrics with polyester; and are more comfortable and exert lower pressure than apparels with spandex.

For common elastic fibers, elasticity depends on the instant recoverability of the length of the fiber on release of the deforming stress. The recoverability in shape-memory fibers depends on the recovery ability of deformed fibers with external stimulus such as heat or chemicals. In this case, the external stimulus is a must.

The main difference between SMP fiber and conventional elastic fibers is the variation of E' in normal using temperature range. For SMP fibers, the variation of E' is very significant. Namely, when the temperature is increased above the transition temperature, E' will sharply decrease and the rubbery state platform will appear and be extended to above 180°C. However, for conventional polyurethane fibers, though in the entire heating scan range, there are some transition areas of E', such as −40°C for Lycra, where the elastic modulus is almost constant and change little with the increase of temperature in room temperature range. Therefore, this point imparts the heating responsive shape-memory properties to the SMP fiber in normal using temperature.

10.4 Main Physical Properties/Benefits of Textiles

10.4.1 Properties of Shape-Memory Fibers

10.4.1.1 Mechanical Properties

Figure 10.1 is the SEM photos of shape-memory fibers in the cross-section and longitudinal views. The cross section of the fiber is irregular oval shape. The longitudinal view shows smooth, rough, or crack according to different spinning conditions.

The tenacity of shape-memory fibers is above 10cN/tex, and the maximum strain at room temperature is in the range of 50%–200% [34], which depends on the molecular structures and the corresponding posttreatment processes. At room temperature, the elastic modulus of SMPU fiber is much higher than Lycra, but lower than the conventional nylon or polyester filament (Figure 10.2).

For the Lycra and XLA fiber, the elongation ratio at break can reach 500%–600% with the sufficient tenacity for knitting process.

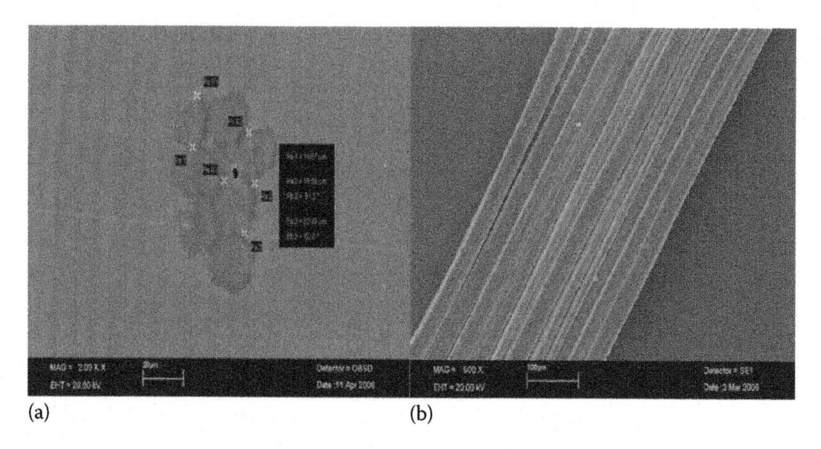

FIGURE 10.1
Cross-section (a) and longitudinal (b) view of shape-memory fibers.

10.4.1.2 Thermomechanical Properties

The elastic modulus of shape-memory fiber varies with heating in the transition area. In the transition area, the shape-memory fiber is able to evolve from glass state to rubbery state, and the difference of elastic modulus is of one or two orders of magnitude. In the normal using temperature range, the elastic modulus of XLA and Lycra remains in the relative fixed value. At room temperature, the XLA and Lycra are located in rubbery state (Figure 10.3).

SMPU fiber shown in Figure 10.4 was heated in the first heating scan from −70°C to 240°C, the transition region can be observed in the range from 0°C to 50°C. And the melting point of crystallization of hard segments was hardly detected. In the subsequent cooling scan, the crystallization exothermal peak at 124°C could be observed, suggesting the stronger crystallizability of hard segments. Correspondingly, the melting peak of hard segments appears in the second heating scan at 155.6°C.

For Lycra, in the cooling scan, the crystallization exothermal peak appears at very low temperature. At usual using temperature, the soft segments are amorphous and located in rubbery state area.

10.4.1.3 Shape-Memory Properties

Shape-memory properties of the fibers were carried out by a tensile tester (Instron 4466). The T_{high} value was set at 70°C and T_{low} at ambient temperature (22°C) [35]. The sample was of 30 mm in length. The cyclic tensile test includes the following steps: Heat the fibers to 70°C and then stretch them to 100% strain at a speed of 10 mm/min ①; Cool the fibers to the ambient temperature while the same 100% strain is kept for 15 min ②; Unload the clamps so they can return to their original position at a speed of 40 mm/min ③; Heat the fibers again to 70°C ④; and Begin the second cycle ⑤. The

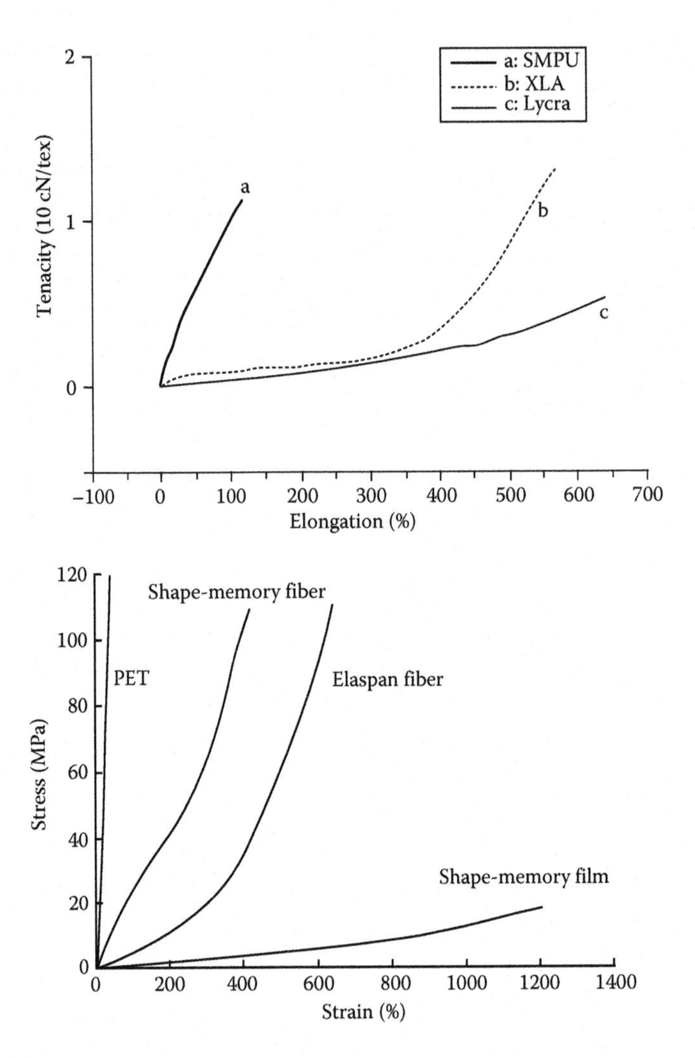

FIGURE 10.2
Stress–strain behaviors of various fibers.

schematic representation of the above cycle is shown in Figure 10.5 [36]. The above cycles were repeated four times and the stress–strain behaviors were recorded for further analysis. Schematic representation of typical stress–strain behavior is shown in Figure 10.6. ε_m is the maximum strain in the cyclic tensile tests, ε_u is the strain after unloading at T_{low}, and $\varepsilon_p(N)$ is the residual strain after recovery in the Nth cycle. The ε_m value is set at 100% strain for the study. The fixity ratio (R_f), and recovery ratio (R_r) are calculated according to following equations [36]:

$$R_f(N) = \varepsilon_u(N);\ R_r(N) = (1 - \varepsilon_p(N)) \times 100\%$$

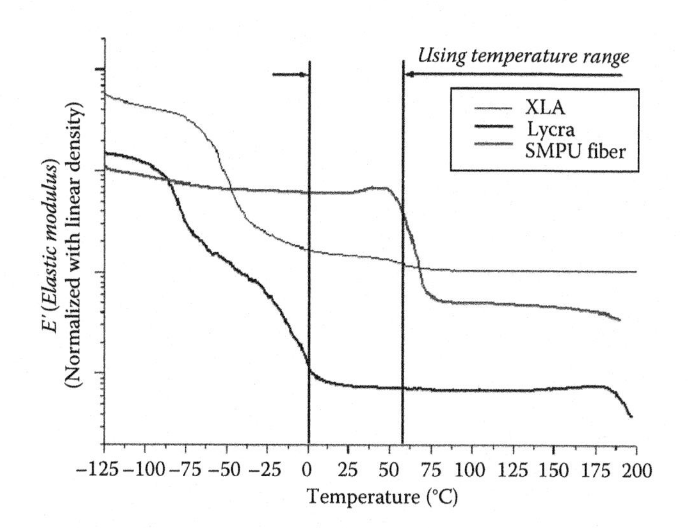

FIGURE 10.3
DSC curves of shape-memory fiber, Lycra, and XLA.

The stress–strain curve of SMPU fibers is located between the high modulus fiber such as nylon and the high elasticity fiber such as Lycra (Figure 10.2).

10.4.2 Shape-Memory Yarns and Fabrics

Shape-memory yarns include core spun yarn, friction yarn, fancy yarn, and other styles. The tenacity of shape-memory yarns is dependent on the difference of yarn structure and outside fibers.

The shape-memory effect of shape-memory yarns also can be tested according to the method for shape-memory fibers. The shape-memory effect is relevant to the shape deformation, shape-memory fiber content. For most kinds of shape-memory yarn, the testing elongation adopts the maximum strain when the outside fiber does not slip when tensile. It is about 8.5%–10%.

According to the shape-memory effects test, the shape-memory fixity and shape-memory recovery are good. The fixity of ring spun core yarn is better than friction core yarn. The shape-memory recoverability is the same. The twist of friction spun yarn is the function of the space of fiber to the core, the hyperbola function. So the inner fiber of friction yarn is tighter than the outer shell fibers. The elastic recovery force is high when tinseled, so that it affects the shape-memory fixity of the core.

Shape-memory fibers and yarns can be woven on common weaving machine. Different shape-memory effects can be inspired and achieved according to fabric structure, yarn configuration, and shape-memory fiber contents. For general fabrics with shape-memory fibers such as plain or twill fabric, an original shape can be set above a hard segment transition temperature, which is higher than the normal use temperature and the switch

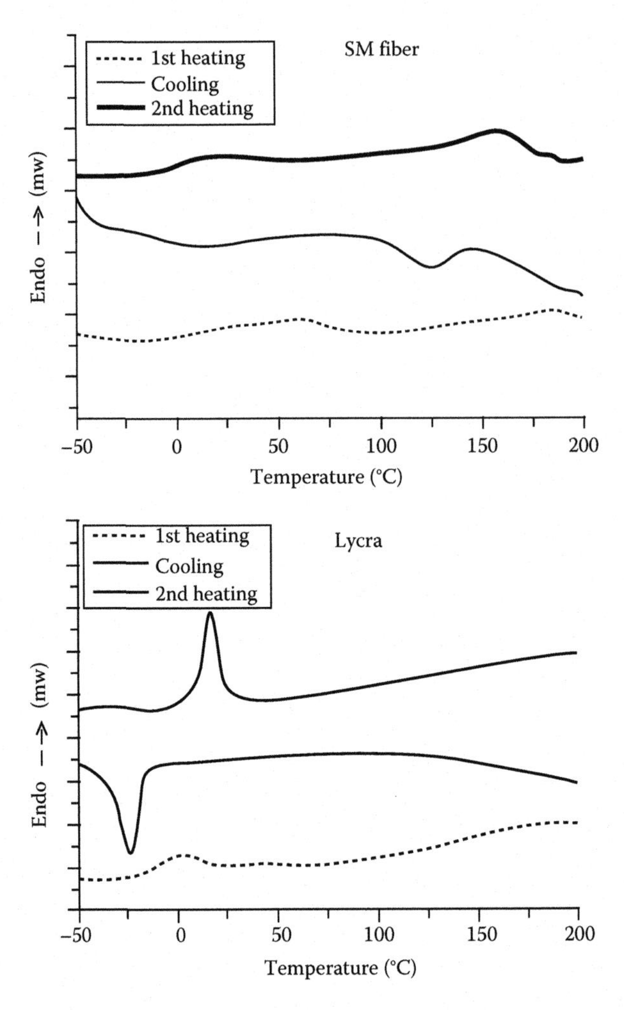

FIGURE 10.4
Differential scanning calorimetric analyses.

temperature, and then cooled to a lower temperature so that the original shape is fixed. In the case where the switch temperature is higher than the normal use temperature, even if the fabric deforms, it easily returns to the original shape it remembers when heated to a temperature higher than the switch temperature. In the case where the switch temperature is lower than the normal use temperature, the article can return to its original shape when the deforming force is removed. A series of fabrics retain their original shape such as flat or bagging, and can be set with an original shape such as crease, at a temperature higher than a hard segment transition temperature, then, cooled to a lower temperature so that the original shape is fixed. Even if the fabrics are deformed, they easily return to their original shape when heated to a temperature higher than the switch temperature.

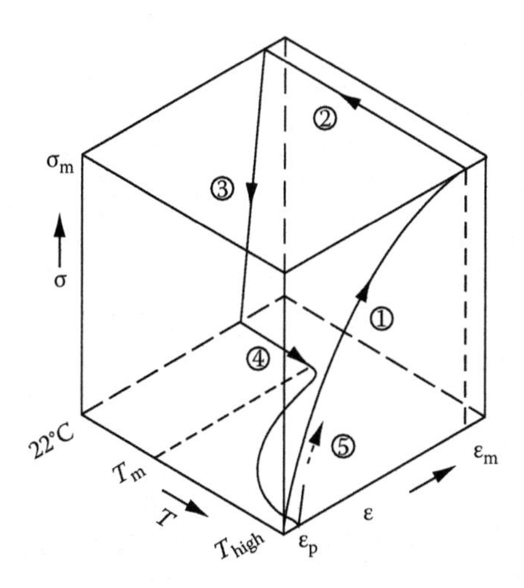

FIGURE 10.5
Cyclic tensile testing path.

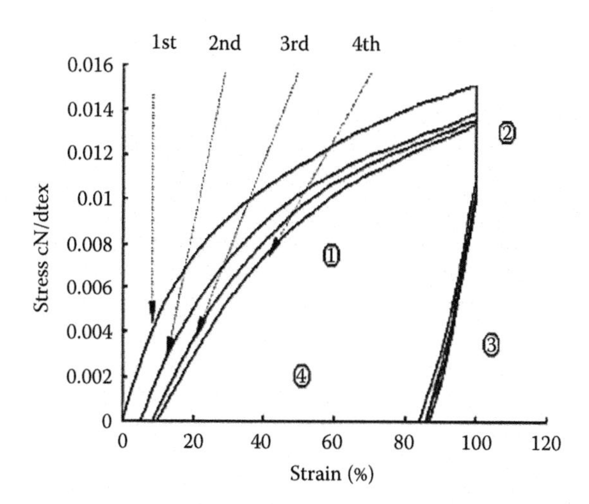

FIGURE 10.6
Cyclic tensile curves of shape-memory fiber.

Shape-memory woven fabric with shape-memory alloy and the shape-memory fabric in the Japanese patent [32] have the following disadvantages:

1. They have hard-hand feel when used below the glass transition of the SMP.
2. The shape-memory effects depend only on the glass transition.
3. They have large shrinkage in hot water and hot air.

4. The nonwoven fabric of SMP tends to be uneven in thickness and in strength and is high in cost owing to the expensive adhesive used.
5. The fibers and fabrics are limited in the structure and in the use of apparel.

The shape-memory fabric developed by Hong Kong Polytechnic University does not have good shape-memory effects but solves the problems above.

Shape-memory knitting fabric can also be knitted on common knitting machine such as flat machine, circular machine, and seamless machine. 3-D textiles, underwear, sportswear, skirt, and gloves have good shape-memory properties and advantages.

1. Better bagging recovery and temperature responsive ability
2. Better capabilities for developing 3-D textiles than spandex and polyester
3. Better elasticity than fabrics with polyester
4. To minimize pressure points and provide a perfect and comfortable fit

Figure 10.7 shows that the bagging height of shape-memory fabric decreases with temperature increasing, but the cotton fabric is little affected by temperature. Figure 10.8 shows the bagging heights recovery with time under different temperatures. It shows that the bagging heights decrease with time and it can recover sharply above 50°C within 3 min.

As compared the bagging recoveries of knitting fabrics with or without shape-memory fibers, Figure 10.9 explains the results of these comparisons. For every tested fabric, with the time increasing, the bagging height decreases and bagging recovery rate increases. Fabrics with shape-memory fibers have better heat sensitivity and distinct bagging recovery rate when heated than other fabrics. The fabrics with shape-memory fibers have lowest begging

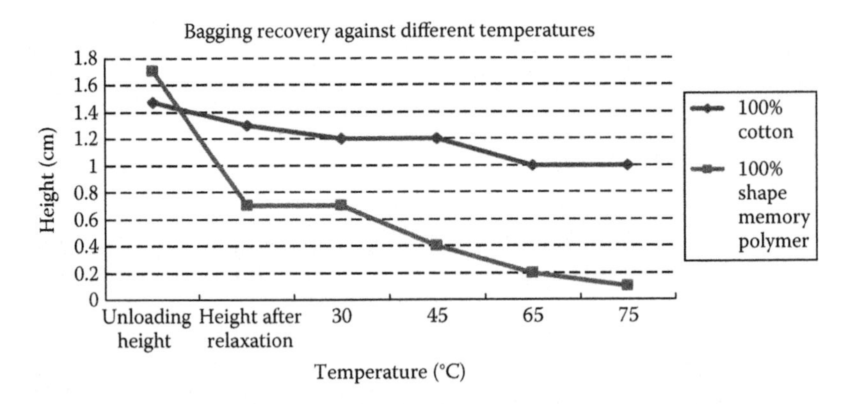

FIGURE 10.7
Bagging heights with temperature increasing.

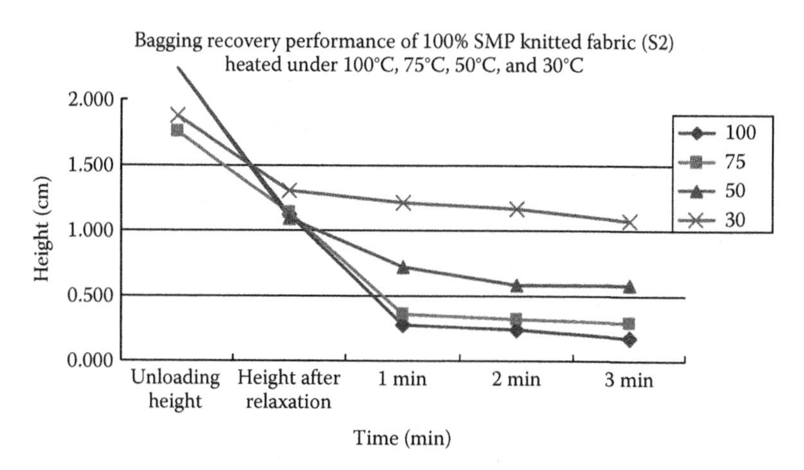

FIGURE 10.8
Bagging heights recovery with time.

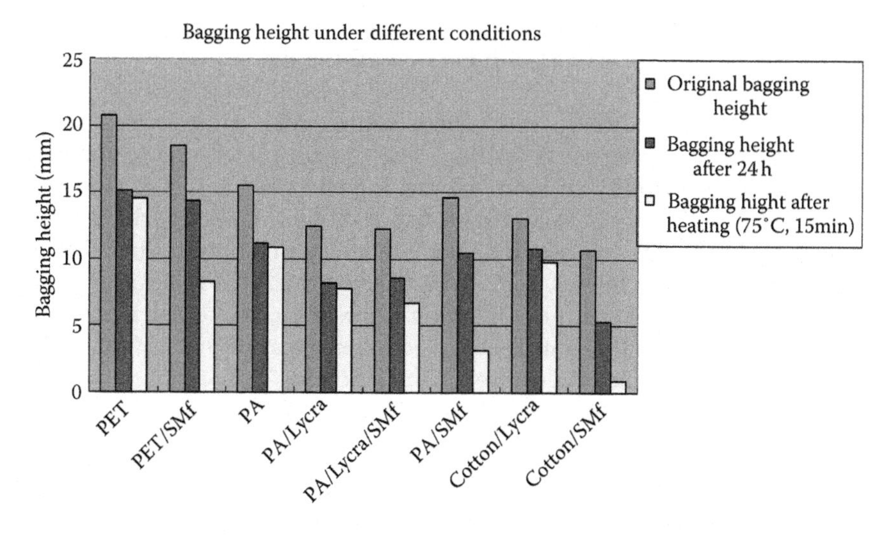

FIGURE 10.9
Bagging recovery of different fabrics.

residual in the same fabric series (PET series; PA series; cotton series). The fabric with shape-memory fibers has lower original bagging height than the fabrics with pure PET fibers or PA fibers or cotton. It means the fabric with shape-memory fibers have some elasticity under common temperature.

10.4.3 Shape-Memory Finishing Textiles

Shape-memory finishing fabric can be acquired with coating shape-memory emulsion or combining shape-memory film. A Japanese patent in [32,33] published a nonwoven fabric combined with SMP and shape-memory adhesive. A U.S. patent in [30,31] published shape-memory fabric adhesives ordinary fabric and SMP powder.

Hong Kong Polytechnic University has studied shape-memory finishing chemicals and technologies for cotton fabrics.

Finished cotton fabrics are flat originally. After washing or storing in closet, the crease and bagging will appear. The switch temperature of shape-memory emulsion is about 60°C. The shape-memory finished fabric feel soft, the crease and wrinkle will disappear, and the fabric recovers the original shape when washed in hot water or dried higher than 60°C.

The advantages of shape-memory finished fabric are as follows:

1. Has good temperature sensitivity
2. Has higher breaking strength
3. Has better performance; has improved wrinkle resistance after drying
4. Has better wash ability and flat appearance, and crease retention increases with increasing number of washing times

Wool fabrics and sweaters have serious felting phenomenon and shape shrinkage due to the oriented friction. The finishing methods to increase the dimensional stability of wool fabrics include resin coating, chlorination shrink proofing, oxidation antifelting, and other methods. Hong Kong Polytechnic University has studied shape-memory finishing chemicals and technologies for wool fabrics and garment finishing. This method is more environmental friendly, simpler finishing technology than other traditional finishing methods, and the fabrics have good temperature sensitivity.

Now this finishing method can reach the following demands:

- Flat appearance
- Bagging recovery
- Dimensional stability
- Fastness to washing

10.5 Applications and Examples

10.5.1 Shape-Memory Yarns

The shape-memory fancy yarn is designed to keep the set shape which is favorable to spinning and weaving at normal temperature. But it recovers to the original shape designed such as curl, helical crimp, etc. Owing to the

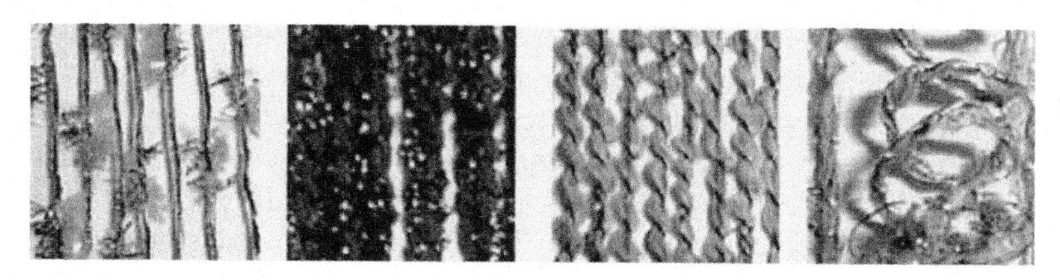

FIGURE 10.10
Shape-memory fancy yarn.

different structure such as the core spun yarn, fancy yarn, etc., the yarn may have different changes and effects. Many styles of the yarn can be obtained depending on the kinds, properties, morphologies of the fibers blended with the shape-memory fibers, and the blending ratio. Figure 10.10 shows different kinds of shape-memory fancy yarns.

A core yarn or friction yarn can be spun from the shape-memory fiber and ordinary cotton fiber, the shape-memory fiber as the core and the cotton fiber as the sheath. After repeated benching or elongation, the core yarn deformed from straight to curl. When put into hot water higher than the switch temperature of the shape-memory fiber, the yarn quickly returned to its original shape.

A loop type yarn was spun from the shape-memory and polyester textured yarn. The so said yarn retained the straight shape at normal temperature. But when heated at a temperature higher than the switch temperature of the shape-memory fiber, the so said fiber deformed and curled, and the polyester textured yarn bulged to form the loops at the surface of the blended yarn. The so said yarn returned to the original straight shape, when stretched to straight and cooled to the temperature below the switch temperature.

10.5.2 Shape-Memory Fabrics

Woven fabrics include denim (Figure 10.11), plain weave etc., which woven by the shape-memory fiber as weft and other ordinary yarn as warp. A small

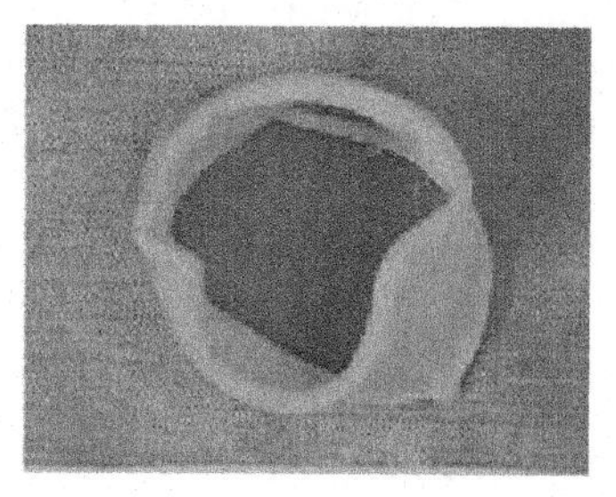

FIGURE 10.11
Fabrics with shape-memory fibers.

dimension annular shape-memory knitting fabric can be knitted and when worn next to the skin, the fabric becomes soft and close to the skin. So these fabrics can be developed in different applications.

A woven corduroy fabric of SMP can be formed by weaving cotton yarn as the warp and weft and a core as the pile weft. After pile cut and hot treatment at the temperature higher than the switch temperature of the shape-memory fiber in the pile weft, the nap of the corduroy formed and set. The so said corduroy has better antireversal properties of the pile. Even if pressured or reversed during usage, the pile can recover erect when heated to a temperature higher than the switch temperature.

A braid was woven by the shape-memory fiber and set at a temperature higher than the switch temperature. The so said braid had better shape stability and softer hand feeling than that woven by other synthetic fiber.

A nonwoven fabric was made by bonding the fiber and cotton fibers. The web of the fiber and cotton fiber were adhered and shaped by the so said shape-memory fiber which had the adhesive property above a temperature when the shape-memory fiber begins to flow. The nonwoven fabric had better hand feeling and had even thickness and strength. The production cost is less than the nonwoven fabric made by short fibers and adhesive of an SMP. (See Japanese Patent Laid—open No. 252353/1986.) Even if it wrinkled or deformed after washing or storage for a long time, it easily returned to its original shape it remembered when heated above the switch temperature. The so said nonwoven fabric can be favorably applied to the pads of collars, cuffs, shoulder, and bras.

A woven fabric was made by the loop type shape-memory yarn. The so said fabric was flat originally. When heated above the switch temperature, the yarn became the loop type yarn and the fabric became puffy and magnificent. When stretched to straight and cooled to a temperature below the switch temperature, the so said fabric recovered the original flat appearance.

A velvet was formed by weaving cotton yarn as the weft and the warp and a core yarn in as the pile warp. After pile cut and hot treatment at a temperature higher than the switch temperature of the shape-memory fiber in the pile weft, the nap of the corduroy formed and set. The so said corduroy has better anti reversal properties of the pile. Even if pressured or reversed during usage, the pile can recover erect when heated at a temperature higher than the switch temperature.

Shape-memory composite fabrics were formed by two layers of shape-memory fabric which included spun-laid webs with a scrim base and needle—punched and scrim compounds.

10.5.3 Shape-Memory Garments

Shape-memory fibers and yarns can be used in knitted garment such as 3-D textiles, underwear, sportswear, skirt, and gloves which have good shape-memory properties and advantages.

A corset was knitted from the friction yarn and set at a temperature higher than the switch temperature to get the original dimension. The corset can be stretched to allow ease of dressing. However, the switch temperature of the so said fabric is below body temperature. Therefore, when dressed on the body, the corset became tighter and tighter and recovered the original shape to achieve a satisfactory tying effect.

A shaped sports bra was knitted by shape-memory fibers or shape-memory yarn. The bra has a switch temperature of about 30°C. The body temperature becomes higher than the switch temperature when people do sports. So the bra became tight and had good protective function to the body. But people did not feel it too tight or uncomfortable.

10.5.4 Shape-Memory Finishing Fabric and Garments

Shape-memory finishing fabric can be acquired with coating shape-memory emulsion or combining shape-memory film.

Finished cotton fabrics are flat originally. After washing or storage in closet, the crease and bagging will appear. The switch temperature of shape-memory emulsion is about 60°C. The shape-memory finished fabric feel soft, the crease and wrinkle will disappear and recover the original shape when washed in hot water or dried higher than 60°C (Figure 10.12).

Shape-memory finishing chemicals and technologies for wool can increase the dimensional stability of wool fabrics and felting. This method is more

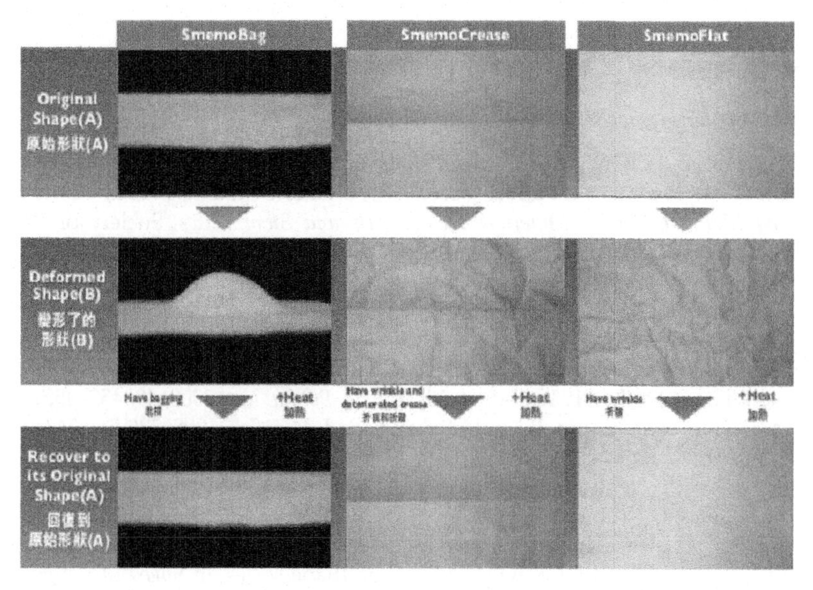

FIGURE 10.12
Shape-memory recovery of cotton fabric

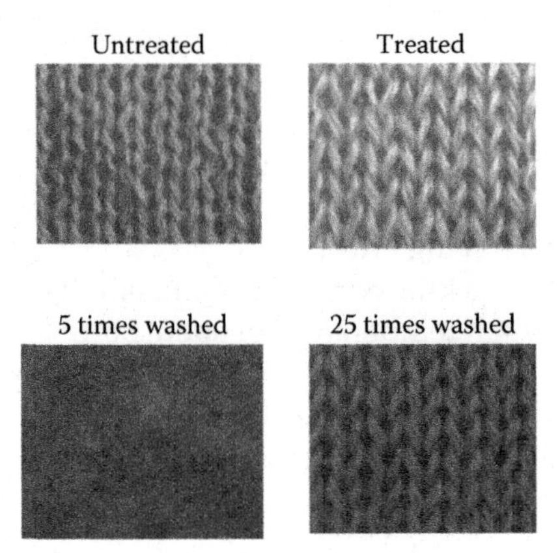

FIGURE 10.13
Texture of untreated and treated wool fabric.

environmental, simpler finishing technology than other traditional finishing methods and the fabrics have good temperature sensitivity.

Figure 10.13 shows the texture of untreated fabric and treated fabric with SMPs. The texture of treated fabrics is clear, but it is distorted in the untreated fabrics. After washing, the texture of treated fabrics is still clear and it can stand 25 times washing and tumble dry according to AATCC wool washing standard, but untreated fabrics has serious felting after 5 times washing.

References

1. Lin, J. R. and Chen, L. W., *J. Appl. Polym. Sci.*, 1998, 69, 1563–1574.
2. Crowson, A., Smart materials based on polymeric system, in *Smart Structures and Materials. Smart Materials Technologies and Biomimetics*, Society of Photo-optical Instrumentation Engineers, Bellingham, WA, San Diego, CA, 1996.
3. Hayashi, S., *Int. Prog. Urethanes*, 1993, 6, 90–115.
4. Liang, C., Rogers, C. A., and Malafeew, E., *J. Intell. Mater. Syst. and Struct.*, 1997, 8(4), 380–386.
5. Du, S. G., *Funct. Mater.*, 1995, 26(2), 107–112.
6. Jeon, H. G., Mather, P. T., and Haddad, T. S., *Polym. Int.*, 2000, 49, 453–457.
7. Wei, Z. G., Sandstroum, R., and Miyazaki, S., *J. Mat. Sci.*, 1998, 33(15), 3743–3762.
8. Tobushi, H., Hayashi, S., and Kojima, S., Mechanical properties of shape memory polymer of polyurethane series, in SEM Spring Conference on Experimental Mechanics, Dearborn, MI, 1993.
9. Tobushi, H., Hara, H., and Yamada, E., Thermomechanical properties in a thin film of shape memory polymer of polyurethane series, in *Smart Structures and Materials: Smart Materials Technologies and Biomimetics*, Society of Photo-optical Instrumentation Engineers, Bellingham, WA., San Diego, CA, 1996.

10. Tobushi, H., Hara, H., Yamada, E., and Hayashi, S., Thermomechanical properties of shape memory polymer of polyurethane series and their applications, in Proceeding of the 3rd International Conference on Intelligent Materials, Lyon, France, 1996.
11. Ding, X. M. and Hu, J. L., *J. China T. R.*, 2000, 21(4), 56–59.
12. Kim, B. K., Shin, Y. J., Cho, S. M., and Jeong, H. M., *J. Poly. Sci. B Polym. Phys.*, 2000, 38, 2652–2657.
13. Li, F. K., Zhang, X., Hou, J. A., Zhu, W., Xu, M., Luo, X. L., and Ma, D. Z., *Acta Polym. Sin.*, 1996, 4, 462–467.
14. Li, F. K., Zhang, X., Hou, J. N., Xu, M., Luo, X. L., Ma, D. ZH., and Kim, B. K., *J. Appl. Polym. Sci.*, 1997, 64, 1511–1516.
15. Lin, J. R. and Chen, L. W., *J. Appl. Polym. Sci.*, 1998, 69, 1575–1586.
16. Adanur, S., Future of industrial textiles, in *Wellington Sears Handbook of Industrial Textiles*, ed. S. Adanur, Technomic Publishing Company, Inc., Lancaster, PA, 1995, pp. 757–759.
17. Ashley, S., *Mech. Eng.*, 1994, 116(2), 36.
18. Bai, Z. W. and Zhang, X. Q., China synthetic rubber industry, *CEPS*, 1999, 22(3), 184–188.
19. Fukuda, T., Ohahima, N., and Hourai, K., Use of shape memory polymer for damping control of fiber composite materials, in Proceedings of the 9th International Conference on Composite Materials, Madrid, Spain, 1993.
20. Gordon, R. F., *Adv. Mater. Process.*, 1994, 3, 9.
21. Kusy, R. P. and Whitley, J. Q., *Thermochim. Acta*, 1994, 243, 253–263.
22. Maycumber, S. G., Fabric that gets cool when you get hot, *Daily News Record*, 1993.
23. Zeng, Y. M., Yan, H. J., and Hu, J. L., *J. China Textil. Univ.* (Chinese edition), 2000, 26(6), 127–130.
24. Luo, X. L., Zhang, X. Y., Wang, M. T., Ma, D. Zh., Xu, M., and Li, F. K, *J. Appl. Polym. Sci.*, 1997, 64, 2433–2442.
25. Li, F. K., Chen, Y., Zhu, W., Zhang, X., and Xu, M., *Polym.*, 1998, 39(26), 6929–6934.
26. Ikematu, T., Kishimoto, Y., and Miyamoto, K., Shape memory polymer resin, resin composition and the shape memorizing molded product thereof, European, 1990. 9. 16.
27. Yagi, T., Tsuda, N., and Tanaka, Y., Polymer material having shape memory characteristics, European, 1990. 9. 5.
28. Hayashi, S., Ishikawa, N., and Giordano, C., *J. Coat. Fabr.*, 1993, 23(7), 74–83.
29. Tobushi, H., Hashimoto, T., Ito, N., Hayashi, S., and Yamada, E., *J. Intell. Mater. Syst. Struct.*, 1998, 9(2), 127–136.
30. US patent no. 5,128,197 7/1992.
31. US patent no. 5098776, 1992.
32. JP patent no. 01-229836, 1989.
33. JP patent no. 01-229879, 1989.
34. Zhu., Y., Hu, J., Yeung, L.-Y., Liu, Y., Ji, F., and Yeung, K.-W., Development of shape memory polyurethane fiber with complete shape recoverability. *Smart Mater. Struct.*, 2006, 15, 1385–1394.
35. Kim, B. K., Lee, S. Y., and Xu, M., Polyurethane having shape memory effect. *Polymer*, 1996, 37, 5781–5793.
36. Lendlein, A. and Kelch, S., Shape-memory polymers. *Angew. Chem. Int. Ed.*, 2002, 41, 2034–2057.
37. Li, F. et al., Shape memory effect of ethylene-vinyl acetate copolymers, *J. Appl. Sci.*, John Wiley & Sons, 1999, 71, 1063–1070.

11

Applications of Shape-Memory Polymers in Biomedicine

Witold M. Sokolowski
Jet Propulsion Laboratory, California Institute of Technology, Pasadena, California

Jinsong Leng
Centre for Composite Materials and Structures, Harbin Institute of Technology, Harbin, P.R. China

CONTENTS

11.1 Introduction

Shape-memory polymers (SMPs) constitute a group of high-performance smart materials that have recently gained widespread attention. They are lightweight, have a high strain/shape recovery ability, are easy to process, and their required properties can be tailored for a variety of applications.

Their potential role in clinical applications has become recognized only in the last 5–6 years. Recently a number of medical applications have been considered and investigated, especially for polyurethane-based SMPs. These SMP materials were found to be biocompatible, nontoxic, and nonmutagenic. Their glass transition temperature (T_g) can be tailored for shape restoration/self-deployment of clinical devices when inserted in the human body. Newly developed SMP foams, together with cold-hibernated elastic memory (CHEM) processing, further broaden their potential biomedical applications. SMP materials can be miniaturized and deformed, inserted in the human body through small catheters, and subsequently recover an original predetermined shape once in a satisfactory position.

In this chapter, SMP materials, their potential, and the already investigated medical applications are described. Their major advantages over other medical materials are delineated. Some simple SMP applications are already in use in clinical settings, whereas others are still in development.

11.2 Characteristics of Shape-Memory Polymers

SMP materials have been under development since the 1980s [1,2]. They offer unique properties for a variety of applications. These materials, especially polyurethane-based SMPs, exhibit large changes in elastic modulus E above and below the glass transition temperature T_g.

The shape-memory function of SMPs allows repeated shape changes and shape retentions [3]. At temperatures above T_g, the material enters a rubbery elastic state where it can be easily deformed into any shape. When the material is cooled below its T_g, the deformation is fixed and the shape remains stable. At this stage, the material lacks its rubbery elasticity, and is rigid. However, the original shape can be recovered simply by heating the material once again to a temperature higher than T_g. This phenomenon is explained on the basis of molecular structure and molecular movements [2]. The molecular chains undergo micro-Brownian movements above the T_g (rubbery state), when the elastic modulus of the polymer material is low. In the rubbery state, the material can be easily deformed by the application of an external force, and the molecular chains oriented in the direction of the tension. When the temperature is lowered below the T_g and the deformation remains constant, the micro-Brownian motion freezes and the chain orientation and the deformation remain fixed. When the material is heated above T_g, the micro-Brownian movement starts again, the molecular chains lose their orientation, and the material recovers its original shape. In this case, the shape recovery function of the material requires cross-linking or partial crystallization.

The basic properties of an SMP material designated as MM-4500 with T_g of 45°C are given in Table 11.1.

The unique properties and major advantages of SMP materials over their predecessors, shape-memory alloys (SMAs), are summarized below.

TABLE 11.1

Basic Properties of MM-4500

Properties	MM-4500
Density (g/cm^3)	1.15
T_g (°C)	45
E below T_g (MPa)	1102
E above T_g (MPa)	3.8
Recovery force (MPa)	Up to 4.1

Source: Hayashi, S. et al., *Plast. Eng.*, 51, 29, 1995.

- SMPs are lightweight. The density is about 1.13–1.25 g cm^{-3} versus 6.4–6.5 g cm^{-3} for nitinol (NiTi).
- The wide range of T_g, from –70°C to +100°C, allows a wide variety of potential applications in different thermal environments.
- Shape recovery up to 400% of plastic strain versus 7%–8% for SMAs.
- Large reversible changes of elastic modulus (as high as 500 times) between the glassy and rubbery states.
- Excellent biocompatibility allowing a wide variety of potential medical applications.
- Easy processing. They are applicable to molding, extrusion, and conventional or computerized numerical control (CNC) machining.
- They are low-cost materials (about 10% of the cost of existing SMA).

However, SMP materials are characterized by low recovery forces, i.e., low actuating forces, and cannot be utilized in high power actuators.

The unique characteristics of SMPs make these materials very attractive to many practical applications. A number of biomedical applications are being considered and investigated for polyurethane-based SMP materials and are described in the following section.

11.3 Biomedical Applications of Shape-Memory Polymers

Polyurethane-based SMPs were found to have excellent biocompatibility. Standard cytotoxicity and mutagenicity tests have been conducted on these materials with excellent results [4]. Another attractive feature is that the T_g can be tailored for shape restoration/self-deployment of various clinical devices when they come in contact with or are implanted in the human body [5]. Minimally invasive surgery has made possible the insertion of small devices and materials by laparoscopes. Alternatively, they can also be inserted via the endovascular route. These new approaches have made essential the miniaturization of devices, designed to pass through small skin apertures.

Hayashi et al. had investigated SMPs for several medical applications [6]. As a result of these investigations, SMP materials will soon be used to manufacture catheters that remain stiff externally for accurate manipulation by the physician, but that become softer and more comfortable inside the human body. With soft catheters there will be fewer arterial wall injuries than with rigid catheters; moreover, they will be easier to manipulate and conduct in tortuous vessels. In addition, polyurethanes are known to be nonthrombogenic, so these catheters will not activate the coagulation cascade. Because of their low thrombogenicity, polyurethanes are widely used in the construction of hematology-related products and devices [7–9].

SMP materials are considered for use in orthopedic braces and splints that can be custom fitted. By heating the shape-memory component above its T_g and then deforming it, a desired fit can be obtained and sustained after cooling. In one of the first applications, a polyurethane-based SMP has been used for a specially designed spoon handle, designated for the physically handicapped [10]. In this application, the spoon handle is heated and deformed to an individual hand shape and then its deformation is fixed at room temperature to provide a comfortable and custom fit.

The use of SMP materials for orthopedic and dental applications is under examination [5]. SMP properties, such as moisture permeability, and energy dissipation and storage, are being considered for bandages and artificial skins. The SMP loss tangent tan δ in the transition region is very similar to that of human skin, providing a natural feel when in contact with or implanted inside the human body.

Recently, Wache et al. conducted a feasibility study and a preliminary development on a polymer vascular stent with shape-memory as a drug delivery system [11] (see Figure 11.1). Samples from a thermoplastic polyurethane-based SMP were manufactured by injection molding. The field of applications of this polymer stent was demonstrated in pretrials. Presently, almost all commercially available stents are made of metallic materials. There are several designs of these minimally invasive vascular stents intended for coronary applications including tubular mesh, slotted tubes, and coils. A common aftereffect of stent implantation is restenosis. The use of the shape-memory polymer stent as a drug delivery system leads to significant reduction of restenosis and thrombosis. An improved biological tolerance in general is expected when using biocompatible SMP materials.

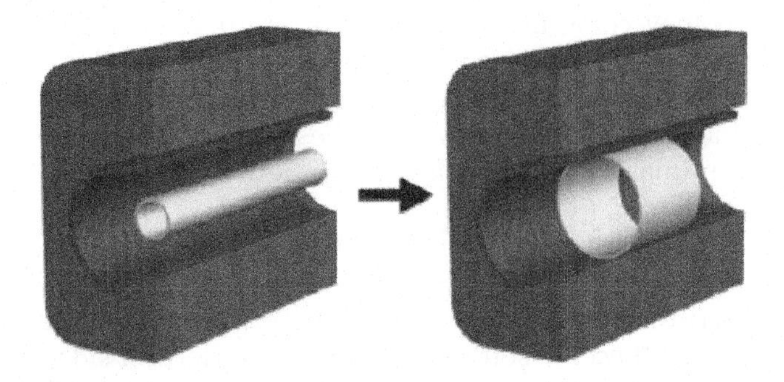

FIGURE 11.1
Schematic of the shape-memory effect prior to application (left) and right after reset (right). (Reproduced from Wache, H.M. et al., *J. Mater. Sci. Mater. Med.*, 14, 109, 2003. With permission.)

The exciting shape-memory polymers can move from one shape to another in response to a stimulus. Thus, SMPs are dual-shape materials. Jeffrey M Bellin et al. [12] report a triple-shape polymer which is able to change from a first shape (A) to a second shape (B) and finally from there to a third shape (C). The triple-shape effect is a general concept that requires the application of a two-step programming process to suitable polymers. The triple-shape SMPs have great potentials for various applications. For example, it may be useful for first, insertion into the body; second, expansion at a target site; and third, removal at a later point in time which may be necessary even with degradable materials (see Figure 11.2).

A group of different implant polymer materials with shape-memory that have been developed for biomedical applications was described by Lendlein and Kelch [13,14]. These implant materials are not a single polymer with the specific composition but polymer systems which allow variation of different macroscopic properties within a wide range through only small changes in the chemical structure. For this reason, a variety of different applications

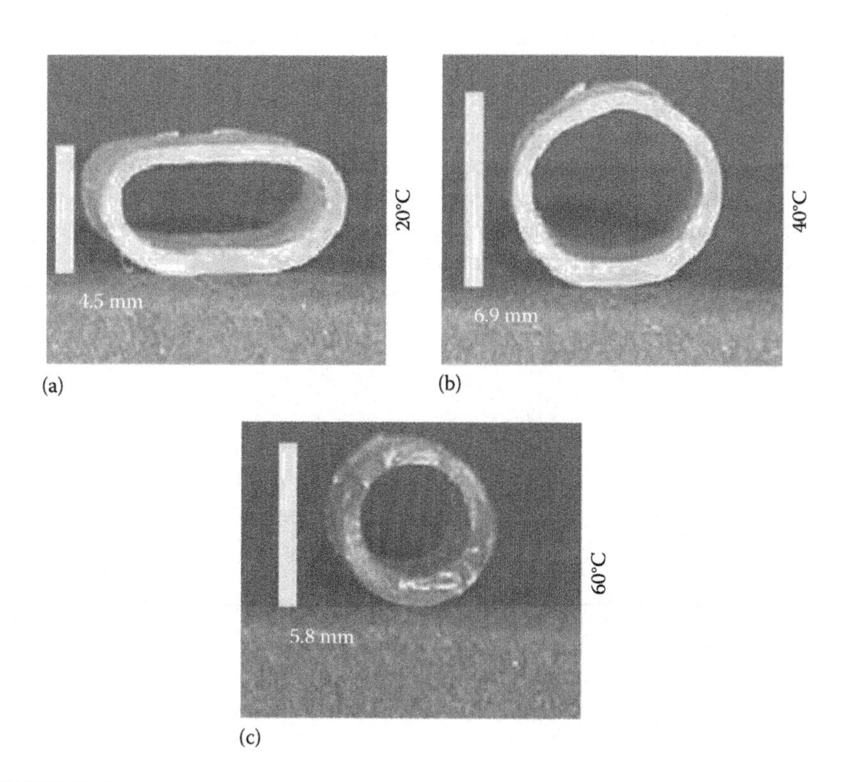

FIGURE 11.2
Through increasing the temperature from 20°C (a) to 40°C (b) an initial shape change was induced followed by a second shape change through increasing the temperature to 60°C (c). (Reproduced from Bellin, I. et al., *Proc. Natl. Acad. Sci. USA*, 103, 18043, 2006; Karp, J.M. and Langer, R., *Curr. Opin. Biotechnol.*, 18, 454, 2007. With permission.)

can be realized with tailor-made polymers of the same family. Both thermoplastic elastomers as well as covalently cross-linked polymer network have been prepared and investigated. Comprehensive in vitro investigation of their tissue compatibility is being performed. The first results were promising.

The thermo/moisture-responsive in polyurethane SMP sheds the light for the possibility to realize micro/nanodevices for surgery/operation at the cellular level. For instance, Ikuta and Coworkers [15] have developed many polymer micromachines, which are about the size of cells or even smaller, and can be triggered for operation by a laser beam outside the cell. However, there are tremendous difficulties to deliver such machines into cells. The thermo/moisture-responsive PU SMP offers a possible solution. As illustrated in Figure 11.3, a piece of originally curved SMP is straightened and then inserted into a living cell. Upon absorbing moisture inside the cell, the SMP recovers its original shape. As the recovery strain in solid or porous SMPs is on an order of hundred percent, it becomes achievable to make cell- or sub-cell-sized machines using the thermo/moisture-responsive SMP and then deliver the machines into living cells for operation controlled by an outside laser beam.

Biodegradability of shape-memory polymers can be achieved by the introduction of weak, hydrolyzable bonds that cleave under physiological conditions. When biodegradable polymers are used in medical devices, the biocompatibility of the solid polymer and its degradation products have to be considered [14,16]. A challenge in endoscopic surgery is the tying of a knot with instruments and sutures to close an incision or open lumen. It is difficult to manipulate the suture so that the wound lips are pressed together

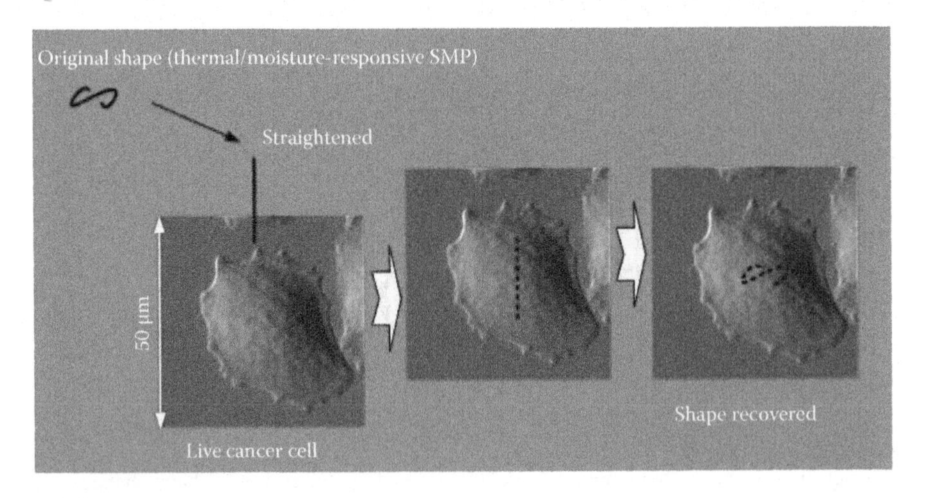

FIGURE 11.3
Concept of delivery of micro/nanodevice into a living cell for cell-level surgery/operation using thermo/moisture-responsive SMP.

under the proper stress [16]. Thus, the usefulness of SMP in wound closure has been investigated recently, and a design of smart surgical suture has been examined, where as the temporary shape is obtained by elongating the fiber under controlled stress (see Figures 11.4 and 11.5) This suture can be applied loosely in its temporary shape under elongated stress. When the

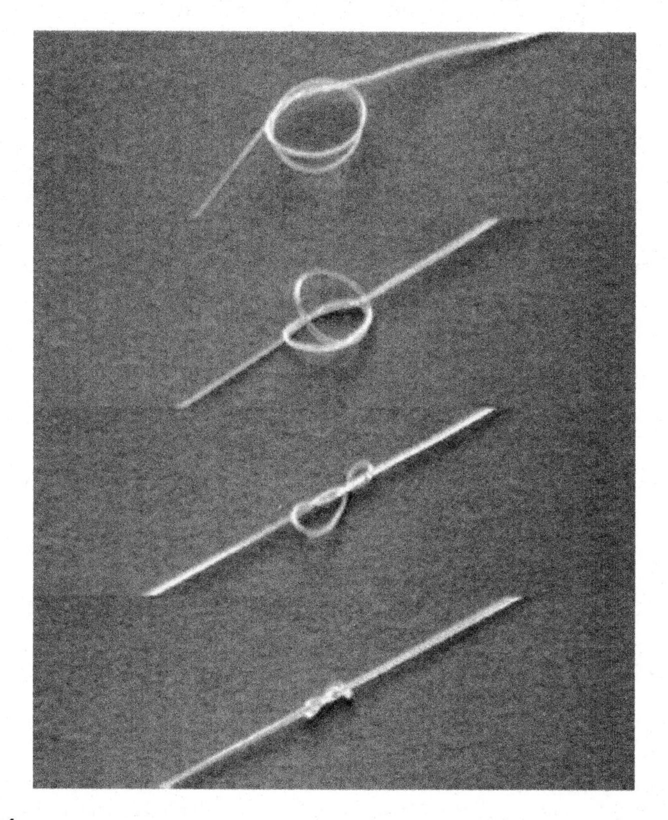

FIGURE 11.4
A fiber of a thermoplastic shape-memory polymer was programmed by stretching about 200%. After forming a loose knot, both ends of the suture were fixed. The photo series shows, from top to bottom, how the knot tightened in 20 s when heated to 40°C. (Reproduced from Lendlein, A. and Langer, R., *Science*, 296, 1673, 2002. With permission.)

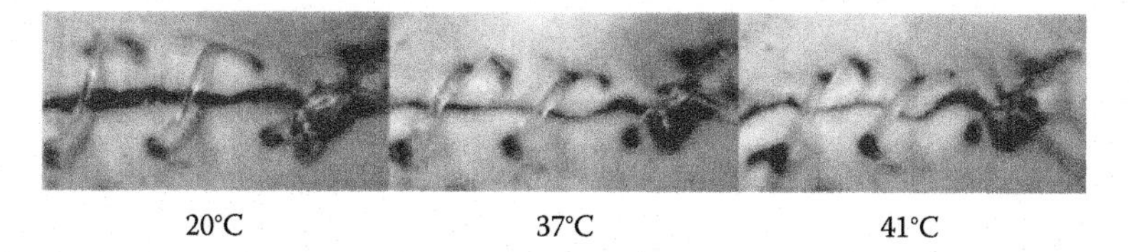

FIGURE 11.5
Degradable shape-memory suture for wound closure. The photo series from an animal experiment shows (left to right) the shrinkage of the fiber while temperature increases. (Reproduced from Lendlein, A. and Langer, R., *Science*, 296, 1673, 2002. With permission.)

temperature is raised above T_g, the suture will shrink and then tighten the knot, in which case it will apply an optimal force.

These sutures should be degradable and show gradual mass loss during degradation. The hydrolyzable ester bonds are introduced into the polymers so that they would cleave under physiological conditions. In this way, the degradation kinetics could be controlled through the composition and relative mass content of the precursor macrodiols. The multiblock copolymers presented show linear mass loss in vitro (Figure 11.6), resulting in a continuous release of degradation products [16].

A blood clot may cause an ischemic stroke, depriving the brain of oxygen and often resulting in permanent disability. As an alternative to conventional clot-dissolving drug treatment, a laser-activated device for the mechanical removal of blood clots is also proposed (Figures 11.7 and 11.8) [17,18]. The SMP microactuator, in its secondary straight rod form, could be inserted by minimally invasive surgery into the vascular occlusion. The microactuator,

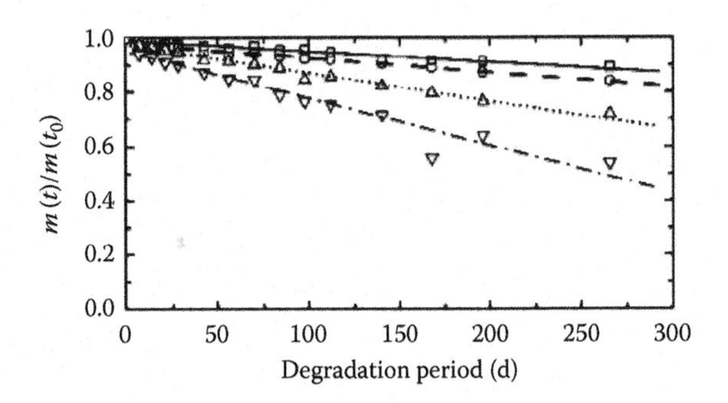

FIGURE 11.6
Hydrolytic degradation of the thermoplastic shape-memory elastomers in aqueous buffer solution (pH 7) at 37°C. The relative mass loss for multiblock copolymers differing in their hard segment content is shown (PDC10, circles; PDC17, squares; PDC31, upward-pointing triangles; PDC42, downward-pointing triangles). Sample mass ($m(t)$) after a degradation period t; $m(t_0)$, original sample mass. (Reproduced from Small IV, W. et al., *Opt. Express*, 13, 8204, 2005. With permission.)

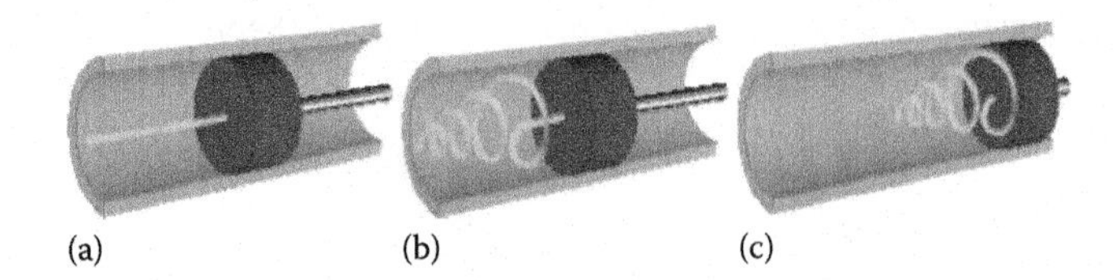

FIGURE 11.7
Depiction of removal of a clot in a blood vessel using the laser-activated shape-memory polymer microactuator coupled to an optical fiber. (a) In its temporary straight rod form, the microactuator is delivered through a catheter distal to the blood clot. (b) The microactuator is then transformed into its permanent corkscrew form by laser heating. (c) The deployed microactuator is retracted to capture the thrombus. (Reproduced from Small IV, W. et al., *Opt. Express*, 13, 8204, 2005; Behl, M. and Lendlein, A., *Mater. Today*, 10, 675, 2007. With permission.)

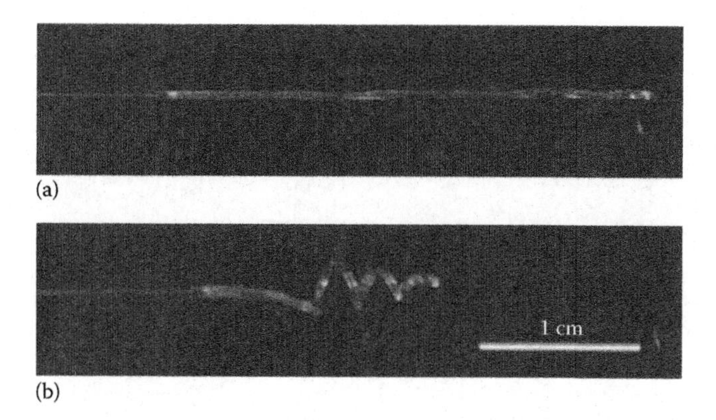

FIGURE 11.8
SMP microactuator coupled to an optical fiber shown in its (a) secondary straight rod and (b) primary corkscrew forms. The maximum diameter of the SMP corkscrew is approximately 3 mm. (Reproduced from Small, IV W. et al., *Opt. Express*, 13, 8204, 2005. With permission.)

which is mounted on the end of an optical fiber, is then transformed into its predeformed straight shape by laser heating. Once deployed into the corkscrew shape, the microactuator is retracted from the captured thrombus, enabling the mechanical removal of the thrombus.

The computer simulation in Figure 11.9 illustrates the light transmission through the SMP microactuator in its corkscrew form. It shows that the total light power drops with distance due to light leakage from the corkscrew turns, where the light paths exceed the critical angle for total internal reflection. According to the simulation results, about 90% of the light is transmitted to the distal end of the corkscrew, and thus the rest of the light is lost at the corkscrew turns. Additionally, the realistic deployment of the light-induced SMP microactuator was demonstrated in 37°C (human body temperature) (Figure 11.10). The laser power was about 4.89 W.

According to the above-mentioned light-induced SMP actuator, the light can only heat the small area around the end of the fiber. The heating area is relatively small and therefore light-inducing velocity is also slow. In order to realize the heating in a large area and fast actuation, a novel laser-activated SMP microactuator (see Figure 11.11) is proposed, where the infrared light can transmit from the side of the fiber [19]. The optical fiber was treated by the aqueous solution of sodium hydroxide; transmission efficiency of the optical fiber increased, as the total contact area between the SMP and the optical fiber increased. An optical fiber was embedded into the SMP for delivery of 3–4 μm laser light for activation. The free strain recovery investigation of a shape-memory polymer thread was performed and the results are shown in Figure 11.12. The original shape of the polymer was beeline, and then it was bent to a deformed shape and then fixed. When laser entered the optical fiber, the shape-memory polymer began to recover. It is possible

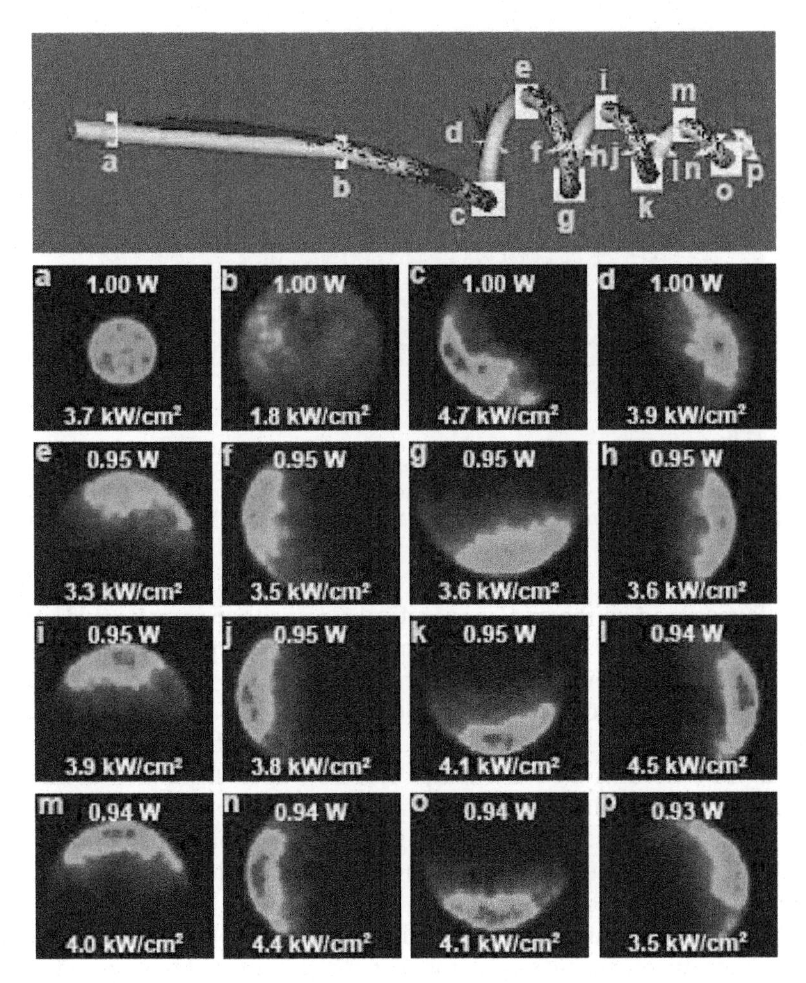

FIGURE 11.9
Computer simulation of laser light propagation through the SMP corkscrew in air. Virtual detectors (a–p) indicate the spatial light distribution (irradiance) at various cross-sectional locations along the corkscrew. (Reproduced from Small IV, W. et al., *Opt. Express*, 13, 8204, 2005. With permission.)

for the thermally activated SMP to initiate an original shape by a touchless and highly selective infrared laser stimulus.

Another example of a medical challenge to be addressed is obesity, which is one of the major health problems in developed countries. In most cases, overeating is the key problem, which can be circumvented by methods for curbing appetite. One solution may be biodegradable intragastric implants that inflate after an approximate predetermined time and provide the patient with a feeling of satiety after only a small amount of food has been eaten. Furthermore, SMP materials could be used in a variety of different medical devices and

FIGURE 11.10
Laser actuation of the SMP device in static water at body temperature. (Reproduced from Small IV, W. et al., *Opt. Express,* 13, 8204, 2005. With permission.)

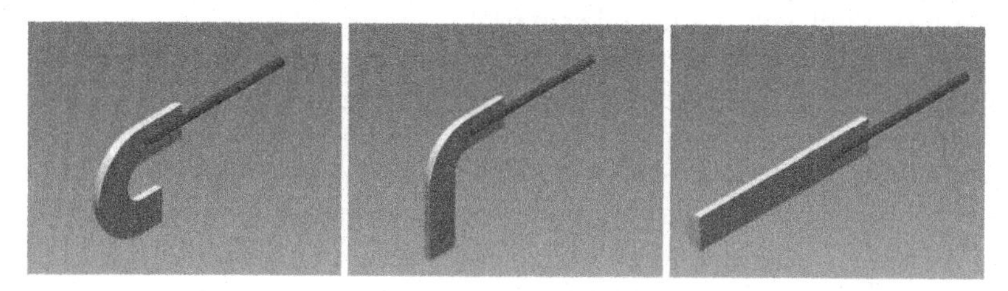

FIGURE 11.11
Illustration of deployment process of SMP actuator induced by infrared light.

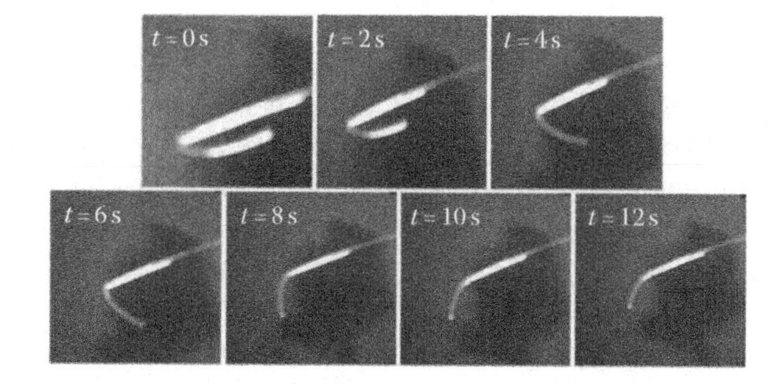

FIGURE 11.12
Shape recovery process of shape-memory polymer. (Reproduced from Zhang, D. et al., Infrared laser-activated shape memory polymer, in *15th SPIE International Conference on Smart Structures/NDE*, San Diego, CA, March 9–13, 2008, 6932.)

diagnostic products as deployable elements of implants from vascular grafts to components of cardiac pacemakers and artificial hearts [20,21].

11.4 CHEM Foam Structures and Medical Applications

Present memory metals such as NiTi are being used as components of different devices and provide a means of inserting a thin, wirelike device contained in a needlelike casing through a small incision. This device can regain a more complex shape once the casing is removed. SMP materials with much higher shape recovery/packaging capability can be inserted through small incisions or in a noninvasive way by catheters and subsequently regain their original large shape/size by the body heat and then stay there to perform a desired function.

Newly developed SMP foams, together with CHEM processing, further broaden their potential biomedical applications and are described in the following sections.

The concept of CHEM has been developed by Sokolowski et al. as a new, simple, ultralight, self-deployable smart structure and is described in details in Chapter 9. The CHEM technology utilizes polyurethane-based SMP in open cellular (foam) structures or sandwich structures made of SMP foam cores and polymeric composite skins [21,22]. The CHEM foam technology takes advantage of the polymer's heat-activated shape-memory in addition to the foam's elastic recovery to deploy a compacted structure. The T_g is tailored to deploy and if needed, rigidize a structure in fully deployed configuration [23,24]. The stages for use of a CHEM processing and foam structure are illustrated in Figure 11.13 and are as follows:

1. Original structure: the original structure is produced/assembled in a room held below T_g.
2. Compaction or rolling: the structure is warmed above T_g to make it flexible and then compacted and/or rolled up for stowing.
3. Hibernation (storage): the compacted/rolled structure is cooled to ambient temperature to achieve the hibernated stowage. As long as the temperature is maintained below T_g, no external forces are needed to keep the structure compressed.
4. Deployment: the compacted/rolled structure is warmed above T_g. The memory forces and the foam's elastic recovery cause the structure to naturally deploy back to its original shape and size.

If needed:

5. Rigidization: the deployed structure is cooled by ambient temperature and becomes rigid.

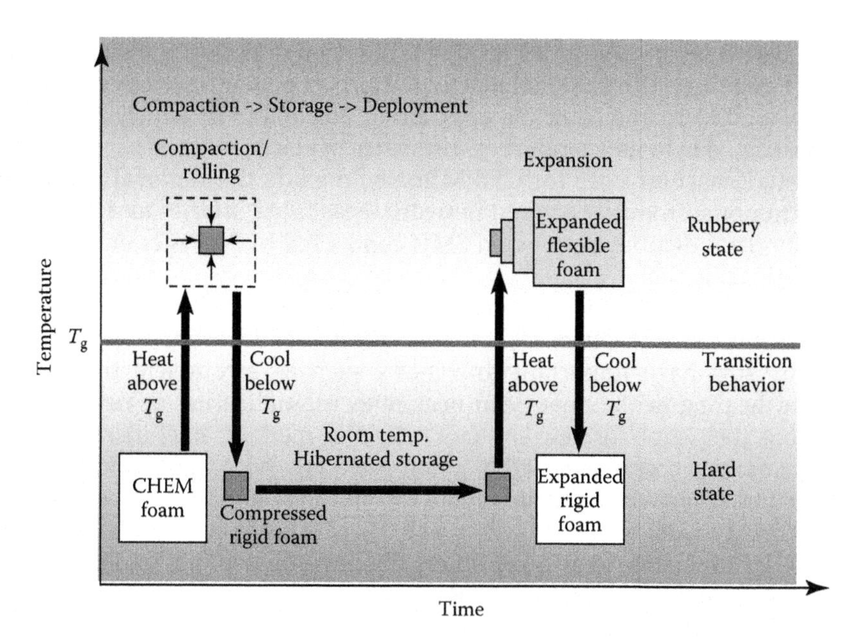

FIGURE 11.13
CHEM processing cycle. (Reproduced from Sokolowski, W. et al., *Biomed. Mater.*, 2, S23, 2007.)

The advantage of this technology is that structures, when compressed and stored below T_g, are a small fraction of their original size and are lightweight (two orders of magnitude lighter than aluminum). The foam configuration ensures lower mass, higher speed of deployment, and higher full/stowed volume ratio when compared with solid SMP materials. Similar to solid SMP, a wide range of T_g can be selected for deployment, including a T_g slightly below the human body temperature of ~37°C [25], thus allowing a wide variety of potential biomedical applications.

Properties that CHEM foams add to SMP properties, including a wide range of porosities, lightweight, high full/stowed volume ratio, and precision of original shape restoration, suggest that they have a potential to be used in functional and deployable elements of different clinical devices.

CHEM foams can be miniaturized, deformed, and inserted in the body through small catheters. Under the body heat, they can be precisely deployed and recover a much larger predetermined required shape when in satisfactory position [5,26]. A CHEM foam porosity can be adjusted to the needs of the application. Also, the CHEM technology provides a simple end-to-end process for stowing and deployment, and avoids the complexities associated with other methods for deployment of medical devices.

CHEM foams may find use in medical applications such as vascular and coronary grafts, orthopedic braces and splints, and medical prosthetics and implants. The CHEM could be used to design artificial grafts for replacing diseased arteries or serve as a scaffold for tissue engineering. It may possibly be used as a three-dimensional matrix to support bone growth in vitro and in vivo. CHEM foams could be appropriate for both soft and hard tissue engineering, due to its T_g and large difference of elastic modulus.

Potential vascular uses for CHEM foams include the removal of blood clot (thrombus) from the arterial network. Feasibility studies and preliminary (in vitro) demonstrations on SMP, conducted by Metzger et al., were optimistic [27].

The intracranial aneurysm represents abnormal outpouching of blood vessels that can cause the strokes. However, present endovascular approaches to aneurysms have important drawbacks such as incomplete treatment, deficient healing at the aneurysm neck, and recanalization or recurrence. Therefore, the search for new and more effective methods of obliterating the aneurysmal sac continues. CHEM foam materials have appealing characteristics for their use as embolic agents and filling material to occlude aneurysms (Figure 11.14).

The CHEM foams were experimentally investigated for endovascular treatment of aneurysm at CHUM Research Center, Notre-Dame Hospital, Montreal, in collaboration with École Polytechnique, Montreal, with encouraging results [4]. Lateral wall venous pouch aneurysms were constructed on both carotid arteries of eight dogs. The aneurysms were occluded peroperatively with CHEM blocks. Internal maxillary arteries were occluded via a 6F transcatheter technique using compressed CHEM blocks. Angiography and pathology were used to study the evolution of the occlusion and neointimal formation at the neck of experimental aneurysms after 3 and 12 weeks. The CHEM extract demonstrated no evidence of cell lysis or cytotoxicity and no mutagenicity. The efficient vascular embolization was confirmed in the aneurysms and good neointimal formation over the neck of treated aneurysms was demonstrated at the CHEM interface. Maxillary arteries embolized with

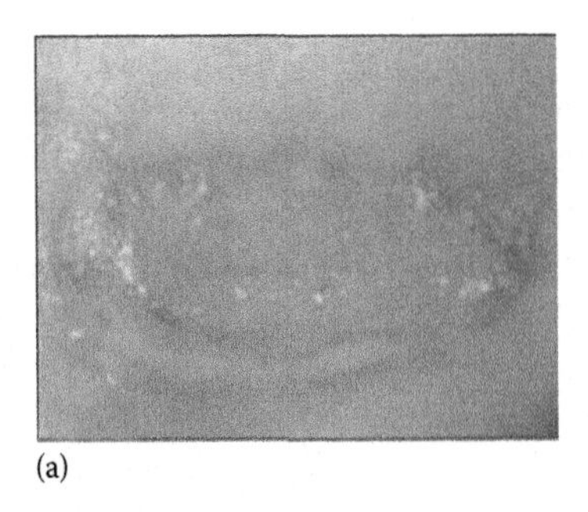

(a)

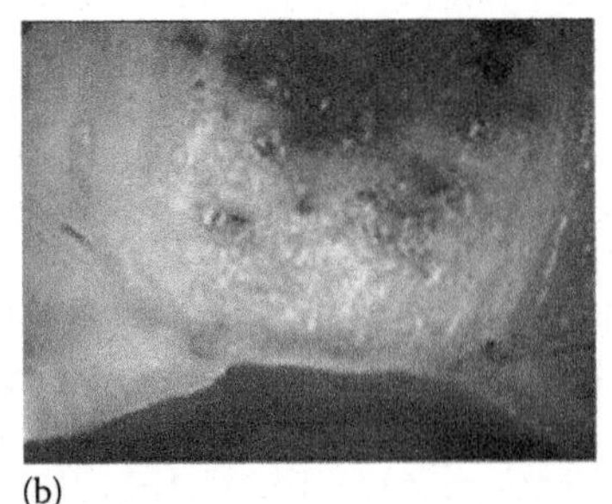

(b)

FIGURE 11.14
Experimental aneurysm occluded with CHEM. Macroscopic photographs showing good neointimal formation over the neck of the aneurysm at the CHEM interface. (a) Aneurysm neck "en face." (b) Axial section of the aneurysm. (From Metcalfe, A. et al., *Biomaterials*, 24, 491, 2003. With permission.)

CHEM foam remained occluded at the time of sacrifice (3 weeks). The major conclusion of the investigation was that the foamy nature of this new embolic agent favors the ingrowth of cells involved in neointima formation and new embolic devices for endovascular interventions could be designed using CHEM's unique physical properties.

The unique attributes of SMP materials like shape-memory effect, biocompatibility, and other properties make them a worthy technology for numerous potential self-deployable medical products. Presently, some SMP applications are already used in the medical world, others are under development.

Recently developed SMP foams together with CHEM processing further widen their potential medical applications. Future progress in the development of different CHEM materials, configurations, and delivery systems will be evaluated and tested in vitro and in vivo. The best systems will be selected for design and development of novel, second generation medical devices including devices for the treatment of aneurysms by endovascular embolization.

The SMP stent represents an innovative alternative to the conventional stent due to less costly manufacturing compared to metal stents. The manufacturing of SMP stents by injection molding, extrusion, or dip-coating technology guarantees an economical production. Compared to the production of conventional metal stents, the production costs are reduced by more than 50%.

It is believed that SMP materials and CHEM foams will significantly and positively impact the medical device industry. Their applications may usher in an era of simple, low-cost self-deployable medical devices. It is thought that the SMP technology will continue to gain attention and will open the door for the design and construction of novel important medical products and devices.

11.5 Summary and Outlook

The new materials will play a key role in the development of new technologies. In the field of medical engineering, a number of new technologies can only be realized if the biocompatible materials required can be developed. In this context, this extraordinary invention of biocompatible and biodegradable polymers with shape-memory properties will play an important role in the "new materials" for the twenty-first century.

References

1. Hayashi, S. and Shirai, Y. (1988). Shape memory polymer: Properties, Mitsubishi Technical Bulletin Number 184.
2. Hayashi, S., Kondo, S., Kapadia, P., and Ushioda, E. (1995). Room-temperature-functional shape-memory polymers, *Plast. Eng.*, 51, 29–31.

3. Hayashi, S. and Ishikawa, N. (1993). High moisture permeability polyurethane for textile applications, *J. Coated Fabrics*, 23, 74–83.
4. Metcalfe, A., Desfaits, A.C., Salazkin, I., Yahia, L., Sokolowski, W.M., and Raymond, J. (2003). Cold hibernated elastic memory foams for endovascular interventions, *Biomaterials*, 24, 491–497.
5. Sokolowski, W., Metcalfe, A., Hayashi, S., Yahia, L., and Raymond, J. (2007). Medical applications of shape memory polymers, *Biomed. Mater.*, 2, S23–S27.
6. Hayashi, S. (1993). Properties and applications of polyurethane series shape memory polymer, *Int. Prog. Urethanes*, 6, 90–115.
7. Robin, J., Martinot, S., Curtil, A., Vedrinne, C., Tronc, F., Franck, M., and Champsaur, G. (1998). Experimental right ventricle to pulmonary artery discontinuity: Outcome of polyurethane valved conduits, *J. Thorac. Cardiovasc. Surg.*, 115, 898–903.
8. Wabers, H.D., Hergenrother, R.W., Coury, A.J., and Cooper, S.L. (1992). Thrombus deposition on polyurethanes designed for biomedical applications, *J. Appl. Biomater.*, 3, 167–176.
9. Wheatley, D.J., Raco, L., Bernacca, G.M., Sim, I., Belcher, P.R., and Boyd, J.S. (2000). Polyurethane: Material for the next generation of heart valve prostheses, *Eur. J. Cardiothorac. Surg.*, 17, 440–448.
10. Tobushi, H.S., Hayashi, S., and Kojima, S. (1992). Mechanical properties of shape memory polymer of polyurethane series, *JSME Int. J.*, 35, 296–302.
11. Wache, H.M., Tartakowska, D.J., Hentrich, A., and Wagner, M.H. (2003). Development of a polymer stent with shape memory effect as a drug delivery system, *J. Mater. Sci. Mater. Med.*, 14, 109–112.
12. Bellin, I., Kelch, S., Langer, R., and Lendlein, A. (2006). Polymeric triple-shape-materials, *Proc. Natl. Acad. Sci. USA*, 103, 18043–18047.
13. Karp, J.M. and Langer, R. (2007). Development and therapeutic applications of advanced biomaterials, *Curr. Opin. Biotechnol.*, 18, 454–459.
14. Lendlein, A. and Kelch, S. (2002). Shape memory polymers, *Angew. Chem. Int. Ed.*, 41, 2034–2057.
15. Maruo, S., Ikuta, K., and Korogi, H. (2003). Optically driven micromanipulation tools fabricated by two-photon microstereolithography, *Mater. Res. Soc. Symp. Proc.*, 739, 269–274.
16. Lendlein, A. and Langer, R. (2002). Biodegradable, elastic shape-memory polymers for potential biomedical applications, *Science*, 296, 1673–1676.
17. Small IV, W., Wilson, T.S., Benett, W.J., Loge, J.M., and Maitland, D.J. (2005). Laser-activated shape memory polymer intravascular thrombectomy device, *Opt. Express*, 13(20), 8204–8213.
18. Behl, M. and Lendlein, A. (2007). Shape memory polymers, *Mater. Today*, 10(4), 675–684.
19. Zhang, D., Liu, Y., and Leng, J.S. (March 9–13, 2008). Infrared laser-activated shape memory polymer. *15th SPIE International Conference on Smart Structures/NDE*, San Diego, CA, SPIE 6932.
20. Kline, J. (1988). *Handbook of Biomedical Engineering*. San Diego, CA: Academic Press.
21. Sokolowski, W. (2004). Cold hibernated elastic memory self-deployable and rigidizable structure and method therefore, U.S. Patent No. 6,702,976 B2, March 9, 2004.

22. Sokolowski, W., Chmielewski, A., Hayashi, S., and Yamada, T. (March 1–5, 1999). Cold hibernated elastic memory (CHEM) self-deployable structures. *Proceeding of SPIE International Symposium on Smart Structures and Materials*, Vol. 3669, Newport Beach, CA, 179–185.
23. Kirkpatrick, E. and Sokolowski, W. (July 7–10, 2003). Heating methods for deployment of CHEM foam structures. Proceeding of International Conference on Environmental Systems (ICES),Vancouver, British, Canada.
24. Sokolowski, W. and Hayashi, S. (March 2–6, 2003). Applications of cold hibernated elastic memory (CHEM) structures. Proceeding of SPIE 10th International Symposium on Smart Structures and Materials, San Diego, CA.
25. Sokolowski, W., Tan, S., and Pryor, M. (April 19–22, 2003). Lightweight shape memory self-deployable structures for Gossamer applications. Proceeding of 45th AIAA Structures, Structural Dynamics & Materials Conference, Palm Springs, CA.
26. Sokolowski, W. (December 12–15, 2004). Potential bio-medical and commercial applications of cold hibernated elastic memory (CHEM) self-deployable foam structures. Proceeding of SPIE International Symposium on Smart Materials, Nano- and Micro-Smart Systems, Sydney, New South Wales, Australia.
27. Metzger, M.F., Wilson, T.S., Schumann, D., Matthews, D.L., and Maitland, D.J. (2002). Mechanical properties of mechanical actuator for treating ischemic stroke, *Biomed. Microdev.*, 4(2), 89–96.

12

Novel Applications and Future of Shape-Memory Polymers

Wei Min Huang
School of Mechanical and Aerospace Engineering, Nanyang Technological University, Singapore, Singapore

CONTENTS

This chapter presents a number of novel applications of two commercially available shape-memory polymers (SMPs), namely, a thermoplastic polyurethane SMP from Mitsubishi Heavy Industry, Japan, and a thermo-set polystyrene SMP from Cornerstone Research Group, United States, that are not well covered in the previous chapters of this book. The basic concepts and prototypes of most of these have been demonstrated in our research group. Also discussed are the major concerns that researchers and engineers should pay attention to in utilizing SMPs for real engineering applications and future research/application directions of SMPs.

12.1 Introduction

Since a rather extensive discussion about the various properties and aspects of shape-memory polymers (SMPs) has been presented in the previous chapters of this book, this chapter will only briefly address some important issues that one, as either an engineer or a researcher, should bear in mind while designing SMP-based devices and/or utilizing this material for novel engineering applications, and then focus on a number of novel applications of SMPs that are not so well discussed in the previous chapters of this book.

Unless otherwise stated, the works discussed in this chapter are from our group at the School of Mechanical and Aerospace Engineering, Nanyang Technological University, Singapore, although a couple of them are inspired by the previous studies of others reported in the literature.

We have used two types of SMPs for different applications, namely, a thermoplastic polyurethane (PU) SMP (same material discussed in Chapter 5) from Mitsubishi Heavy Industry, Japan (MHI), and a thermo-set polystyrene (PS) SMP from Cornerstone Research Group, United States (CRG). While the latter is thermo-responsive, the former is both thermo-responsive and moisture-responsive as revealed in Chapter 5. Given the excellent biocompatible nature of PUs, the PU SMP turns out to be a good choice for bio-related applications. On the other hand, the PS SMP is more suitable for the structural part of a device in which high stiffness and excellent durability are required.

Future directions in developing new SMPs and potential applications are discussed at the end of this chapter before conclusions.

12.2 Concerns in Materials Selection and Design

SMPs have the ability to recover their original shapes even after being severely distorted. As shown in Figure 12.1, two pieces of PU SMP plates (from MHI) are partially cut and bent in the in-plane direction for 180° at room temperature (about 23°C). Subsequently, they are placed atop a hot plate for thermally induced shape recovery. At the end of the heating process, as we can see, both plates almost fully recover the original straight shape. The high shape recovery

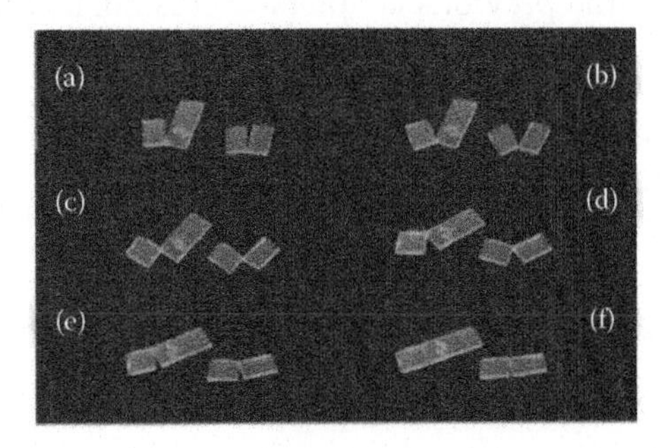

FIGURE 12.1
Shape recovery in severely pre-distorted PU SMP plates upon heating.

capability is because the recoverable strain in SMPs is normally well over 100%, which is at least an order higher than that of shape-memory alloys (SMAs) [1].

At present, there are a number of different types of SMPs that have been developed [2–4] and even made commercially available in the market, namely, the PU SMP from MHI and the PS SMP from CRG. More SMPs are being explored in order to meet different application requirements. Such efforts include the development of SMP composites blended with a variety of types of fillers for improved performance, enhanced properties and/or additional actuation approaches, etc. [5]. The selection of a right SMP among a number of candidates is an important task for the eventual success of a particular application. The actuating requirement (stimulus) should be the first concern, while whether the polymer is thermoplastic or thermo-set is another.

Size and shape are most likely the next issues for consideration. Compared with other shape-memory materials (e.g., SMAs), apart from those much lower in material cost, the SMP is much easier and cheaper in processing by conventional polymer processing technologies (e.g., injection molding, film casting, fiber spinning, profile extrusion, and foaming, etc.) into different sizes and even complicated shapes. Although bulk SMP is more or less the current focus in terms of not only scientific research but also applications, recently we found that the shape-memory phenomenon persists in 300 nm thick PU SMP (MHI) films, which are produced by water casting (Figure 12.2). This finding provides solid evidence of the possibility of developing applications of SMPs in micron/nanodevices.

Although the recoverable tensile strain in SMP solids is very high, SMP foams are more suitable for applications in which a high recoverable compressive strain is required. As reported in [6], a PU SMP foam (from MHI) can fully recover a pre-compression strain up to 94% (Figure 12.3).

Thermomechanical properties of an SMP are required even in the early design stage. Different SMPs may have remarkably different properties. Although many properties of most new SMPs have yet to be fully characterized, the properties of some commercial SMPs are well documented after many years of intensive studies (e.g., the PU SMP from MHI mentioned in Chapter 5). However, the variation in properties due to different processing

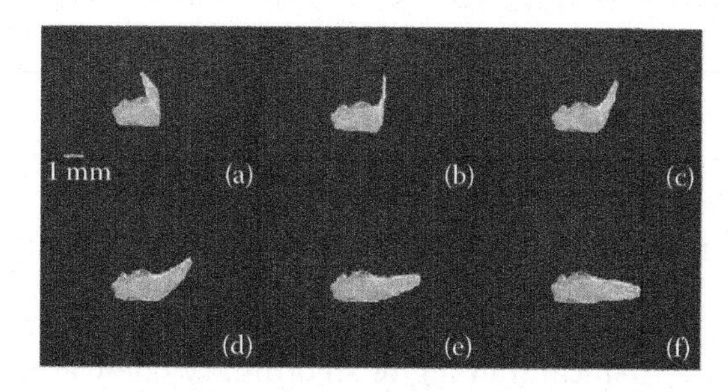

FIGURE 12.2
Shape recovery in a 300 nm thick PU SMP films upon heating.

FIGURE 12.3
PU SMP foam. Left: compressed; right: after shape recovery.

techniques and procedures provides us a convenient way for tailoring a material for better performance, but at the same time it brings a challenge and/or an uncertainty in design. As such, first-hand experiments are most likely required to get reliable data for design engineers.

Since pre-deformation is normally required for SMPs, we should bear in mind the difference in pre-deforming a piece of SMP at low and high temperatures, which correspond to the glass and rubber states of a material, respectively. Take uniaxial tension as an example, at low temperatures (i.e., below the glass transition temperature, T_g), PU SMPs, which are normally relatively ductile, go through a procedure of elastic deformation, necking, transition and front propagation, and finally fracture (Figure 12.4a). As such, at a low pre-strain level, the pre-deformation is not uniform. However, for brittle ones, such as PS SMPs, materials often break suddenly right after necking. On the contrary, since at the above T_g SMPs are in a rubber state, they can be stretched easily and in a more or less uniform manner without any propagation phenomenon. As such, pre-deformation at high temperature is normally preferred. On the other hand, upon heating, the shape recovery is approximately uniform everywhere even in the severely deformed part in a partially deformed sample (Figure 12.4b).

It should be pointed out that different from that of SMAs, which can be used for cyclic operations by means of the two-way shape-memory in terms of either the intrinsic two-way shape-memory effect through a proper thermo and/or mechanical training of an ordinary one-way shape-memory SMA or the mechanical two-way shape-memory by a bias resetting system [1], SMPs are more likely for one-time operation only. This is because most

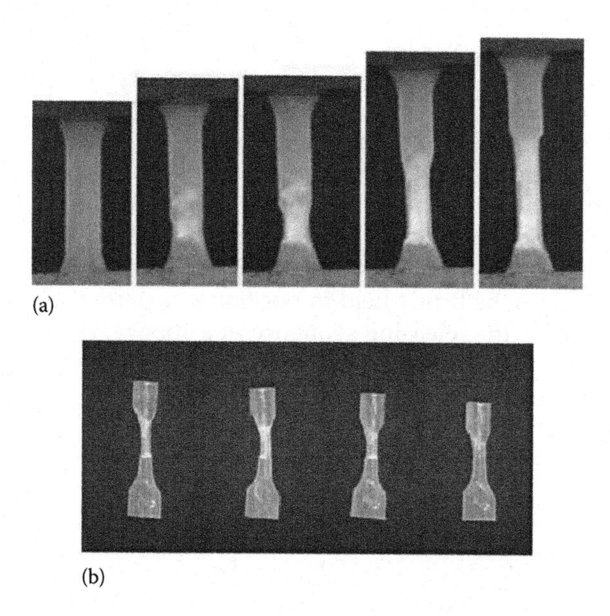

FIGURE 12.4
(a) Uniaxial tension at low temperature and (b) shape recovery upon heating.

SMPs become much softer upon applying the corresponding stimuli, which is a phenomenon opposite to that in SMAs. Thus, an automatic resetting system cannot be achieved for cyclic operation. The SMP recently reported in [7] seemingly has provided a possible solution to overcome this limit.

12.3 Applications

In this section, we present a number of novel applications of SMPs that we have worked out in the past years in our group. Based on the type of SMP used, these applications are split into two groups, i.e., PU SMP (MHI) (and its composites) and PS SMP (CRG). As mentioned above, the PU SMP from MHI is soft and ductile at room temperature and biocompatible, thus, more suitable for bio applications; while the PS SMP from CRG is stiff and brittle and more applicable as a structural material. Blending with various types of fillers can modify the performance and properties of SMPs as discussed in detail in Chapter 7 of this book.

12.3.1 PU SMP and Composites

The prototypes of a few devices have been worked out using PU SMPs with a nominal transition temperature of 35°C or 65°C depending on the application requirement. This SMP was found to be thermo- and moisture-responsive [8], i.e., the actuation of this SMP can be triggered not only by heat, but also upon absorbing moisture/water.

Figure 12.5 shows a live ant with a SMP micro-tag mounted on one of its legs. First, a SMP tube was made from a piece of SMP wire. Subsequently, the hole was enlarged at 50°C which is above the T_g, 35°C, of this SMP. After cooling back to room temperature, the SMP tube was mounted on one leg of a live ant. Upon heating to 35°C, that is the transition temperature of this SMP, the tube shrinks and can be firmly held in position as a permanent tag.

Figure 12.6 shows the releasing sequence of a sponge, which was initially held by a SMP wrapper. The wrapper was heated for recovering its original straight plate shape. Consequently, the sponge was released. Similar devices can be designed as a switcher or control element for triggering a reaction upon imposing the right stimulus.

FIGURE 12.5
SMP micro-tag. (From Huang, W.M. et al., *J. Intell. Mater. Syst. Struct.*, 17, 753, 2006. With permission.)

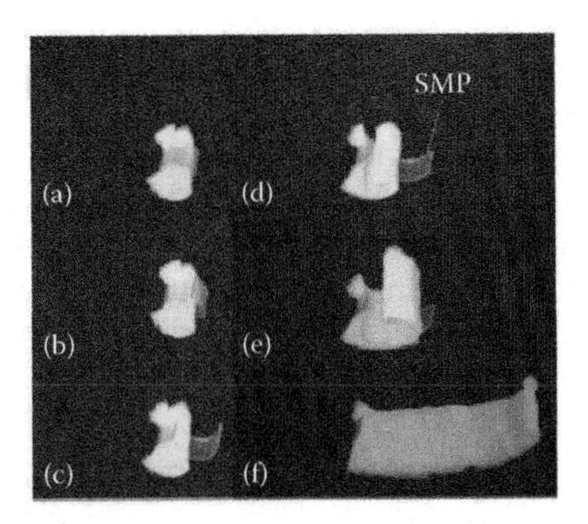

FIGURE 12.6
Opening of a SMP wrapper upon heating. A piece of folded sponge was then released.

At present, obsolete electrical devices, such as hand-phones, have become a threat to our environment. Recycling is a good solution to stop the trend. However, from the practical point of view, we face the problem of high labor cost in manure disassembling and sorting in recycling. Active disassembly using shape-memory material (both alloys and polymers) has been proposed [10–12]. This concept is illustrated by the following example. A piece of SMP rod can be easily deformed into a screw like shape at high temperature (Figure 12.7a). After that, it can be used as an ordinary plastic screw in any device as shown in Figure 12.7b. Upon heating, the thread on the SMP screw disappears, so that active disassembly can be realized by slightly shaking the device without any tedious manure unscrewing work involved. On the other hand, SMP can be used for screw-free assembly. For instance, different-sized screws can be replaced by pre-stretched SMP rods with one single size. These SMP rods can be placed inside the holes with/without thread. Upon heating, SMP rods recover their original shape and so fully fill the holes to achieve screw-free assembly (Figure 12.8).

Braille is a writing system which enables blind and partially sighted people to read and write through touching. It consists of patterns of raised dots arranged in cells of up to six dots in a 3×2 configuration. Currently, it is difficult to correct any mistake typed either manually (using a stylus and slate, Figure 12.9a) or automatically (using a thermo-set typing machine). Using

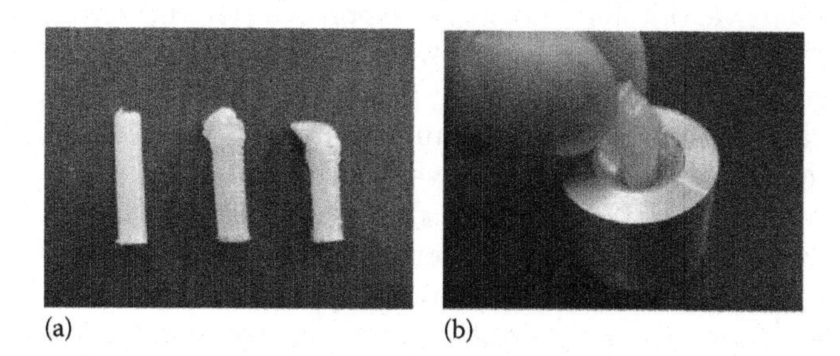

FIGURE 12.7
SMP for active disassembly. (a) From left to right: initial shape, deformed shape (screw), and recovered shape of the SMP rod; (b) SMP screw in working position.

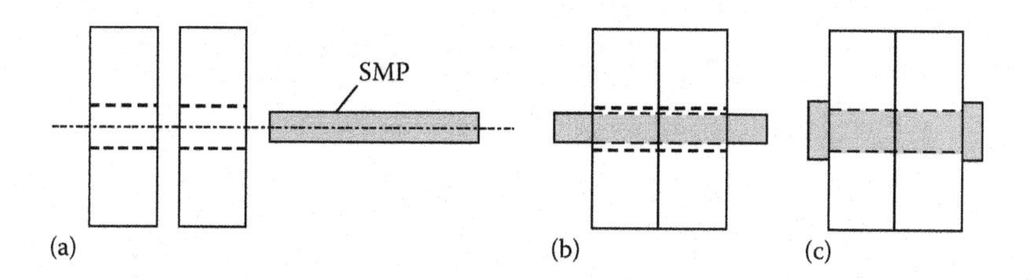

FIGURE 12.8
Screw-free assembly using SMP. (a) To be assembled in pieces and predeformed SMP; (b) SMP in position; (c) after heating SMP.

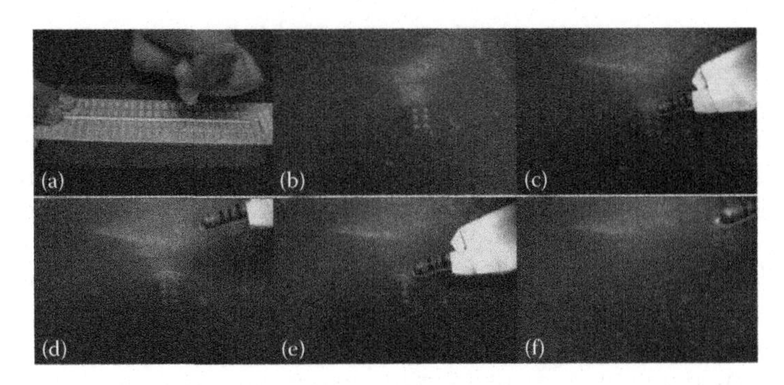

FIGURE 12.9
SMP Braille paper. (a) Writing; (b)–(f) removing dots using a point heater in a one-by-one manner.

SMP instead of normal Braille paper could potentially help the blind. A point heater has been designed to allow precise and accurate correction of any mistakes written on a piece of SMP. The point heater is able to make specific portions of a SMP paper revert back to its original shape, thus correcting any mistake, as demonstrated in (Figure 12.9b through f).

Thermo-responsive SMP sutures have been proposed for self-tightening of knots for minimally invasive surgery [13]. Since the PU SMP is not only thermo-responsive, but also moisture-responsive [14], the actuating of pre-stretched PU SMP sutures can be easily achieved upon immersing into room temperature water without involving any troublesome heating process for self-tightening (Figure 12.10) or self-unraveling (Figure 12.11). Figure 12.12 further demonstrates the concept of water-driven SMP suture, in which of a piece of 0.4 mm diameter PU SMP, which was gently wrapped around a piece of sponge, was immersed into room temperature water for self-tightening.

Further utilizing the thermo/moisture-responsive feature of this PU SMP, we can develop stents which can shrink for easy removal late on. As shown

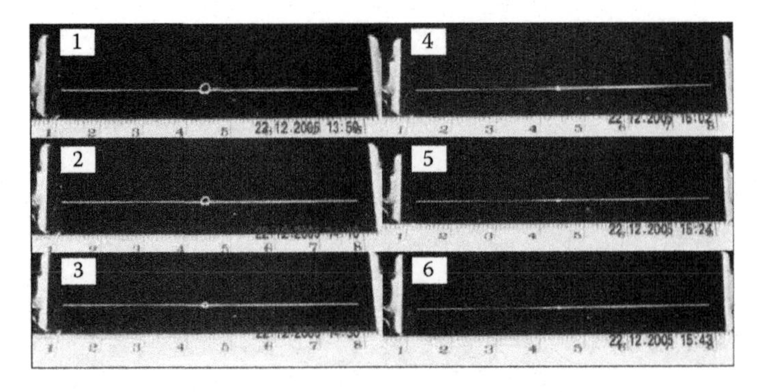

FIGURE 12.10
Self-tightening of a SMP wire upon immersing into room temperature water.

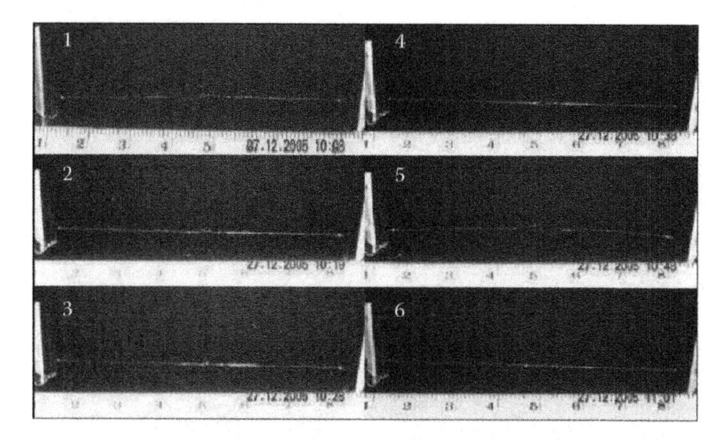

FIGURE 12.11
Self-unraveling of a SMP wire upon immersing into room temperature water.

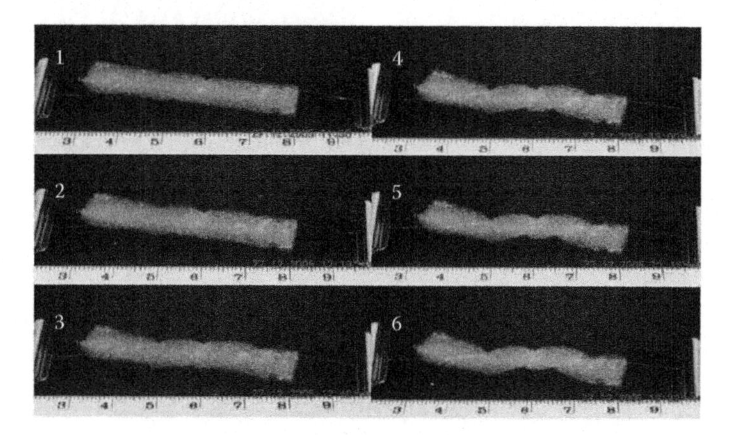

FIGURE 12.12
Self-tightening of a SMP wire wrapped around a sponge upon immersing into room temperature water.

in Figure 12.13a, a PU SMP thin film was pre-stretched and formed into a tube shape. The tube can be mechanically deformed into a star shape with a much smaller diameter (Figure 12.13b) and then mechanically deployed in water (Figure 12.13c), just like the deployment of a stent inside a blood vessel. Upon immersing into water, the SMP absorbs water, which causes its T_g to drop continuously. When the T_g is below the ambient temperature, the SMP starts to shrink (Figure 12.13d), which makes the task of taking it out from the inner side of a blood vessel easier.

The drop of T_g upon immersing into water can also be utilized to make a piece of SMP with different transition temperatures at different parts, i.e., a piece of SMP with functionally gradient T_g at different parts. In Figure 12.14,

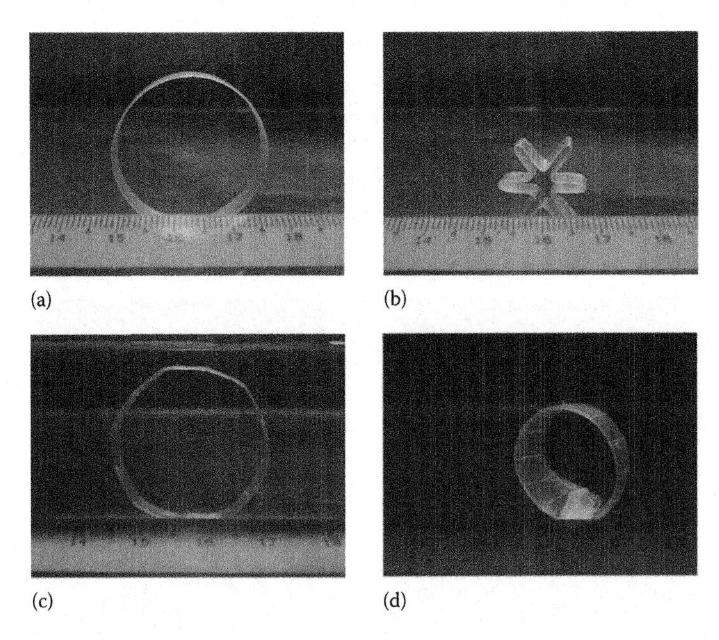

FIGURE 12.13
Retraction of a pre-deformed SMP in water. (a) After pre-stretching. (b) After folding. (c) After deployment in water. (d) After retraction in water.

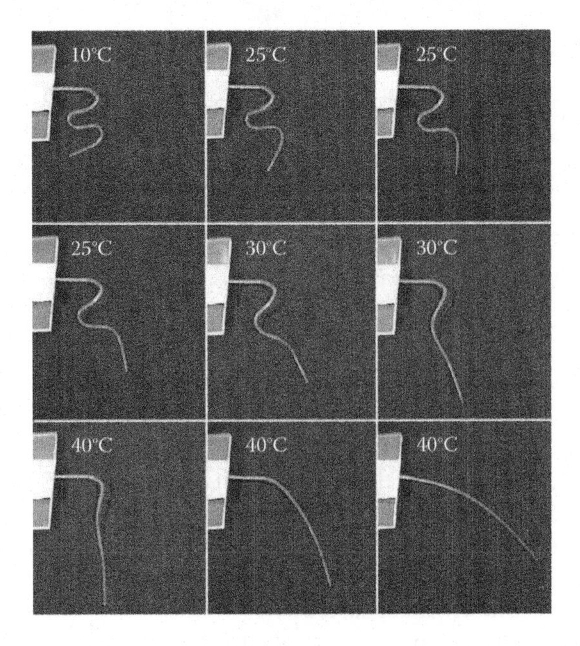

FIGURE 12.14
Recovery in a programmable manner in a SMP wire upon heating. (Reprinted from Huang, W.M. et al., *Appl. Phys. Lett.*, 86, 114105, 2005. With permission.)

a piece of 1 mm diameter SMP wire has different T_g at different segments. The bottom segment has the lowest T_g as it is immersed into water for the longest time; while the top segment is kept dry all the time. Hence, upon heating, at 25°C, the bottom segment recovers the straight shape; at 30°C, the middle segment recovers, and finally at 40°C, the top segment becomes straight. That is to say, the shape recovery can be controlled in a programmable manner. A combination of moisture-responsive and functionally gradient T_g features of this SMP results in the programmable shape recovery upon immersing into water as shown in Figure 12.15.

Since we have proved that at a submicron-size scale, this SMP still has excellent shape-memory (refer to Figure 12.2), it might be possible to make tiny actuators only a few microns or less in size and even nanomachines. Utilizing the moisture-responsive feature of this SMP, one can deliver a tiny device made of SMP into a live cell. The device recovers its original shape after it is inside a cell upon absorbing water as illustrated in Figure 12.16. It is

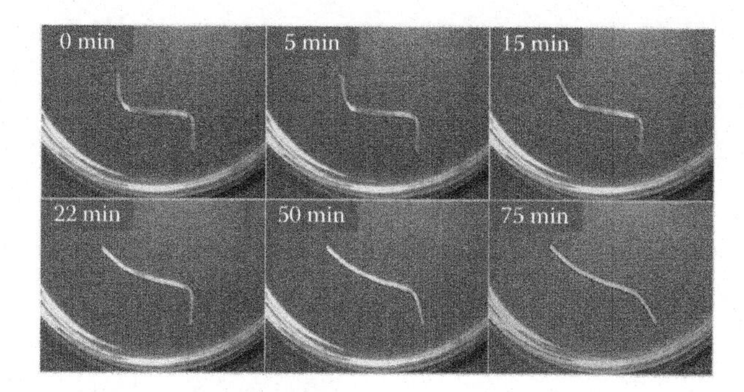

FIGURE 12.15
Recovery of a 1 mm diameter polyurethane SMP wire in water in a sequence. The wire was produced by extrusion. The top-half wire was placed in water for a lower T_g, while the bottom-half was kept dry. The wire was then bent into a Z-shape. Upon immersing into room temperature water, the top-half of the wire recovered first and the bottom-half started to recover later. (Reprinted from Huang, W.M. et al., *Appl. Phys. Lett.*, 86, 114105, 2005. With permission.)

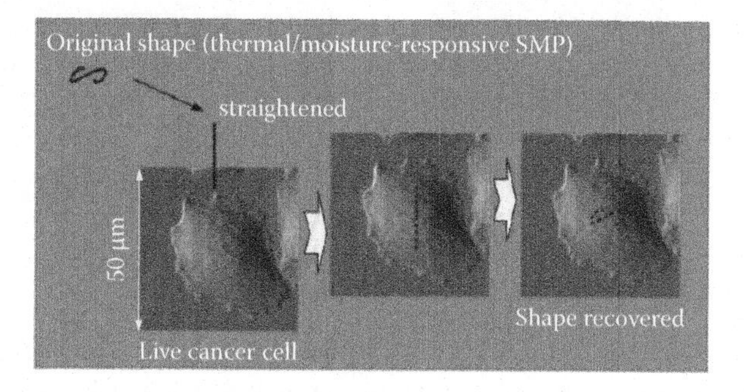

FIGURE 12.16
Delivery of a piece of S-shape thermo-moisture-responsive SMP wire into a living cell (illustration).

possible to have more complicated shape devices and machines made of this SMP. Subsequently, we should be able to have surgery inside a cell or a small machine operating inside a cell (e.g., carrying out mechanical test on DNA strings inside a live cell). The basic concept of delivery of a piece of coiled SMP into a hole is further demonstrated at millimeter scale in Figure 12.17, in which a piece of hydrogel is used to represent a cell.

SMPs are electrically nonconductive in nature, which limit the techniques for activating them. In addition, PU SMPs are relatively soft. As such, they are unsuitable in applications where a high stiffness is required. Upon blending with different types of fillers, the thermo electro mechanical properties of the SMP composites can be tailored.

We mixed the SMP with carbon black, which is electrically conductive in nature, to produce electrically conductive SMPs, which can be easily Joule heated by passing an electrical current, just like that of SMAs. Figure 12.18a

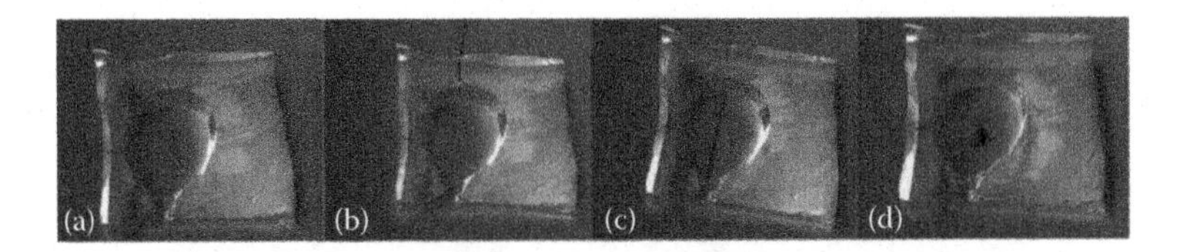

FIGURE 12.17
Demonstration of concept of delivering a coiled SMP wire (0.4 mm diameter) into a hole inside a hydrogen gel. (a) A piece of gel with a hole in room temperature water; (b–c) delivery of a piece of straightened SMP wire into the hole inside gel; (d) recovery of SMP wire inside the hole.

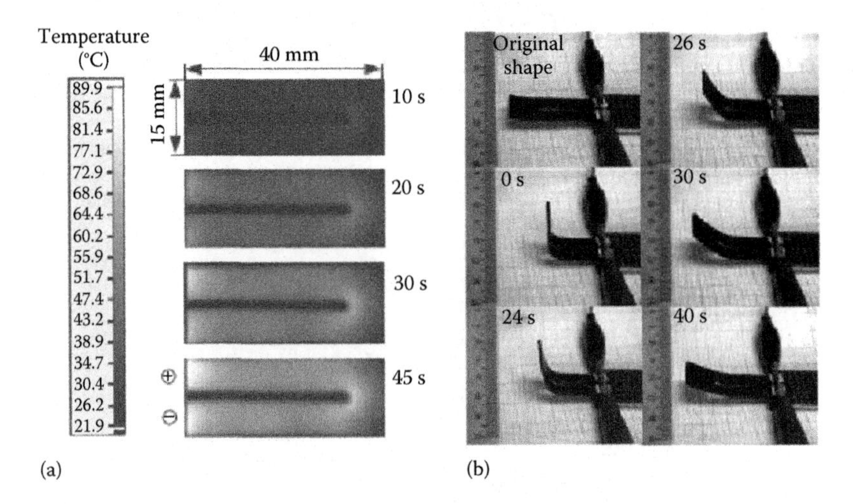

FIGURE 12.18
Shape recovery in a conductive SMP composite (mixed with 13% volume fraction of carbon black) upon passing an electrical current. (a) Temperature distribution captured by an infrared camera; (b) shape recovery. (From Yang, B. et al., *Eur. Polym. J.*, 41, 1123, 2005. With permission.)

reveals the temperature distribution of a piece of SMP composite (mixed with 13% volume fraction of carbon black), in which the highest temperature is over 90°C when a 15 V of current is applied. Figure 12.18b shows the sequence of shape recovery (from bent shape to straight in 40 s). Such conductive SMPs can be used to drive the motion of toys (Figure 12.19) or even for morphing wings of airplanes (Figure 12.20). It should be pointed out that the conductive SMP still can be triggered by moisture as revealed in Figure 12.21. Thus, such SMPs are thermo/electrical/moisture-responsive. This feature provides great flexibility for actuating.

In addition to mixing with carbon black, we have developed thin film SMPs with magnetic chains inside using Fe_3O_4 micro-powders [17]. The chains were formed automatically inside SMP solution upon applying

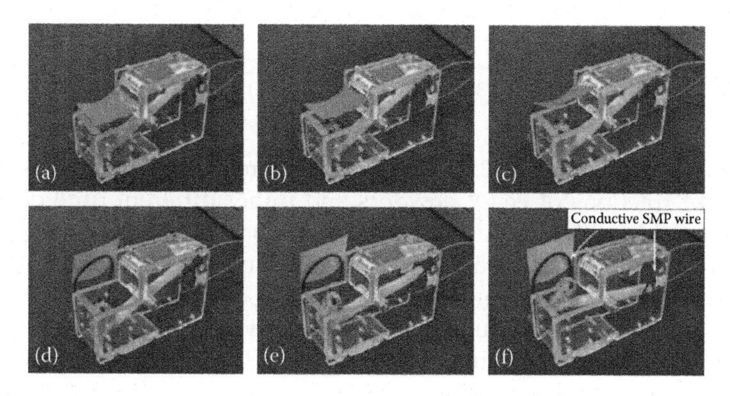

FIGURE 12.19
Electrical conductive SMP wires for actuation of a toy. (a–d) Opening the cover; (d–f) pushing the toy tiger out.

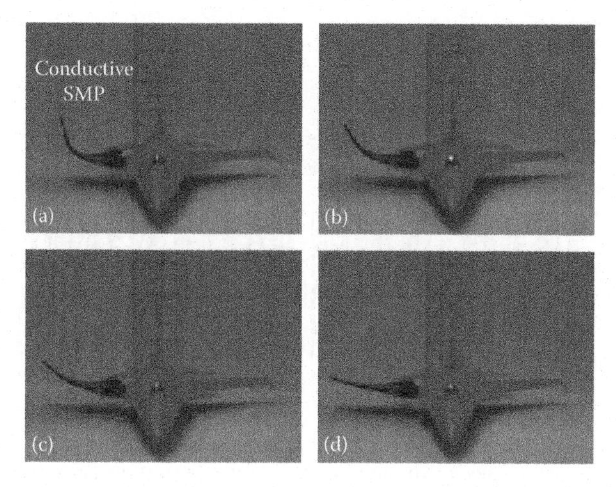

FIGURE 12.20
Conductive SMP for morphing wing.

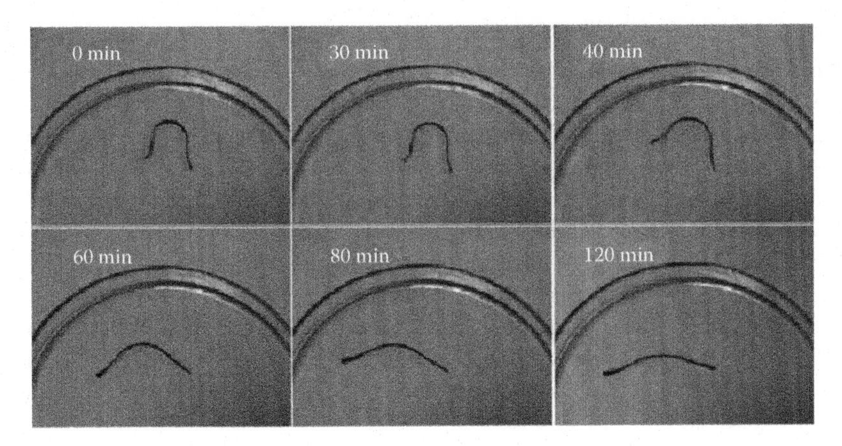

FIGURE 12.21
Shape recovery of a piece of conductive PU SMP actuated by room temperature water. (From Yang, B. et al., *Eur. Polym. J.*, 41, 1123, 2005. With permission.)

a magnetic field. Figure 12.22 reveals the formation of chains at different weight fractions of Fe_3O_4 powders and under different time period of magnetic field. With the increase in powder content and time of applying magnetic field, the chains become continuous and thicker. Upon heating to 180°C, which is above the melting temperature of this polymer, the chains in the thin film switch gradually by 90° if a vertical magnetic field, which is perpendicular to the previous setup, is applied (Figure 12.23). Replacing the

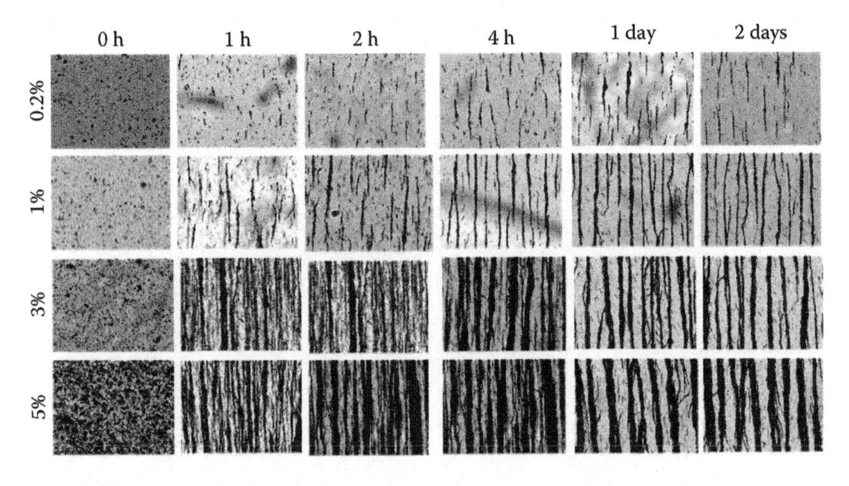

FIGURE 12.22
Formation of magnetic chains. (From Lan, X. et al., Electrically conductive shape memory polymer with anisotropic electro-thermo-mechanical properties, in *Polymers in Defence and Aerospace Applications*, September 18–19, Toulouse, France, Smithers Rapra Technology Ltd, Billingham, Cleveland, U.K., 2007, ISBN 978-1-84735-019-0, paper 9. With permission.)

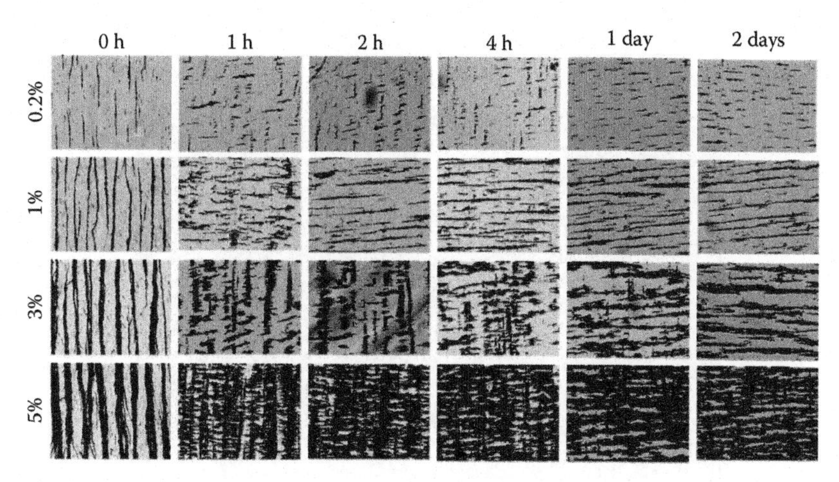

FIGURE 12.23
Switching of magnetic chains. (From Lan, X. et al., Electrically conductive shape memory polymer with anisotropic electro-thermo-mechanical properties, in *Polymers in Defence and Aerospace Applications*, September 18–19, Toulouse, France, Smithers Rapra Technology Ltd, Billingham, Cleveland, U.K., 2007, ISBN 978-1-84735-019-0, paper 9. With permission.)

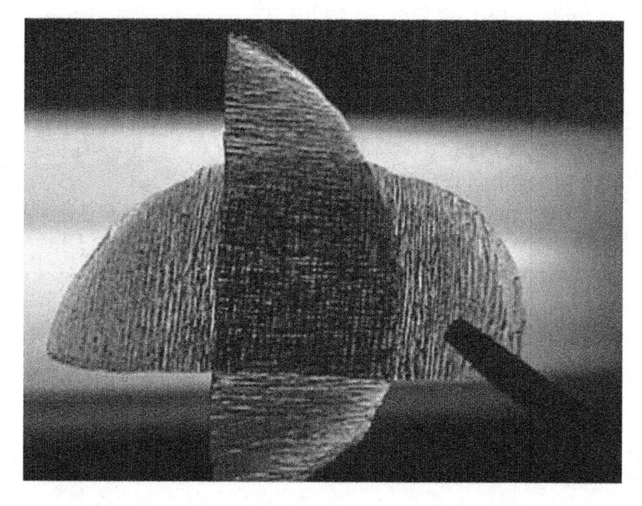

FIGURE 12.24
A double-layered composite. (From Lan, X. et al., Electrically conductive shape memory polymer with anisotropic electro-thermo-mechanical properties, in *Polymers in Defence and Aerospace Applications*, September 18–19, Toulouse, France, Smithers Rapra Technology Ltd, Billingham, Cleveland, U.K., 2007, ISBN 978-1-84735-019-0, paper 9. With permission.)

Fe_3O_4 powders (which is electrically nonconductive) by nickel micro-powders (which is electrically conductive), we have produced conductive SMP thin films with well-aligned nickel chains inside [18]. Figure 12.24 shows a composite with two pieces of such thin films perpendicularly bonded together [17].

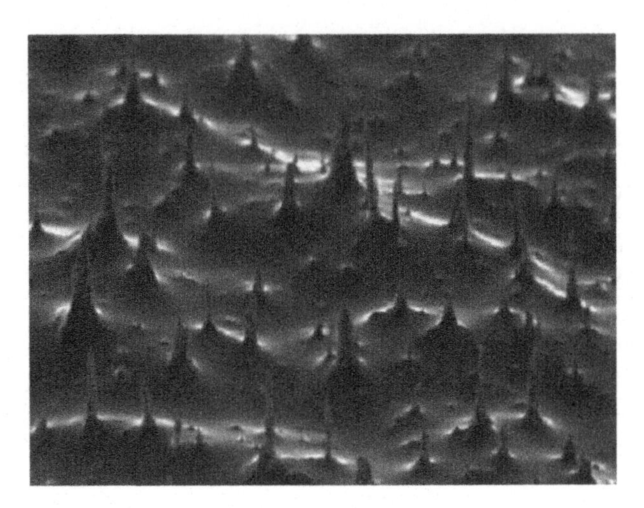

FIGURE 12.25
Micro-chains atop SMP (at 1% volume fraction of nickel particles).

Instead of forming in-plane chains, vertical chains can be produced in a similar way as presented in Figure 12.25. These vertical chains can be flattened mechanically or by applying a magnetic field at high temperature. The deformation of these chains can be well preserved at room temperature. However, upon heating, they can bounce back to their original shape as in Figure 12.26. This technique provides a convenient approach to dramatically alter the surface morphology of a material.

Since SMP foams can be severely compressed, they are an excellent candidate for deployable structures [19]. Figure 12.27 shows that a toy car is compressed into a compact shape, and then recovers its full size/shape after heating. Despite of the simplicity, this example reveals the advantages of SMP forms in the potential applications of space missions, in which deployable structures are widely used to save space and minimize weight during rocket launching.

FIGURE 12.26
Reversible protrusive micro-chain. (a) Original, (b) after deformation, and (c) after heat.

(a)

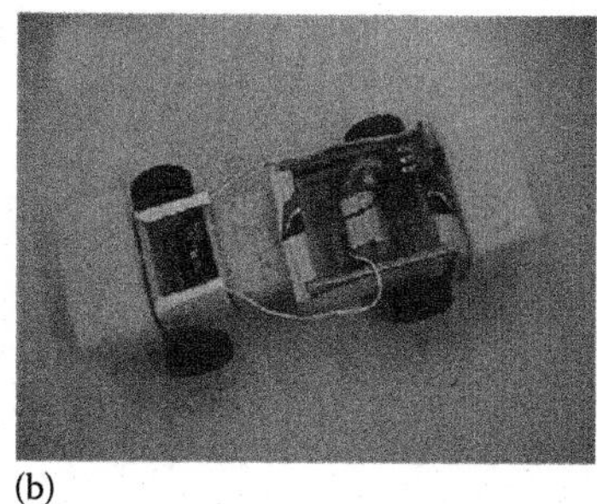
(b)

FIGURE 12.27
SMP transformer. (a) Packed shape; (b) shape restored. (From Huang, W.M. et al., *J. Intell. Mater. Syst. Struct.*, 17, 753, 2006. With permission.)

Porous polymers are very important in many applications such as tissue engineering where they are applied as a scaffold for cellular attachment and tissue development. Porous PU SMPs have been proposed for a few biomedical devices [20,21]. The common agents used for polyurethane to develop porous or foaming structures are organic solvents. The residues of these agents remaining in the material may be harmful to cell and tissue [22]. As the PU SMP absorbs water easily, we have developed porous PU SMPs using water as a nontoxic agent. The fabrication procedures are as follows [23]. SMP thin sheets are immersed into room temperature water for different hours to absorb various amounts of water. After that, the SMP sheets are heated to different high temperatures, namely, 110°C, 120°C, 130°C, and 150°C, for three minutes and then cooled in air to room temperature. Figure 12.28 presents the optical images of these samples, in which closed pores are observed. In general, with the increase of heating temperature, the size of pores increases remarkably.

In another set of SMP samples, which are immersed in water for 1, 2, 6, and 48 h, respectively, they are heated to 120°C for 3 min and then quenched in air to room temperature. Figure 12.29 reveals the images obtained by an optical microscope. It shows that there are some pores developed after one-hour immersion in water. With the increase of immersion time more moisture is absorbed in SMP and more pores can be built upon heating to over 100°C. In general, at a lower ratio of moisture the individual closed pores are formed, whereas if more moisture is absorbed, the pores expand and coalesce. This results in larger pores with open cells.

An interesting finding in the course of forming pores inside the PU SMPs is the reversible bubbles. In one sample, which is soaked in water for two hours, we heat the sample to 102°C and then quench it in air to room temperature. After heating again to 80°C, the bubble reduces its size significantly (Figure 12.30). This finding might be utilized to realize reversible micro-bubbles and even channels, for instance, for drug release upon immersing into water (owing to the moisture-responsive feature of this SMP).

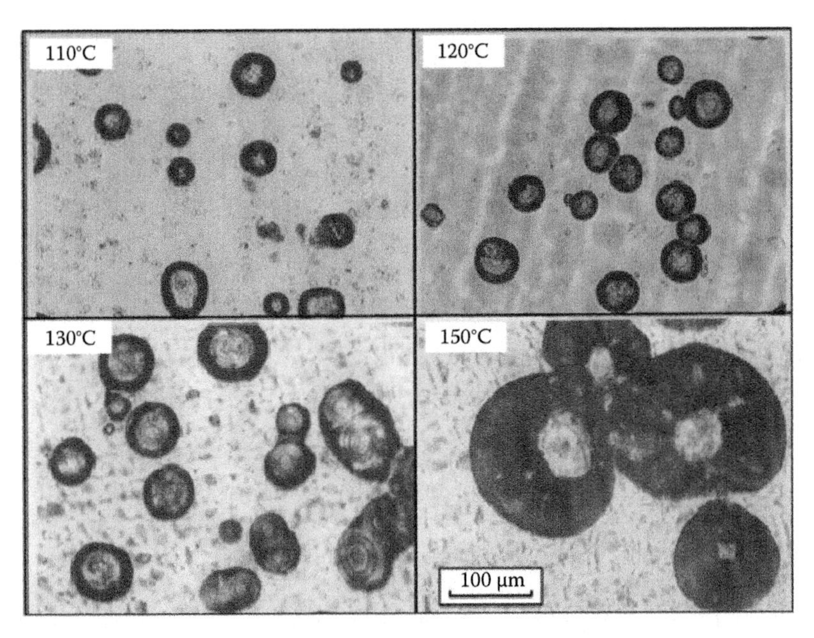

FIGURE 12.28
SMP foams using water as nontoxic agent: influence of heating temperature.

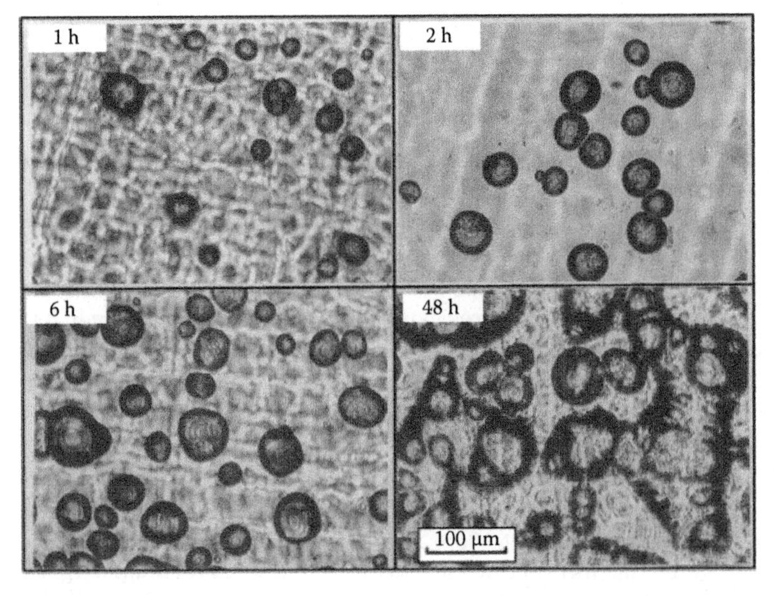

FIGURE 12.29
SMP foams using water as nontoxic agent: influence of water content.

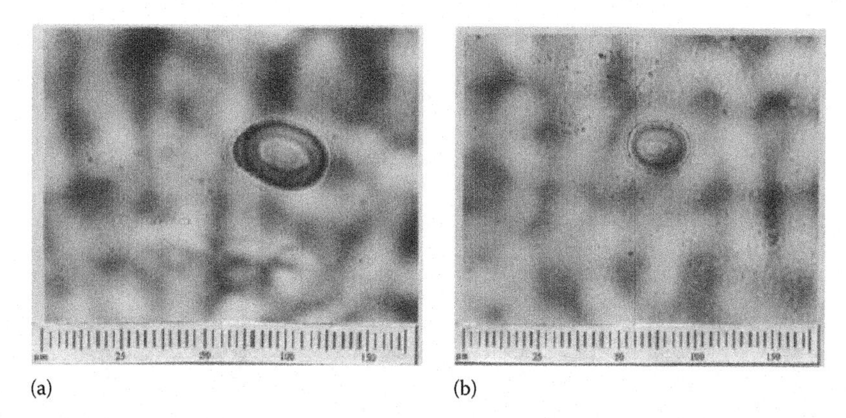

FIGURE 12.30
Reversible bubble. (a) Initial bubble and (b) after subsequent heating.

12.3.2 PS SMP

The PS SMP from CRG is much stiffer at room temperature as compared with the above PU SMP from MHI. As such, we can polish the surface of the PS SMP to achieve an average surface roughness which is only a few nanometers. SMPs with such smooth surface can be utilized for many interesting applications.

If one stretches a piece of well-polished PU SMP to over 50% strain at high temperature (well above its T_g), after cooling back to the room temperature, many butterfly-shaped features (with sizes from 6 to 90 μm) can be found on the sample surface upon surface scanning using, for instance, a WYKO interferometer (Figure 12.31). If subsequently the sample is slightly polished, all these butterflies will switch by 90° after heating for shape recovery. The underlying mechanism is associated with the breaking of the hard segments upon stretching and shape recovery at high temperature/upon heating [24]. The occurrence of these butterflies reveals the damage of the material, which affects the shape recovery ability of the SMP. Hence, this finding provides an experimental approach to evaluate quantitatively not only the shape recovery ratio in SMPs but also the evolution of shape recovery in SMPs under cyclic actuation.

1. A typical *garden* of micro-butterfly of various sizes (from about 6 to 90 μm). Black parts (wings of big butterflies) are areas that the WYKO interferometer failed to collect data at this scanning scale.
2. A typical micro-butterfly. Left: 2-D view; right: 3-D view.

If we coat a thin layer of metal, for example , nickel or gold, atop a piece of pre-stretched SMP, which has been well polished, after heating for shape

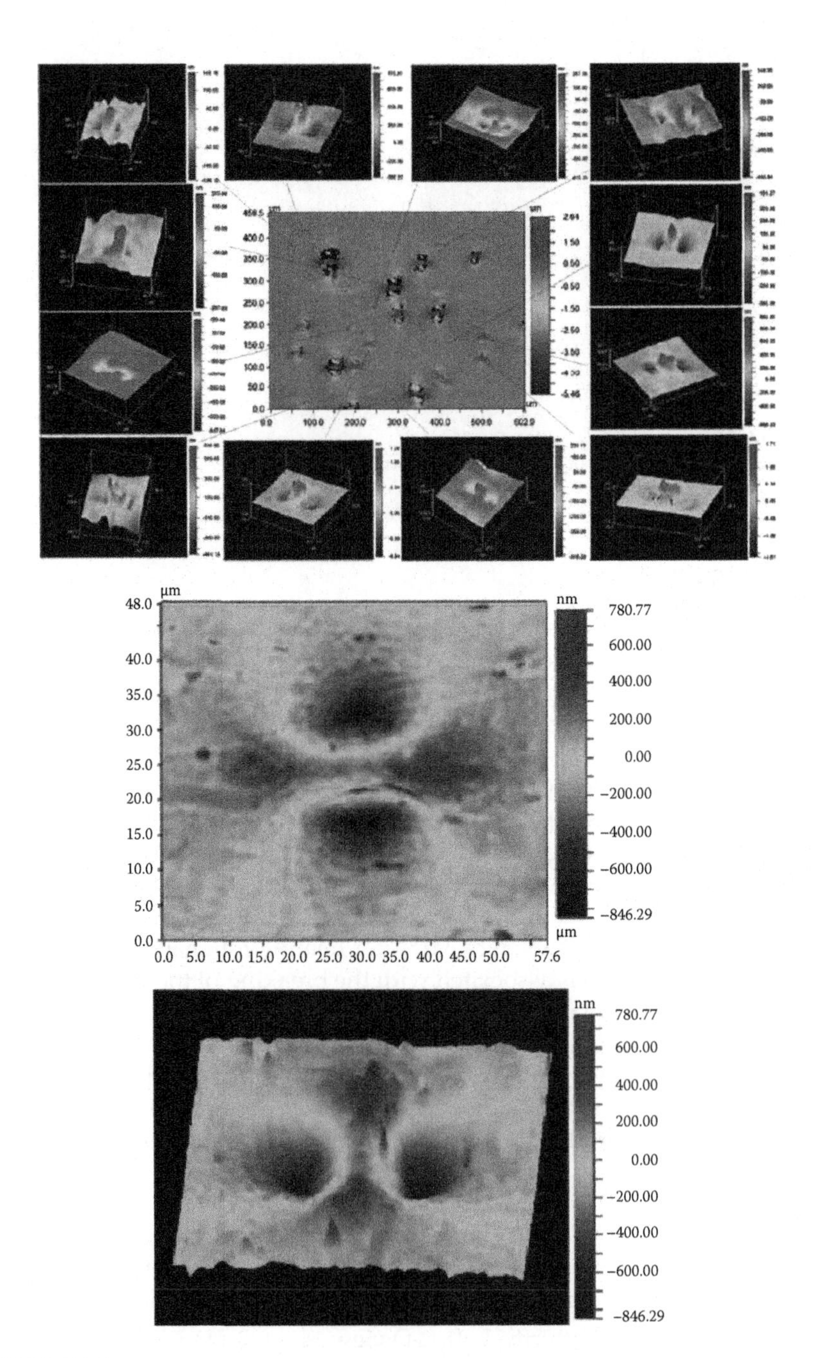

FIGURE 12.31
Micro-butterflies. (From Liu, N. et al., *Surf. Rev. Lett.*, 14, 1187, 2007. With permission.)

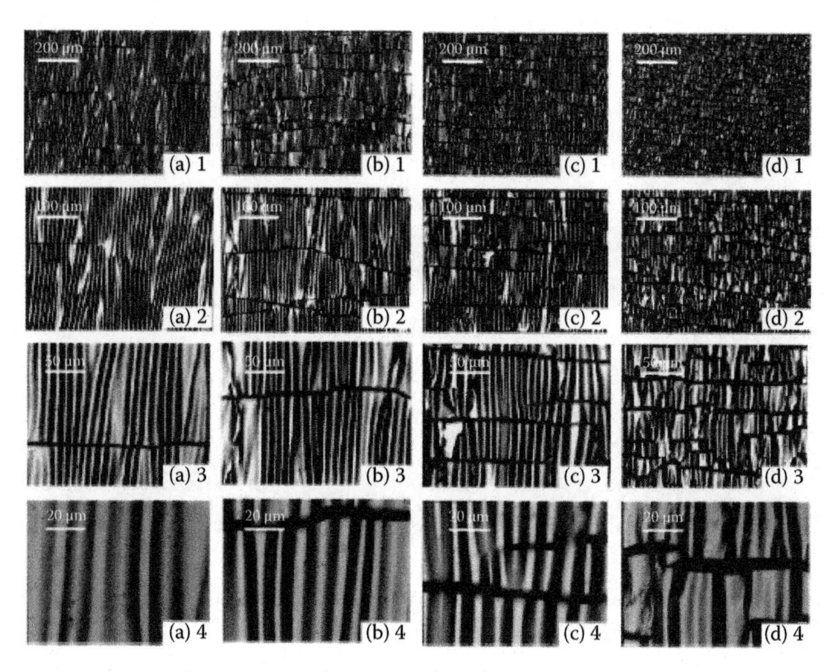

FIGURE 12.32
Strip-shaped wrinkles formed atop a pre-stretched SMP (coated with a thin layer of 50nm nickel). (a) 5% pre-strained; (b) 10% pre-strained; (c) 20% pre-strained; and (d) 40% pre-strained.

recovery, strip-shaped wrinkles can be observed as shown in Figure 12.32. These wrinkles result by the buckling of the elastic metal layer during the shape recovery of the SMP substrate underneath (which is pretty soft during this period of time as well). If the coated metal is brittle, such as nickel, hairline cracks can be found along the transverse direction due to the expansion in this direction during shape recovery (Figure 12.33). The more pre-stretching in the SMP, the more cracks are resulted. The number of cracks per unit length is about a linear function of the amount of pre-strain. Without pre-stretching, labyrinth-shaped wrinkles can be found upon heating the coated SMP to over 150°C. Zoom-in view of these two types of wrinkles, one is more or less anisotropic, while the other is isotropic, is presented in Figure 12.34 for comparison. Droplet test reveals that the pattern of wrinkles can significantly affect the contact angle and shape of a water droplet deposited atop the surface.

In addition to the above approach for surface patterning, which can only produce limited types of patterns at micro-scale, we have developed two other techniques to produce various kinds of protrusive features (which are more difficult to produce than dimples/indents) and surface patterns atop PS SMP at a scale from millimeter to nanometer.

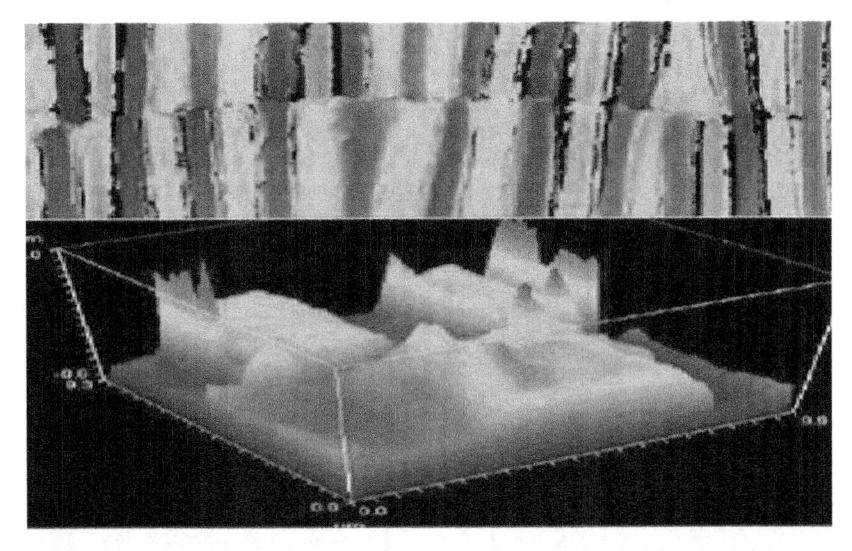

FIGURE 12.33
Enlarged image (top view and 3-D view, by WYKO scanning) of a fine hairline crack that causes slight shear of the wrinkles.

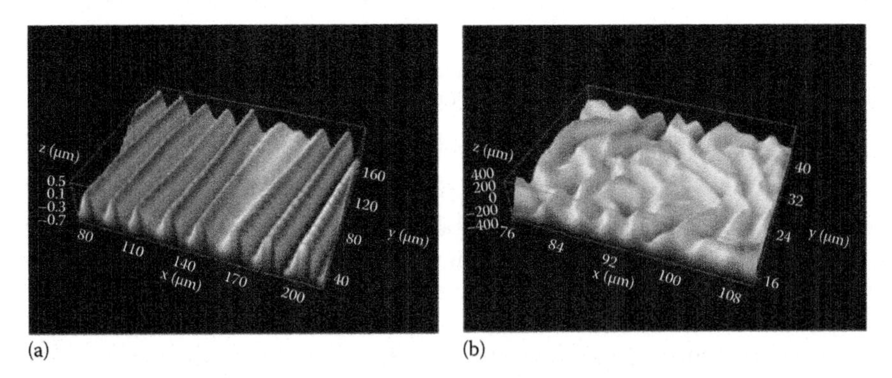

FIGURE 12.34
Wrinkles atop SMP (50 nm thick gold-coated). (a) Strip (anisotropic) wrinkle; (b) labyrinth (isotropic) wrinkle.

Packing a layer of hard balls of the same size atop a SMP and then compressing the balls hard at either room temperature or above T_g produces an array of indents. While compressing at low temperature results in sink-in indents, at high temperature, the indents are pile-up, subsequently polishing the SMP and then heating it to above T_g for shape recovery. Eventually, an array of features can be obtained. Depending on the polishing depth, the resulted feature could be circular-shaped trench, flat-top circular protrusion, or crown [25]. Figure 12.35 shows typical micro-protrusive arrays with flat and sharp heads (3-D view, top view, and cross-sectional view). Steel balls (1 mm diameter) are used to make the indents.

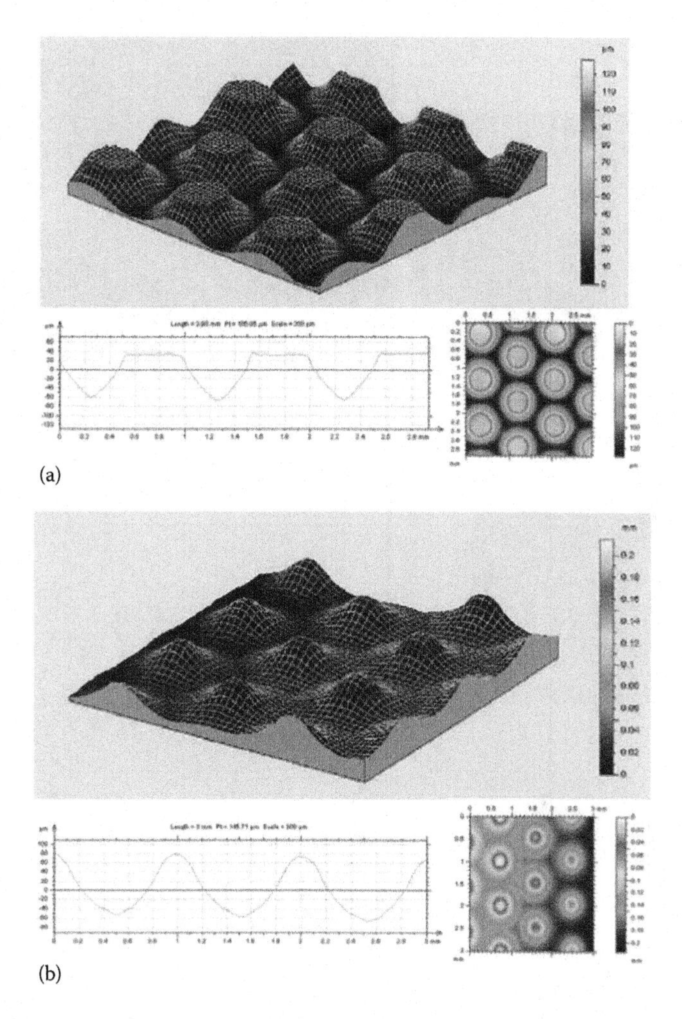

FIGURE 12.35
(a) Macro-protrusion arrays with flat head and (b) sharp (crown-shaped) head.

Using 30 μm diameter silica balls, micron-sized protrusion array can be produced as shown in Figure 12.36. The height and base size of the protrusion are only about 500 nm and 15 μm in diameter, respectively. If the SMP is pre-stretched before indented, elliptical-shaped protrusion array can be produced (Figure 12.37).

Instead of hard balls, different-shaped protrusions can be produced depending on the shape of the indenter. Figure 12.38a is a micron-sized pyramid which is produced using a nano Berkovich indenter. Protrusions down to less than 100 nm in height have been produced (Figure 12.38b).

In order to provide a more cost-effective approach to produce crown-shaped protrusion arrays, one can pre-compress a SMP in the thickness

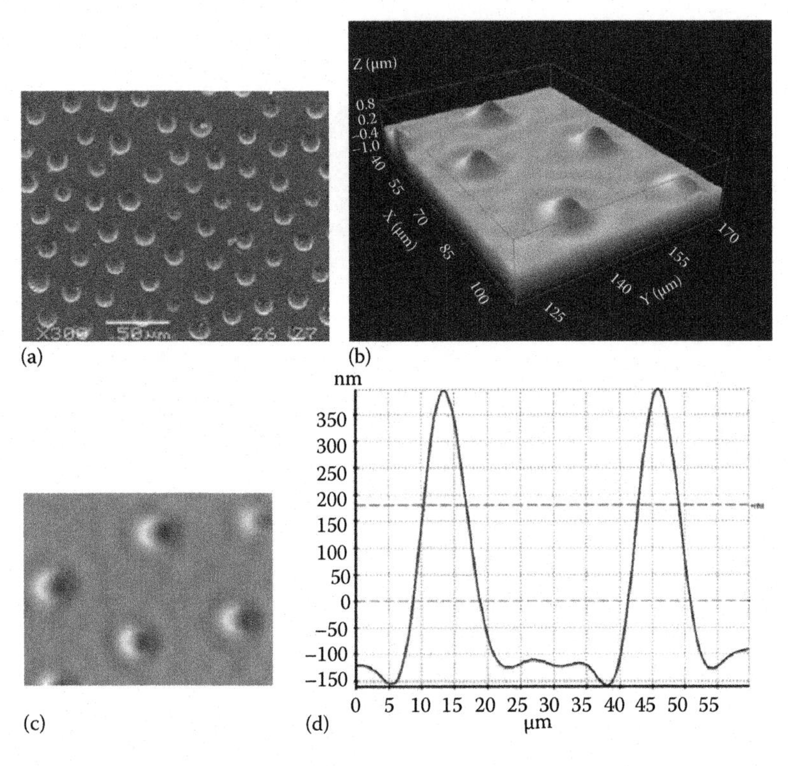

FIGURE 12.36
Details of a micro-protrusion array. (a) SEM image showing the surface fully covered by micro-dots; (b) 3-D profile (a zoom-in view); (c) top view; (d) cross-sectional view.

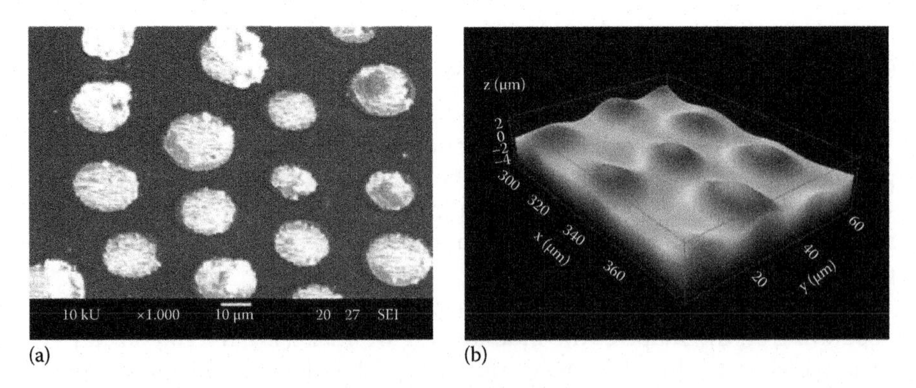

FIGURE 12.37
Elliptical-shaped protrusion array. (a) SEM image and (b) 3-D confocal scanning.

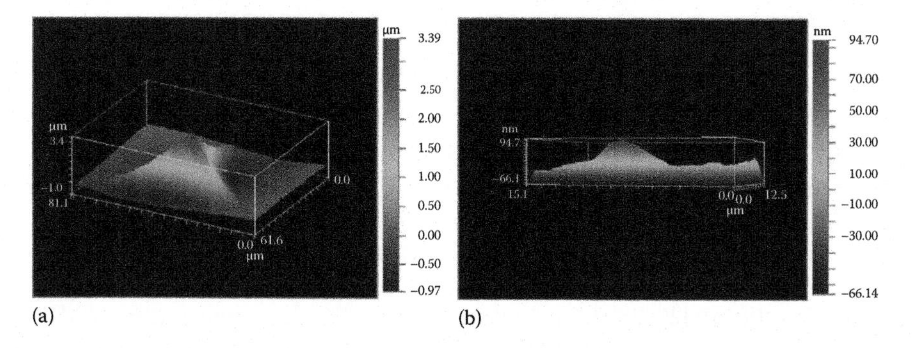

FIGURE 12.38
(a) Micron-sized pyramid atop PS SMP realized using nano Berkovich indenter and (b) micron-sized particle.

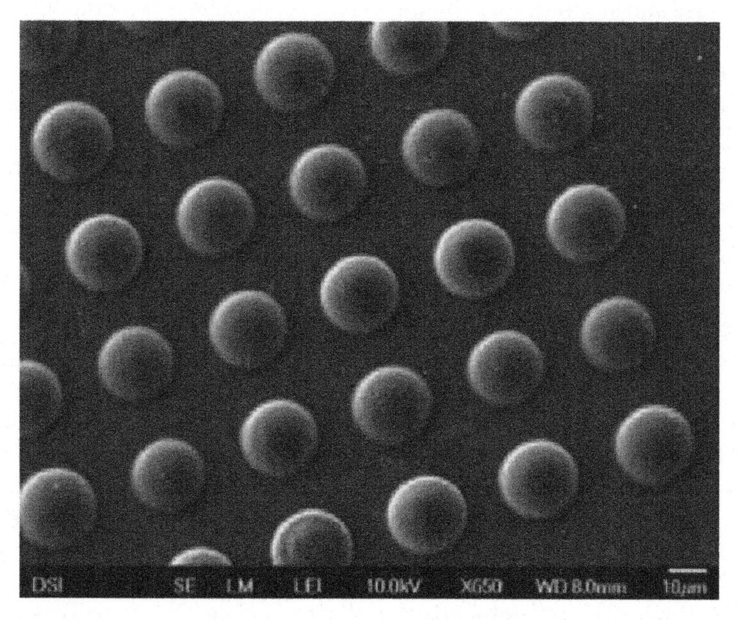

FIGURE 12.39
Micro-protrusive array produced by laser local heating atop a precompressed PS SMP.

direction and then use a laser to heat locally the top surface of the SMP in a point-by-point manner or use a micro-lens array, which results in an array of points to be heated simultaneously, for massive fabrication. Figure 12.39 shows an array of protrusions produced by laser local heating atop a pre-compressed SMP.

12.4 Future Directions in Developing New SMPs and Applications

Despite many different types of SMPs having been developed [2–5], new SMPs are always highly in demand in order to meet different and new requirements or achieve better performance.

Alternative stimulus, for example, electrical field, magnetic field, light (in particular, visible light), pressure, sound, mechanical stress, chemicals, or molecular stimuli [26–28], for actuation is one of the directions for new SMPs. Chemical-responsive SMPs can be designed to detect the change in environmental conditions and even take reactions automatically. One possible application is in automatic water cleaning after harmful chemicals are detected. A SMP can be developed to combine the detecting and cleaning functions together. In terms of bio-related applications, SMPs are useful in automatic drug release and synthesis of protein-polymer conjugates for therapeutic applications when a certain biological environmental condition is sensed [28,29].

Apart from the biomedical applications, SMPs, in particular thermo-responsive SMPs, have been proposed for smooth and dramatic shape change from one configuration to another in, for instance, morphing wings of airplanes (Figure 12.20), hood/seat assembly and tunable automotive brackets in vehicles [30–32], and airflow control [33].

However, almost all SMPs developed so far are only capable of one-time operation. SMPs which can be used in cyclic actuations are needed for continuous and periodical operations just like that of SMAs. There are many potential applications for such reversible SMPs. For instance, they can be used as a new cost-effective approach for producing mechanical energy from, for instance, solar energy (light-responsive SMP [34]), chemical energy (chemical-responsive SMP), or waste heat (thermal-responsive SMP). Surfaces with different types of reversible dimples or protrusions made of SMPs can change their many surface properties (e.g., surface tensile, sound/light reflection, etc.) remarkably. Such surfaces can also alter the drag for optimal performance in vehicles (airplanes, submarines, and ships) in different missions.

A SMP, which can respond to a few different types of stimuli, can provide novel solutions in many engineering applications, thus significantly widening the applications of SMPs in future. A unique feature of such multi-stimuli-responsive SMPs is that their actuation can be better controlled for shape recovery in a step-by-step manner than the current ones, which are achieved by having either two T_g for different transitions [35,36] or functionally gradient T_g at different locations [15].

In addition to long durability, high actuation force, and fast actuation speed in the new SMPs, excellent biocompatibility and/or biodegradability (even with adjustable degradation rate) are necessary for bio-related applications [37,38]. In thermo-responsive SMPs, while a broadened glass transition temperature could be our interest [39], a narrow transition temperature range, down to a few degrees only, could dramatically increase the actuation speed, while an improved thermal conductivity, which can be achieved by, for instance, mixing with fillers with high thermal conductivity, could reduce the cooling time remarkably. In addition to SMPs with a T_g around our body temperature in bio-related applications, high transition temperature SMPs (T_g about 150°C and higher) are also useful in certain applications (e.g., reusable mandrels [40]).

SMP composites and nanocomposites can result in high performance SMPs with tailored properties for a particular application. Carbon nano-tubes have been used as fillers to produce electrically conductive SMP with improved strength [5]. Cheap fillers, such as some clays (e.g., attapulgite, which is a kind of nano-sized fibrous clay mineral), provide an alternative approach for strengthening SMPs in a cost-effective way [41]. This study is still in the early stages [42].

Regardless of type and size, fillers, if formed in a pattern inside a polymer matrix, can result in better performance along certain directions than those with randomly distributed fillers [43,44]. By forming nickel particle chains, enhanced electrical conductivity has been achieved in a thermo-responsive PU SMP with/without carbon black, so that they can be easily actuated by Joule heating [18,45]. One can easily align nickel powders into chains by applying a magnetic filed. However, this approach is not applicable to nonmagnetic fillers. Generic and convenient approaches for forming patterns of fillers inside SMPs are required for SMP composites with better performance.

In addition to Joule heating (in electrically conductive SMP composites) and light heating (by means of laser, inferred light, radiation, etc.), hysteresis heating has been used for actuating thermo-responsive SMPs filled with magnetic fillers under a high-frequency alternating magnetic field [46]. So the composites turn to be kind of magnetic-responsive. This approach is wireless and remote-controllable, and very much suitable for no-hole surgery inside a human body. Similarly, other types of fillers, which have great hysteresis, that generate enough heat, upon applying a certain alternating stimulus, can be used for heating a thermo-responsive SMP. For instance, ferroelectrical and pizeoelectrial materials can be used as fillers for electro-responsive SMP composites.

Built-in temperature sensors have been integrated into SMPs for temperature monitoring [47]. SMPs with other types of sensing capabilities should be very much useful as well.

12.5 Conclusions

At present, a few SMPs are commercially available in the market. A number of novel applications have been realized. The development of new materials not only accelerates but also widens the applications of SMPs. As compared with other shape-memory materials, such as shape-memory alloys, SMPs are much cheaper and more convenient in processing, so that they can be used in from macro- to nano-sized applications. More importantly, SMPs can be designed to have tailored characteristics to meet the requirements (different stimuli and various properties).

As a relatively new type of shape-memory material, to most researchers and engineers, the development and application of SMPs seemingly lag behind other shape-memory materials at present. However, given their advantages and multifunctional nature, we can expect SMPs to become a leading player in a variety of fields very soon.

Acknowledgments

The contributions from previous and current students in our research group are gratefully acknowledged. The PU SMP foams were kindly provided by the Nagoya R&D Center, MHI through Dr. WM Sokolowski at JPL, United States. All PS SMP sheets were from CRG. Financial supports include an ACRF (RG 16/00) and a BPE Cluster SEED Grant (SFP011-2005) both from Nanyang Technological University, Singapore. The laser used for patterning atop SMPs was kindly provided by Dr. Q Xie from DSI, Singapore, through a collaborative project.

References

1. Huang, W. M. 2002. On the selection of shape memory alloys for actuators. *Materials and Design* 23:11–19.
2. Liu, C., Qin, H., and Mather, P. T. 2007. Review of progress in shape-memory polymers. *Journal of Materials Chemistry* 17:1543–1558.
3. Ratna, D. and Karger-Kocsis, J. 2008. Recent advances in shape memory polymers and composites: A review. *Journal of Materials Science* 43:254–269.
4. Behl, M. and Lendlein, A. 2007. Actively moving polymers. *Soft Matter* 3:58–67.
5. Gunes, I. S. and Jana, S. C. 2008. Shape memory polymers and their nanocomposites: A review of science and technology of new multifunctional materials. *Journal of Nanoscience and Nanotechnology* 8:1616–1637.
6. Tey, S. J., Huang, W. M., and Sokolwski, W. M. 2001. On the effects of long term storage in cold hibernated elastic memory (CHEM) polyurethane foam. *Smart Materials and Structures* 10:321–325.
7. Chung, T., Romo-Uribe, A., and Mather, P. T. 2008. Two-way reversible shape memory in a semicrystalline network. *Macromolecules* 41:184–192.

8. Yang, B., Huang, W. M., Li, C., Lee, C. M., and Li, L. 2004. On the effects of moisture in a polyurethane shape memory polymer. *Smart Materials and Structures* 13:191–195.
9. Huang, W. M., Lee, C. W., and Teo, H. P. 2006. Thermomechanical behavior of a polyurethane shape memory polymer foam. *Journal of Intelligent Material Systems and Structure* 17:753–760.
10. Chiodo, J. D., Harrison, D. J., and Billett, E. H. 2001. An initial investigation into active disassembly using shape memory polymers. *Proceedings of Institute of Mechanical Engineers* 215:733–741.
11. Chiodo, J. D., Jones, N., Billett, E. H., and Harrison, D. J. 2002. Shape memory alloy actuators for active disassembly using 'smart' materials of consumer electronic products. *Materials and Design* 23:471–478.
12. Chiodo, J. D. and Boks, C. 2002. Assessment of end-of-life strategies with active disassembly using smart materials. *The Journal of Sustainable Product Design* 2:69–82.
13. Lendlein, A. and Langer, R. 2002. Biodegradable, elastic shape-memory polymers for potential biomedical applications. *Science* 296:1673–1676.
14. Yang, B., Huang, W. M., Li, C., and Li, L. 2006. Effect of moisture on the thermomechanical properties of a polyurethane shape memory polymer. *Polymers* 47:1348–1356.
15. Huang, W. M., Yang, B., An, L., Li, C., and Chan, Y. S. 2005. Water-driven programmable polyurethane shape memory polymer: Demonstration and mechanism. *Applied Physics Letters* 86:114105.
16. Yang, B., Huang, W. M., Li, C., and Chor, J. H. 2005. Effects of moisture on the glass transition temperature of polyurethane shape memory polymer filled with nano carbon powder. *European Polymer Journal* 41:1123–1128.
17. Lan, X., Huang, W. M., Leng, J. S., Liu, N., Phee, S. J., and Yuan, Q. 2007. Electrically conductive shape-memory polymer with anisotropic electro-thermo-mechanical properties, in *Polymers in Defence and Aerospace Applications*, September 18–19, Toulouse, France, Smithers Rapra Technology Ltd, Billingham, Cleveland, U.K., 2007, ISBN 978-1-84735-019-0, paper 9.
18. Leng, J. S., Lan, X., Liu, Y. J., Du, S. Y., Huang, W.M., Liu, N., Phee, S. J., and Yuan, Q. 2008. Electrical conductivity of thermo-responsive shape-memory polymer with embedded micron sized Ni powder chains. *Applied Physics Letters* 92:014104.
19. Sokolowski, W. M., Chmielewski, A. B., Hayashi, S., and Yamada, T. 1998. Cold hibernated elastic memory (CHEM) self-deployable structures. *Proceedings of SPIE* 3669:179–185.
20. Metcalfe, A., Desfaits, A. C., Salazkin, I., Yahia, L., Sokolowski, W. M., and Raymond, J. 2003. Cold hibernated elastic memory foams for endovascular interventions. *Biomaterials* 24:491–497.
21. Sokolowski, W., Metcalfe, A., Hayashi, S., Yahia, L., and Raymond, J. 2007. Medical applications of shape memory polymers. *Biomedical Materials* 2:S23–S27.
22. Mooney, D. J., Baldwin, D. F., Suh, N. P., Vacanti, J.P., and Langer, R. 1996. Novel approach to fabricate porous sponges of poly(D,L-lactic-co-glycolic acid) without the use of organic solvents. *Biomaterials* 17:1417–1422.
23. Huang, W. M., Yang, B., Wooi, L. H., Mukherjee, S., Su, J., and Tai, Z. M. 2007. Formation and adjustment of bubbles in a polyurethane shape memory polymer, in *Materials Science Research Horizons*. Nova Science Publishers, Hauppauge, NY, pp. 235–250.

24. Liu, N., Huang, W. M., and Phee, S. J. 2007. A secret garden of micro butterflies: Phenomenon and mechanism. *Surface Review and Letters* 14:1187–1190.
25. Liu, N., Huang, W. M., Phee, S. J., Fan, H., and Chew, K. L. A generic approach for producing various protrusive shapes on different size scales using shape-memory polymer. *Smart Materials and Structures* 16:N47–N50.
26. Capadona, J. R., Shanmuganathan, K., Tyler, D. J., Rowan, S. J., and Weder, C. 2008. Stimuli responsive polymer nanocomposites inspired by the sea cucumber dermis. *Science* 319:1370–1374.
27. Kim, B. K. 2008. Editorial corner—A personal view shape memory polymers and their future development. *eXPRESS Polymer Letters* 2:614–620.
28. Nelson, A. 2008. Stimuli-responsive polymers: Engineering interactions. *Nature Materials* 7:523–525.
29. Farokhzad, O. C., Dimitrakov, J. D., Karp, J. M., Khademhosseini, A., Freeman, M. R., and Langer, R. 2006. Drug delivery systems in urology-getting "smarter." *Urology* 68:463–469.
30. Browne, A. L. and Johnson, N. L. Shape memory polymer seat assemblies. United States Patent 7309104.
31. Browne, A. L. and Johnson, N. L. Hood assembly utilizing active materials based mechanisms. United States Patent 7392876.
32. Alexander, P. W., Browne, A. L., Johnson, N. L., Mankame, N. D., Muhammad, H., and Wanke, T. Active material based tunable property automotive brackets. United States Patent 7401845.
33. Browne, A. L. and Johnson, N. L. Method for controlling airflow. United States Patent 7178859.
34. Jiang, H., Kelch, S., and Lendlein, A. 2006. Polymers move in response to light. *Advanced Materials* 18:1471–1475.
35. Liu, G., Ding, X., Cao, Y., Zheng, Z., and Peng, Y. 2005. Novel shape-memory polymer with two transition temperatures. *Macromolecular Rapid Communications* 26:649–652.
36. Bellin, I., Kelch, S., Langer, R., and Lendlein, A. 2006. Polymeric triple-shape materials. *Proceedings of National Academic Society* 103:18043–18047.
37. Behl, M. and Lendlein, A. 2007. Shape-memory polymers. *Materials Today* 10:20–28.
38. Kelch, S., Steuer, S., Schmidt, A. M., and Lendlein, A. 2007. Shape-memory polymer networks from oligo [(ε-hydroxycaproate)-co-glycolate] dimethacrylates and butyl acrylate with adjustable hydrolytic degradation rate. *Biomacromolecules* 8, 2007:1018–1027.
39. Miaudet, P., Derre, A., Maugey, M., Zakri, C., Piccione, P. M., Inoubli, R., and Poulin, P. 2007. Shape and temperature memory of nanocomposites with broadened glass transition. *Science* 318:1294–1296.
40. Everhart, M. C., Nickerson, D. M., and Hreha, R. D. 2006. High-temperature reusable shape memory polymer mandrels. *Proceedings of SPIE* 6171:61710K.
41. Pan, G. H., Huang, W. M., Ng, Z. C., Liu, N., and Phee, S. J. 2008. Glass transition temperature of polyurethane shape memory polymer reinforced with treated/non-treated attapulgite (playgorskite) clay in dry and wet conditions. *Smart Materials and Structures* 17:045007.
42. Cao, F. and Jana, S. C. 2007. Nanoclay-tethered shape memory polyurethane nanocomposites. *Polymer* 48:3790–3800.

43. Balazs, A., Emrick, T., and Russell, T. P. 2006. Nanoparticle polymer composites: Where two small worlds meet. *Science* 314:1107–1110.
44. Manias, E. 2007. Nanocomposites: Stiffer by design. *Nature Materials* 6:9–11.
45. Leng, J. S., Huang, W. M., Lan, X., Liu, Y. J., and Du, S. Y. 2008. Significantly reducing electrical resistivity by forming conductive Ni chains in a polyurethane shapememory polymer/carbon-black composite. *Applied Physics Letters* 92:204101.
46. Mohr, R., Kratz, K., Weigel, T., Lucka-Gabor, M., Moneke, M., and Lendlein, A. 2006. Initiation of shape-memory effect by inductive heating of magnetic nanoparticles in thermoplastic polymers. *Proceedings of National Academic Society* 103:3540–3545.
47. Kunzelman, J., Chung, T., Mather, P. T., and Weder, C. 2008. Shape memory polymers with built-in threshold temperature sensors. *Journal of Materials Chemistry* 18:1082–1086.

Index

N

O

P

R

S

T

V

W

Y